Nonlinear Dynamics in Human Behavior

Studies of Nonlinear Phenomena in Life Sciences – Vol. 5

Nonlinear Dynamics in Human Behavior

editors

W. Sulis
McMaster University
Canada

A Combs
University of North Carolina
USA

World Scientific
Singapore • New Jersey • London • Hong Kong

Published by

World Scientific Publishing Co. Pte. Ltd.

P O Box 128, Farrer Road, Singapore 912805

USA office: Suite 1B, 1060 Main Street, River Edge, NJ 07661

UK office: 57 Shelton Street, Covent Garden, London WC2H 9HE

British Library Cataloguing-in-Publication Data
A catalogue record for this book is available from the British Library.

NONLINEAR DYNAMICS IN HUMAN BEHAVIOR

ISBN 981-02-27426

Printed in Singapore.

CHAOS EDGE

KARL H. PRIBRAM
Professor Emeritus, Stanford University
James P. and Anna King University Professor
and Eminent Scholar, Commonwealth of Virginia
BRAIN Center
Radford University
432 Russell Hall
Radford VA 24142

The current interest in chaos theory is matched only by the number of pronouncements on consciousness and cognition. The scientific climate in the human sciences has shifted totally from mid-century when radical behaviorism held center-stage. It is my belief that with each turn of the wheel of fashion new insights into universal problems are achieved, that knowledge accrues along a helical path, that the essentials that are intuited at the beginning of each turn are tested and gradually made explicit. Once explication has achieved a plateau, interest wanes, some new shift takes place. But a generation or two later, the issues are revisited and additional insights are achieved.

Nonlinear Dynamics in Human Behaviour captures chaos studies in mid-cycle. This accounts for two rather different flavors to the papers: about one third develop the tools of the theory, rigorous arguments based on data and analyses; the other two thirds use their intuitions based on the theory's vocabulary, to provide metaphors. The application by Zausner of chaos to the analysis of painting is of this nature: the dictionary definition rather than strict nonlinear development of a Lyapunov-guided function is involved.

A virtue of the collection is the transparency of the writing that, for the most part, allows one to readily realize into which category each contribution belongs. As can be expected when such a range of interests is represented, different readers will be turned on by different manuscripts. For my part, here is what intrigues me about chaos theory in general and this book in particular:

As noted by Christian Scheier and Wolfgang Tschacher, all nonlinear dynamics is not chaos theory and all chaos theoretical concepts are not equally appropriate to all problems. Thus when "chaos theory" is used as a metaphor the user needs to distinguish between the various modes of nonlinearity he is coming from. Further, as in the specific example Scheier and Tschacher are dealing with, time series analysis, the problem of distinguishing stochasticity from deterministic chaos is fraught with difficulties. In our laboratory, for instance, we have found that nerve impulse spike trains obtained under anesthesia from parietal and hippocampal cortices are better characterized as stochastic.[1] In fact, most analyses of spike trains have come up with "stochastic resonance with noise" models (see Pribram[2] for several reviews).

As indicated by Frederick Abraham, even ordinary chaos theory has a split personality. When the path described is formed by a point or a limit cycle attractor a relatively stable geometry results. If, however, the attractor is truly chaotic and there are no countervailing constraints, the resulting bifurcations lead to progressive dissipation and instability. It would have been so much more accurate and less confusing (and therefore less exciting?) had the theory been named "turbulence theory". So, once again, metaphoricians need to indicate which form of the theory they have in mind. But would Abraham then be able to come up with the very apt idea that Greek tragedy is indeed modelled by chaos theory: the flaw in a person's personality, his positive Liapunov coefficient, leads ultimately to his dissipation.

Along these same lines, Jeff Goldstein distinguishes between diversification and dissipation and between these and the creation of local stabilities in turbulent streams. Goldstein also emphasizes the role of causal factors as releasers of potentials. Actualizers, as he calls them, are prompts which criticalize parameter values. Turning on a light switch is such an actualization. His use of the word "prompt" immediately made me want to apply his way of thinking to the memory retrieval process. But before this could happen, I was swamped with other fascinating ideas: a return to Ashby's dictum of requisite variety in the guise of requisite complexity (does Goldstein mean requisite diversity?); the difference between efficient causes as local (shoulder to the wheel) vs. final causes as global emergents; only to be faced by Newton as efficient and Einstein as final and geometrical; the new science is geometrical not causal–except as Stengers has pointed out Descartes had states that "all of my physics is geometry".

Goldstein's arguments regarding causal factors as releasers of a potential reminded me of my own interpretation of entropy as a measure of a potential to be actualized by choice, the change to be measured as information in Shannon's terms (a conception consonant with Shannon's own measures on the reduction of uncertainty).

The potential in chaos theory is not just the Liapunov function. I have argued for years that in addition to initial conditions, concurrent constraints determine the path of a chaotic system. Goldstein recognizes this and notes that a small change in parameters near a bifurcation can lead, not only to a quantitative change but to a qualitative jump to a new realm of dynamical behaviour in the system.

Pekka Aula takes a different stance. She claims that chaos is the new paradigm of communications research. At the risk of seeming sexist, I have experienced this type of communication for years with my women friends and lovers. But I disagree with the statement that "life is nonlinear". Rather, biological processes are often characteristically piecewise linear. Nor have I seen any experimental or formal evidence for the kind of integrative (constructive) communication in nonlinear systems that Aula discusses and diagrams– although I heartily agree with the statement that dissipative communication is destructive.

I find Shannon's and Gabor's contributions to communication measurement theory more compatible with formulating my experience in brain research. Stimulus sampling theory, response- reinforcement theory, as well as carefully chosen applications of nonlinear models need all be used in their appropriate place– propriety is usually determined by the level of inquiry being pursued.

Alan Stein's paper is an example of an attempt to locate nonlinear dynamics (expressed in self-organization far from equilibrium) within a psychoanalytic framework. To some extent he succeeds admirably although he misrepresents Freud's project on the basis of Sullaway's misinterpretations of that remarkable document. I wonder if either of them have ever read it. Freud, of course, did not have the insights Prigogine has provided us, but he was privy to Boltzman's thermodynamics from which he formulated the basis of the difference between primary and secondary processes. Secondary processes were negentropic by way of reality testing, and thus were self organizing. As Merton Gill and I[3] have carefully annotated (taking 15 years, during 5 of which we met weekly), the Project and the neurobiology of the turn of the century Vienna, were not as primitive as they are made out to be: in the case of Freud it was the miserable translations by people not versed in neurophysiology that obfuscated meaning.

Harry Erwin's excellent paper shows what can be done with sophisticated nonlinear modeling of social phenomena. His choice of subject matter: the origins of agriculture and the collapse of complex cultures is brilliant, because in these major societal developments, the concurrent constraints that protect a particular path fall away, leaving development to the mercy of its built-in conditions.

Finally I will address Bill Sulis' and Allan Combs' rich presentations. Sulis contributes a Prelude which centers on brain function, and Combs clarifies an all important aspect of consciousness. Both use nonlinear dynamics in the strict sense to develop their theses.

Sulis' modeling of neural pattern coding in terms of a dynamical automaton in which the code resembles a language is, to my mind, "right on". His

TIGoRS, using Hamming distance to determine the amount of discordance between patterns, appears to be a useful metric. His demonstration that the ability to induce TIGoRS is a function not only of the number of inputs but the symmetry class of the automaton is superb.

But, Sulis felt that this model was unsatisfactory because of biological implausibility, and he went on to develop a model based on cellular automata. This works remarkably well; for instance, TIGoRS can recognize 30^{100} patterns whereas cell assemblies are limited to a maximum of 2^{100} patterns.

Even cellular automata have shortcomings, however, according to Sulis; I will counter these "shortcomings" by suggesting that (1) binding is global within brain subsystems; (2) test apply the model on subsystems; and (3) the dynamics of TIGoRS are like those of the connection web of synapses and dendrites, though not of neurons. I will take up these issues in reverse order.

Regarding item (3): For almost four decades, I have been urging, in concert with a few others neuroscientists, that the synapse and the dendritic web in which it operates is "the unit" of computation in the brain.[4,5,6,7] The operations of the connection web fit much better the computational needs of PDP type simulations than do those of neural circuits.[6]

Regarding item (2): The brain operates with respect to psychological categories as a set of subsystems. The role of each subsystem is different, and its mode of operation is different, although overall similarities may obtain. Some of the primary sensory systems appear to operate piecewise linearly; perisensory systems operate in terms of symmetry groups to define objects; higher order systems such as those in the temporal lobe function in terms of attractors. The evidence in terms of data and models for these statements can be found in Pribram.[6]

Regarding item (1): The so-called binding problem is a chimera. There are no feature detectors in the cortex. Conjoined in every receptive (that is, dendritic) field of each neuron in the visual cortex are sensitivities to several properties of the input such as orientation of gratings; luminance change; direction and speed of motion; endstopping (hypercomplex) and phase sensitivity (simple) or non-phase sensitivity (complex); spatial frequency; and often color and opponens color sensitivities are there as well. Each neuron's receptive field has different properties conjoined–much as each person has different "properties": some are blond, female, wear glasses; others are blond, male; others are brunette, female, and tall, etc. Sometimes groupings of neurons with receptive fields displaying similar conjunctions occur, but always changes in output can be produced by a variety of changes in input.

Thus, there is no way that the axonal output of a single neuron can, by itself, define a feature for the next station in processing. The basic "pattern" of such activity is stochastic[1] and any code that operates must operate on a base of stochastic resonance with noise.[2] It is the spatial pattern of neuronal activity–such as population vector coding that is the most likely agent of computation and communication in the brain.

TIGoRS and piecewise linear reciprocal transformations (such as Gabor wavelets) therefore are excellent candidates for accomplishing what needs to be accomplished. Sulis' prelude is a welcome introduction to this fascinating field of research.

Allan Combs' coda on consciousness should be read in conjunction with the prelude. The operations of the connection web of synaptodendritic processes lead readily to higher order, more holistic, ecological and context sensitive views since the neural processes are themselves of this nature. In fact, the processes in the connection web are autopoetic (in the sense of self-reorganizing) and are constituted, not of nerve impulses but of interference patterns created by hyperpolarizations and depolarizations. All we see is constituted of such fluctuating (ripples upon ripples) polarization. There are no propagated nerve impulses to be found in the retina until the last layer (the ganglion cell layer) of processing is reached. Then when long distance communication is at stake, nerve impulses, resistant to further processing, are generated.

Yes, remembering as well as moods, thoughts and dreams exhibit periods of constant rhythmic activity and other periods of chaotic change, as Combs notes. This observation gives me the opportunity to indicate the relationship between my own work using holographic and holonomic models. Fourier and Gabor functions are linear, invertible, and can be used to construct continuous symmetry (Lie) groups. Nonlinear dynamics is the basis for theories that involve deterministic chaos. But linearity is not the issue. Quantum holography (based on holonomic theory) can be nonlinear and the dynamics can be linear. What is important is that all of holographic-like modeling is time symmetrical: it captures the "moment". Deterministic chaos theory, on the other hand, is time asymmetric– it describes a path taken. When a path is chosen and undertaken, time symmetry can be broken. Prigogine has detailed the precise conditions under which this is achieved.[8,2,9]

If a photographer wishes to capture "the decisive moment", we can model his neural and psychological processes at that moment in quantum holographic-like, Gabor wavelet processes. If, however, we want to map the photographer's past– and possible future path–to and from that moment, deterministic chaos affords the appropriate tool.

In summary, Sulis' prelude and Combs' coda provide the tools for understanding such processes as remembering and thinking and their neural entailments as well. We are soon entering the 21st Century–our children will be looking back at this decade as the transition between 20th and 21st Century thinking about neuropsychological issues. *Nonlinear Dynamics in Human Behaviour* heralds this transition.

Thus, for me, this volume was an exciting one to read. I have not mentioned the works on creativity; the idea that we live on the edge of criticality when we are about to fall in love (love is a strange attractor); and the tests of nonlinear models to handle simple memory and choice tasks–the results indicated that whether chaotic or non-chaotic formulations were more appropriate depended on the values of the control parameters used (which supports my intuition that, in biological systems, a range of formalisms, from piecewise linear to wholly nonlinear needs to be explored for each set of parameters operative in a situation). There is real substance here, whether speculative or data driven, I feel honored to have been asked to participate in the venture.

References

1. M. Xie, K.H. Pribram and J. King in *Origins: Brain and Self Organization*, ed. K. Pribram, (Lawrence Erlbaum, New Jersey, 1994).
2. K. Pribram, *Origins: Brains and Self Organization*, (Lawrence Erlbaum, New Jersey, 1994).
3. K.H. Pribram and M.m. Gill, *Freud's Project Re- Assessed: Preface ot Contemporary Cognitive Theory and Neuropsychology*, (Basic Books, New York, 1976).
4. K.H. Pribram, *Ann. Rev. Psych.* **11**, *1-40, (1960).*
5. *K.H. Pribram, Languages of the Brain: Experimental Paradoxes and Principles in Neuropsychology, (Prentice-Hall, New Jersey, 1971).*
6. *K.H. Pribram, Brain and Perception: Holonomy and Structure in Figural Processing, (Lawrence Erlbaum, New Jersey, 1991).*
7. *K.H. Pribram, in Science of the Mind, The 21st Century, ed. R. Solso*
8. *I. Prigogine, in Origins: Brain and Self Organization, ed. K.H. Pribram, (Lawrence Erlbaum, New Jersey, 1994).*
9. *A. Combs, in Nonlinear Dynamics in Human Behaviour, eds. W. Sulis and A. Combs, (World Scientific, New Jersey, 1996).*

INTRODUCTION

This volume offers a selection of the best papers presented during the Fourth Annual Conference of the Society for Chaos Theory in Psychology and the Life Sciences, held at Johns Hopkins University, Baltimore, MD, June 24-27, 1994. The Society for Chaos Theory in Psychology and the Life Sciences was constituted in 1991, when a group of far sighted individuals gathered together at the Saybrook Institute in California, to ponder the role that the new sciences of nonlinear dynamics and complex systems theory might play in furthering our understanding in the social sciences. It has long been felt that the social sciences could benefit from the adoption of quantitative modeling strategies, but until recently models based upon linear, reductionist approaches have proven to be woefully inadequate to deal with the complexity inherent in social systems. With the spread of nonlinear dynamics beyond the narrow confines of mathematics and physics and into popular culture, it was recognized that many aspects of complex social systems bore a striking similarity to phenomena observed in nonlinear dynamical systems. Unfortunately, few social scientists possessed the mathematical or physical background necessary to use these new ideas effectively. Moreover few mathematicians and physicists appeared interested to learn the experimental literature in the social sciences in order to construct ecologically valid models. Each side persisted in the belief that the other had little to offer.

Like the Santa Fe Institute, the Society was formed in order to bring together scientists from all disciplines who shared a common interest in the study of complex social systems. The Society would serve as a facilitator, enabling dialogue between representatives of these diverse disciplines so as to build a common language and a consensus of ideas and models. The original Society consisted mostly of psychologists but in its short life has expanded to include mathematicians, physicists, biologists, economists, sociologists, engineers, physicians, physiologists, anthropologists. It now boasts a worldwide membership and in 1997 will launch a new journal, *Nonlinear Dynamics in Psychology and the Life Sciences* which will spearhead this emerging interdisciplinary discipline and provide a permanent vehicle for communicating the fruits of this labour.

This volume fairly reflects the Society and its mandate, to bring together many different individuals, ideas and approaches in order to share, to dialog, to interact, and ultimately to further our understanding of the complex social systems with which our lives are so intimately intertwined. Understanding and solutions to the problems of warfare, genocide, racism, poverty, physical and mental abuse, family breakup, mental illness are foremost in the minds of people today. For too long science has turned its back on such issues, choosing instead to focus upon isolated, controllable systems. That approach has been spectacularly successful but if science is to have any meaning in our lives beyond the mere production of technology, it must find ways to deal with complex, nonisolatable and uncontrollable social systems. It is the goal of the Society to foster the search for such means of understanding and to communicate these to the broader constituency of researchers across all disciplines.

The current volume is a testament to the diversity of approaches represented within the Society. Papers range from the mathematical to the philosophical, and straddle topics from the motion of the extraocular muscles through to warfare and creativity. The reader, whether scientist, social scientist, philosopher or clinician should find much of interest here.

Acknowledgement

I would like to thank all of the individuals who took part in this conference, the organizers and the presenters, especially those who graciously agreed to have their work presented in this volume. In particular I would like to thank Robin Robertson, who shepherded this work through its tentative, early days and who encouraged me to join the project and to help bring it to fruition. I should also thank my wife Ruth and children Kara, Tegan and Alex, who put up with me over the many months required to make this a reality.
William Sulis
McMaster University
April 30, 1996

CONTENTS

PRELUDE
TIGORS AND NEURAL CODES

WILLIAM SULIS
Departments of Psychiatry and Psychology,
McMaster University
Hamilton, Ontario, Canada L8S 4K1

There are at present two major diametrically opposed theories of neural encoding. The first asserts that neurons encode information in their average firing rates while behaving as stochastic units. The second asserts that neurons encode information in precise (spatio) temporal patterns of spike trains while behaving as deterministic (chaotic) units. Both approaches assume that the correspondence between the neuronal response and the stimulus is stable. Current research has failed to support either approach satisfactorily. Recent research on place encoding in the rat hippocampus raises doubts about the stability of neural representations. A new approach, based upon transient induced global response synchronization (TIGoRS) in complex systems is proposed. This approach avoids the dichotomy of current models as well as the local stability criterion while still providing for a stable response of the system to a stimulus at the macroscopic level

1 Introduction

The search for an understanding of the relationship between psychological processes and the neurophysiological processes subserving them has been a major goal of neuroscience research during the past decade. A variety of novel theoretical and experimental tools have been developed during this time. The introduction of large multi-unit recording devices has extended the scale of observation from the recording of single neural units to the simultaneous recording of over 100 units. The development of sophisticated imaging techniques such as functional MRI, MEEG, QEEG, SPECT and PET scanning has permitted the observation of functional correlations between psychological processes and entire regions of the brian. Theoretical tools such as neural networks, cellular automata and coupled map lattices have provided fresh insights into the dynamics of weakly coupled systems with simple dynamical agents.

The concept of computation underlies much of the current research. The usage of the term computation varies significantly among researchers. On the one hand computation serves merely to indicate the use of computer simulations to help model and explore the phenomenon. This is the sense in which it is used in computational neuroethology [1]. On the other hand, computation may be used to indicate that the phenomenon itself is to be understood as a computation, usually in the strong sense of being algorithmic. This is the sense in which it is used in computational neuroscience [2] and frequently in artificial intelligence and artificial life [3].

In order to assert that the phenomena of interest are computations, it is necessary to establish some correspondence between the salient features of the process underlying the phenomenon and some abstract computational scheme so that the succession of features observed in the process can be seen to correspond to a succession of computational transformations of the symbolic system. Such a correspondence is more readily established if a very direct connection can be made between the various physical states of the process and the symbolic representation. In the case of neural systems, such a correspondence amounts to asserting the existence of a neural code. This means that certain patterns of neural activity can be put into a one-one correspondence with an abstract symbol system in such a way that manipulation of the symbol system via its rules directly corresponds to the observed succession of activity patterns recorded from these neurons.

The search for such a correspondence dates back to Semon and Lashley and the search for the engram. The modern search has shifted from the study of memory to that of perception and in particular, the binding problem. This refers to understanding the mechanisms by which the activities of a disparate collection of feature detectors, such as for colour, shape and motion become bound together to form a unitary percept. The major impetus for this shift was the discovery of neurons having firing rates which appeared to be selectively tuned to specific features of stimuli such as auditory frequency [4] or, in the case of vision, line orientation or direction of motion [5]. The discovery of such cells suggested that they were functioning as feature detectors, that is, responding maximally only when a specific instance of a specific feature was present in a stimulus. The presence of such cells gave support to the concept of the individual cell acting as a computational device, abstracting out the instance of the feature and encoding this information in its average firing rate. These cells generally occur quite early in the sensory pathways from sensory cell to association cortex. This anatomical arrangement seemed to fit a computational paradigm. However there was a problem with this model. If information about a stimulus passed first through a set of such feature detectors prior to passage to those cortical areas which presumably subserve conscious awareness, then how is it that in consciousness one perceives a unitary stimulus? In other words, how does the information acquired by these individual feature detectors become subsequently bound together in order to give rise to a unitary conscious experience? Attempts to answer this question have brought about a resurgence of interest in neural representations and neural codes.

The problems of neural representation become apparent when actual neural response patterns to stimuli are observed. Shadlen and Newsome have demonstrated this dramatically [6]. They obtained a series of 210 spike train recordings from a single cortical neuron in extrastriate visual cortex in a macaque monkey in response to repeated presentations of an identical pattern of dynamic random dots. The resulting raster diagram reveals a clear pattern of systematic variation of response manifest as an impression of vertical contours throughout the raster. Although the instantaneous firing rates varied reliably over a range of 80-100 Hz, the actual pattern of spike trains generated in any given run provided only a very noisy estimate of the rate pattern. Moreover the timing of individual spikes appeared to be random, having the characteristics of a non-homogeneous Poisson point process. The only reliable correlate to the stimulus was the pattern of instantaneous rate.

The earliest and most durable proposal for neural coding is that of rate coding. Since each tunable neuron varied its average firing rate in response to specific instances of a specific feature (such as the specific orientation of a line), it seemed plausible to assume that the average firing rate encodes the information output by that neuron about the instance of the feature which it detected. In order to bind these outputs together into a single percept, one merely requires that the individual neurons which encode the features involved in the percept send collateral axonal branches to converge on a single neuron (the 'grandmother cell') or a small population of neurons (the 'cell assembly') which then, in turn, will fire maximally if all of the requisite features are present and minimally otherwise. Thus this cell or cell assembly will rate encode the presence or absence of the complete percept.

Although robust and durable, the rate encoding hypothesis has met with considerable criticism. Simple combinatorial arguments preclude solutions such as the grandmother cell. Membership issues and bandwidth problems plague simple cell assembly models. Although robust against neuronal loss and noise, a noisy rate code is limited in the amount of information which it can convey.

A precise temporal code in which the precise timings of individual spikes would encode information would be preferable since the information carrying capacity would be significantly enhanced. Such an interpretation meets serious opposition unless the apparent noise in the spike trains of individual neurons is reinterpreted as part of the information being carried. This has led to an alternative proposal in which each individual spike train is presumed to arise, not as a result of random variation, but instead as a result of complex interactions between the individual neuron sampled and a host of other neural systems. The underlying dynamics is presumed to exhibit deterministic chaos so that the variation in spike trains is a result of subtle variations in initial conditions as well as concurrent processes during the presentation of the stimulus. Softky and Koch[7] have been the major proponents of this point of view.

In this account, feature binding arises as a result of the synchronized activity of a cell assembly. Two cells belong to the assembly if their spike trains are synchronized over time. Membership is assigned on functional grounds alone and cells may take part in several different assemblies over time. Considerable empirical evidence has been amassed during the past ten years which had demonstrated this form of synchronization in behaving animals[8]. In some cases what has been observed is a phase matching of periodic outputs of neural oscillators[9]. In others cases what has been observed is a synchronization of complexly varying spike trains[10]. More recently what has been observed is merely a complex pattern of correlation between spike trains of neurons in response to a stimulus[11]. In addition there is evidence for robust temporal patterning in at least some neural systems, for example in auditory cortex[4].

Unfortunately there has been some theoretical work which suggest significant limitations on the ability of interconnected neural networks to produce populations of synchronized neurons of the kind required by this model[12]. In addition several studies demonstrate that the most robust correlations between stimulus and neural response are between stimulus and average firing rate. Moreover a problem which has not been sufficiently addressed in the literature is how to ensure that disparate feature detectors, distributed spatially in cortex and presumably utilizing different strategies in order to detect and respond to individual features in a stimulus, can nevertheless produce spike trains which are temporally synchronized to one another. It is conceivable that detectors for the same feature might be able to respond in this manner but it is less conceivable that detectors for disparate features should do so. It is even less conceivable that detectors for features in different sensory modalities should do so, yet we clearly are capable of experiencing a unitary percept across many different sensory modalities.

The argument has recently shifted to explore the dynamical response characteristics of individual neurons in order to determine what kinds of encoding schemes are physically plausible. This problem was actually first addressed over 30 years ago by Gerstein and Mandelbrot[13] who studied the statistics of neuronal interspike intervals. They demonstrated that the distribution of interspike intervals is well modelled as a Poisson process and that a random walk model for the somatic membrane potential could adequately reproduce these statistics. Softky and Koch[7] took this further and addressed whether the current models of neuronal dynamics were capable of reproducing the statistics of in vivo cortical neuronal spike trains. They compared these for the standard integrate and fire models with and without a refractory period, for a leaky integrate and fire model, for models with variable EPSP magnitudes and durations as well as with adaptation and hyperpolarizing currents, and as well for a detailed compartment model simulation of a cortical pyramidal neuron with conventional and barely plausible parameters. Finally they considered a compartment model with active simulation of the dendritic tree. They were able to demonstrate that the only model which produced spike trains comparable to those naturally observed was the model with active dendritic simulation in which the dendrites exhibited strong nonlinearities in their dynamics and submillisecond time constants. Softy [14] later demonstrated that such submillisecond dendritic responses would enable a neuron to function as a coincidence detector which would enable the neuron to support a finely detailed temporal code. A major problem with this model however is that it requires dendritic membrane time constants which are one tenth to one hundredth that of the measured somatic membrane time constant. Until dendritic measurements can be obtained, the question remains open.

Some experimental data in support of the Softky-Koch conjecture has recently been published by Mainen and Sejnowski [15] who have demonstrated that cortical neurons in vitro in can produce fairly reliable, temporally complex spike train patterns in response to Poisson-like stimuli. However the stimuli used were not biologically plausible and it remains to be seen whether cortical neurons in vivo are capable of such finely scaled temporal discriminations.

Shadlen and Newsome [6] revisited the Gerstein and Mandelbrot model and have argued that the pattern of interspike intervals observed in intact animals can be reproduced if the soma of the neuron acts as an integrate and fire device in which the somatic potential varies like a random walk under more or less equally balanced excitatory and inhibitory potentials. Such a model could only support a rate code. A major difficulty with this model is that, in cortex at least, the excitatory synapse outnumber the inhibitory synapses by a factor of almost 6:1. However it is unknown whether the functional contributions of these two systems follow their anatomical distribution.

Both of these approaches address only the question of the vehicle by which information could be encoded by a neuron. They do not address the problem of reading the code. The current literature leaves one with the impression of two diametrically opposed models of neural encoding and a sparsity of conflicting theoretical and experimental evidence in support of either. It is important to note that although theoreticians present these views as being exclusive, the evidence to date fully supports neither, suggesting that the vehicle for a neural code might be some combination of these two approaches or perhaps something altogether different. Indeed it may well be the case that different neural systems utilize different vehicles for encoding information and so the idea of a single comprehensive coding scheme for the brain may be just a pipe dream.

2 The Stability Problem

A neural code, regardless of its nature, must meet several other conditions in order to be plausible as an encoding mechanism to be utilized by biological neural systems. Such encoding much be robust against synaptic noise and random neural drop out. Synaptic transmission is a noisy process. The reliability of synaptic transmission appears to be variable and mediated by such factors as LTP and LTD and quantal neurotransmitter release[16]. The response of the postsynaptic membrane in terms of the MEEP induced by a single quantum of neurotransmitter can vary as much as 40 fold[17]. Neurons die at a steady rate throughout life yet little impact is noted at the behavioral level until such neuronal loss is well advanced either through advanced age or illness.

It is reasonable to assume that there should be a stable correspondence between the encoding of a stimulus and the stimulus itself. This appears to be the case with rate encoding while in the case of temporal encoding one must generalize the 'stimulus' to include the totality of the external and internal environments of the organism and not just the stimulus applied by the experimenter. This poses a formidable problem for those other neural subsystems which receive as input the output of the system under consideration. Consideration must be paid to how neural systems further downstream can read or interpret the encodings of neurons in primary sensory areas. Moreover, if learning occurs then how are prior encodings to be preserved in the face of subsequent learning so that the original encoding will be accessed if presented with the original stimulus?

Recent research into the behaviour of rat hippocampal place cells raises serious concerns about the nature of any encoding scheme. Briefly, place cells [18] are neurons located in the CA1 region of the rat hippocampus which show tuning of their firing rates in response to the animal being located in specific locations in the physical environment which it inhabits. The existence of such cells provided the foundation for the view that the hippocampal region is significant for spatial navigation and has, as one of its function, that of maintaining a cognitive map of the physical environment[19]. This view was strongly supported in a paper by Wilson and McNaughton[20] in which a computer system analyzed the simultaneously recorded spike train patterns of a collection of 40-80 hippocampal neurons and used the firing patterns of identified place cells to locate the position of the animal in the spatial field. In their work, a simple rate code was sufficient to enable the researcher to reconstruct portions of a presumed cognitive map of the rat.

However subsequent research has demonstrated that the situation is much more subtle than previously imagined. This work has been summarized in a recent paper by McNaughton and his group at the University of Arizona [21]. They have been able to permanently implant electrodes into the CA1 region of the hippocampus of freely behaving animals and to sample the spike trains of some 80-100 cells over time under different experimental arrangements. In this way they were able to directly explore questions of the stability over time of these representations.

For example, Jung and McNaughton [22] observed the place fields of rats moving in a radial maze. Well defined place fields could be detected which were stable over several trials in the maze. Occasionally however the complete set of fields recorded on a given trial appeared but in a different orientation (i.e. different arms of the maze) relative to prominent external visual landmarks. All cells recorded maintained the same relative rotation. Quirk [23] noted that rats introduced into an environment in darkness sometimes exhibited a new distribution of place fields, completely different from that observed during previous sessions in the same environment under illumination. Moreover the new fields often persisted after the lights were turned on.

Bostock [24] studied rats trained in a cylindrical environment in which a cue card was used to break the rotational symmetry. When the cue card was changed from white to black, the place fields underwent a random rotation on the first trial, and then were completely different on subsequent trials. An even more striking example [21] involved animals trained in a particular track for more than one week, who were then exposed to the familiar environment, then exposed for 1 hour to a non familiar environment and then returned to the original environment. In many cases the place fields upon return were completely rearranged.

Most interesting is the observation that if the animal is restrained, the place cells do not fire. If, however, the animal is passively placed in a place field and is able to move but does not move, then the place cell fires. In other words, there seems to be a necessity for some component of intentionality in order for the place cell system to fire [25].

The hippocampal place cell system exhibits a striking metastability. It does not provide a stable, context free cognitive map of the environment but instead its activity depends strongly on immediate past and present context. McNaughton and his group argue [21] that the hippocampal place cell system is a component of a path integration system and that there exist preconfigured idiothetic reference frames which can become activated in response to a novel environment. Failure to activate a preexisting frame, already correlated with prior experience with a particular environment could account for the metastability exhibited by this system. A given place cell may be a functional component of several different reference frames and activate in response to differing regions of space relative to the animal. As a result there is no context free interpretation of its behaviour.

Even if the hippocampal place cell system can be understood as a computational system, and that the spike trains of individual neurons can be understood as an encoding of information which is being manipulated by this system, whether through rate or some other representation, it is still not possible to understand what information is being represented by that code without a knowledge of the immediate past and present context in which the animal is situated. This presents a formidable problem for the experimenter interested in understanding the meaning, if any, which can be attached to the output of a given neuron. It provides an even greater problem for the animal since it is not at all clear how downstream systems are to know what meaning to attach to the output of such a neuron.

A stronger interpretation than that of McNaughton may be made. It appears as though the hippocampal place cell system behaves more like a scratch pad for carrying out computations and there is no intrinsic meaning attributable to the actions of a given neuron or collection of neurons beyond that imparted to them by the context of the moment. It may be possible to constrain the function of the place cell system to providing a scratch pad for calculations involving path integration, but not much more can be said about any particular use of such a system independent of context. The system may use a fixed coding vehicle to represent information, just as one might only use English letters on a scratch pad, but the key required to decode any such information is created with each new context, just as it is with a scratch pad. It would be misleading to associate an individual neuron with a specific computation in an invariant manner. It would appear that, at best, such a attribution can be made only at the level of the hippocampus or perhaps even only at the level of the animal itself.

A deeper problem for spatial memory ensues. If we presume that the hippocampal place cell does play a functional role in the establishment of spatial memories, and that relevant spatial information is encoded within the place cell system, then how do stable, coherent spatial memories become established? Presumably spatial memories become established as a result of neural activation of remote cortical areas such as parietal cortex. If such activation follows upon activation of hippocampal place cells, then how do these remote areas encoded this activity into coherent representations when the encoding at the level of the hippocampal is metastable so that the information conveyed by the activation of a given place cell varies according to context? Although neurophysiological evidence is lacking, it is conceivable that the metastability of the hippocampal place cell system extends to other cortical systems as well, including memory. In that case, memory itself would be a metastable process, context dependent. Memories would not be stored in an invariant form but would be created on the spot given a specific context. That this is likely has been suggested by research into the nature of autobiographical memory [26].

A major distinction must be made between the presumed vehicle which supports some neural code and the interpretation of an instance of that code as conveying specific information about the computational process currently being undertaken by that neural system. It is clear from the hippocampal place cell experiments that the correct interpretation of the code requires knowledge outside of the system per se, including both the past history of the animal, its current behaviour as well as details of its environment. The vehicle by which information may be represented by the system, i.e. as a pattern of instantaneous firing rates or of temporal spike train patterns or whatever, may be consistent throughout the history of the animal but knowledge of the vehicle alone is insufficient to permit the current code to be interpreted outside of a knowledge of the system as a whole. Written language provides an accessible analogy. A paragraph extracted at random from a book may be recognized as having meaning (that is properly encoded) but that meaning may not be fully comprehended without knowledge of the context, that is, the book.

This suggests that the proper setting in which to study neural representation and coding is, paradoxically, that of the whole animal-environment context. It is only at this level that meaning can be attached to neural events. It is not a priori necessary that the metaphor of computation at the animal-environment level be extended to the neuronal level. Any neural process which facilitates an adaptive relationship between the animal and its environment warrants consideration whether computational or otherwise. Indeed our excessive emphasis on the interpretation of neural processes as computations may blind us to other dynamical features of their behaviour which are important in maintaining the adaptive relationship between the animal and its environment. Considerations of this sort lead to the concept of computational competence.

3 Computational Competence and Transient Languages

In several papers [27,28,29] I have argued that in order to understand the nature of computation in complex adaptive dynamical systems such as the brain, it is first necessary to make a distinction between computational competence and computational performance. Computational competence refers to the ability or capability of the system under scrutiny to carry out the computation under consideration. In order to address this question it is first necessary to specify, in advance, the environmental context in which the computation is to take place, the specific features of the environment which will be manipulated by the observer and comprise the initial data as it were, the behaviours to be exhibited by the animal which indicate that the required computation did indeed take place, and the duration within which this task is to be performed. Since it is never possible to fully constrain an environment and no animal is ever so constrained that it can exactly reproduce a given behavioral sequence, it is necessary to specify the behavioral constraints on the environment and the animal separately as sets of behaviours, each behaviour having a predetermined duration. The experiment then consists of specifying some set of initial conditions and the environmental constraints to be induced by the experimenter and then allowing the animal-environment system to evolve over the allotted time and finally observing whether or not the subsequent behaviours of both the environment and the animal lie within the previously assigned sets. If they do, then it is possible to assert that the requisite computation took place. If not, not.

The question of computational performance is a secondary consideration. It refers to the actual mechanism by which the computation is carried out. For example, one might ask whether the computation is carried out through some sequence of internal states which can be interpreted as the acting out in material form of some algorithm, and if so which algorithm, or whether a non-algorithmic process is involved. In order to consider questions of computational performance it is first of all necessary that the animal satisfy the conditions of computational competence for the computation in question. A major problem with theory in the neurosciences today is the wholesale acceptance of the computation as algorithm metaphor. This insistence upon algorithmic means of performance blinds many researchers to looking at the deeper question of computational competence and to considering as a result other mechanisms by which computations in the sense described above could be carried out. It is this emphasis on the computation as algorithm (and neuron as computer) which underlies the continued search for a neural code and impedes consideration of the broader view of neural systems as complex adaptive dynamical systems. It is my belief that a consideration of neural systems from this latter perspective, with due attention to questions of computational competence, is more likely to lead to an understanding of the neural mechanisms of computational performance than the slavish imposition of a specific metaphor of computational performance drawn from computer science. There is no a priori reason to presume that the two must be related except perhaps at a level of abstraction beyond that being considered here.

My approach to these problems is based on the transient language formulation of a dynamical system[27,28,29]. The standard formulation of a dynamical system does not readily yield information about its computational capability under natural ecological constraints. The transient language/dynamical automaton formulation of a dynamical system was introduced to rectify this problem. In the standard approach to a dynamical system, the set of all instantaneous states in which the system can exist forms a state space S. For any time t, and any initial state x, the dynamic yields a new state x_t which is the state in which the system is found at time t having begun in state x at time 0. We can express this concisely by defining a mapping $f_t : S \longrightarrow S$ such that $f_t(x) = x_t$. Each map which takes S to itself is termed a selfmap of S. The set of selfmaps of S is denoted S^S. The full dynamic is thus given as a map $f : T \longrightarrow S^S$ where T is some temporal set. A trajectory from an initial point x is just the set $\{f_t(x) \mid t \in T\}$ which corresponds to a continuous path in the state space. The standard formulation readily yields information about the asymptotic behaviour of the system, in particular whether or not it possesses attractors. However it does not readily yield information about the short duration transients of the system and it is precisely these transients which correspond to our usual concept of behaviour. A different approach is required in order that information about these transients can be derived.

In the transient language formulation, the set P of all short duration transients forms the state space and the trajectory of a dynamical system is represented as a sequence of finite duration transients. The dynamic is constructed from the set of selfmaps of P as in the standard formulation. These short duration transients can be thought of as letters in an alphabet so that a trajectory becomes a sequence of letters, much like a word. In this way changes in the system such as those due to external inputs give rise to transformations on these 'words'. These transformations in turn are generated by a 'grammar' and in this way a computational representation of the dynamical system is obtained, a dynamical automaton.

A complete formulation is given elsewhere [30]. A simplified version suffices here. A transient language (S, T, d) consists of a semigroup S, a positive abelian semigroup T and a mapping (duration) from $S \xrightarrow{d} T$ such that $d(ss') = d(s) + d(s')$. As a prototypical example, consider the set of all maps of the form $[0, a) \xrightarrow{f} \mathcal{R}$ where $a \in \mathcal{R}$, the real line. Given $[0, b) \xrightarrow{g} \mathcal{R}$, define fg as $[0, a+b) \xrightarrow{fg} \mathcal{R}$ where $fg(x) = f(x)$ if $x \in [0, a)$ and $fg(x) = g(x)$ if $x \in [a, a+b)$. Each such map represents a transient of a dynamical system, hence the name. A dynamical automaton $(\mathcal{S}, \mathcal{E}, T, \Delta)$ consists of two transient languages $\mathcal{S} = (S, T, d)$ and $\mathcal{E} = (E, T, d')$ and a dynamic Δ such that given $\psi \in S$ and $\eta \in E$, we have $\Delta(\psi, \eta) = \psi'\psi_\eta$ where $d(\psi) = d(\psi')$ and $d(\psi_\eta) = d'(\eta)$ and $\Delta(\psi, \eta\eta') = \Delta(\Delta(\psi, \eta), \eta')$. Define $P(\psi, \eta) = \psi_\eta$.

Every dynamical system gives rise to a corresponding dynamical automaton.

It has been argued elsewhere that the dynamical automaton formulation is the natural one in which to address computational issues in naturally occurring computational systems [31]. In a typical experiment there will be an open set of stimuli Q reflecting the actions of the experimenter and the effects of the environment, an open set R of responses deemed representative of the appropriate response or correct computation, and some set Σ of possible initial histories. The system is competent to carry out the required behaviour or computation if $P(\Sigma, Q) \subset R$. In the case of a perceptual stimulus, the percept becomes associated with some subset R of responses and so corresponds to a distribution of spatiotemporal patterns.

The pairing (Q, R) effects the computation but merely establishes a high level relationship between stimulus and response. In [34] it was argued that such a high level relationship provided a solution to the binding problem if a) the pairing is unique, b) the pairing is robust against internal noise, and c) the pairing is robust over time, that is, repeat presentations of the stimulus produce the same association. The low level dynamics which gives rise to such high level correlations is the domain of computational performance questions. One mechanism which can effect such a pairing is transient induced global response synchronization (TIGoRS).

TIGoRS occurs when a transient stimulus applied to a dynamical system produces a set of responses which cluster closely in pattern space. Assume that a suitable metric ρ has been imposed on the transient language. In order to avoid an erroneous attribution of clustering to TIGoRS when actually due to mere statistical coincidence, we say that a stimulus η induces TIGoRS if, given any two initial histories ψ, ψ', it follows that $\rho(\psi_\eta, \psi'_\eta) < 1/2\rho(|rand(\psi_\eta)|, |rand(\psi'_\eta)|)$ where $rand(\psi_\eta)$ and $rand(\psi'_\eta)$ are randomly generated patterns of same norm as ψ_η, ψ'_η.

TIGoRS has been detected in a variety of complex systems models including tempered neural networks[32], coupled map lattices with input[33] and driven cellular automata[29].

4 Simulation Results

A series of simulations have been carried out in order to assess the robustness of the TIGoRS phenomenon. The results with cellular automata are most extensive and will be discussed here. A 2 state, 3 neighbor cellular automata consists of a set of cells arranged around a circular lattice. Each cell can exist in one of 2 states, 0 or 1. At time step n, the cell determines its state at time step n+1 by entering its state value and those of its adjacent neighbors into a function f. This means that $a_i^{n+1} = f(a_{i-1}^n, a_i^n, a_{i+1}^n)$. The function is the same for all cells and all of the cells are updated simultaneously. An input pattern was generated either randomly or by a second cellular automaton, in either case on a lattice having the same size as that under consideration. A fixed duration of 450 time steps was chosen since most TIGoRS occurred within this bound. The pattern was sampled randomly. If the sampled pattern cell had value 1, then the corresponding cell in the automaton had its state set to 1 regardless of the value determined for that cell via the update rule.

The metric chosen for these studies is the Hamming distance. Two finite duration transients are compared cell for cell at each time step and the number of sites at which the two patterns disagree is summed. The total number of discordant sites is the Hamming distance between the two patterns.

It was demonstrated that the ability to induce TIGoRS was a function of the number of cells receiving input and the symmetry class of the automaton to which it was applied. The symmetry class reflected the dominant symmetry present in the autonomous patterns produced by the automaton. The classes are uniform, linear, complex and chaotic[29] and were shown to be distinct from Wolfram's now classical classification scheme.

This study proved unsatisfactory for three major reasons. First, the input mode was unduly restrictive and the automata seemed unable to generalize their responses. Second, homogeneity and updating synchrony are implausible characteristics for any biological organism. Moreover there was the strong possibility that the TIGoRS observed was an artifact of the underlying temporal synchronization provided by the updating scheme. Finally, the automata were non-adaptive. This too is biologically implausible.

To address these problems, a further study was undertaken using the cocktail party automaton[34]. This is an adaptive cellular automaton which can be controlled for the degree of adaptive response, inhomogeneity and asynchrony. Each cell is provided with both a state and a rule. Updating can be done either synchronously, or via a fixed asynchronous scheme, or via a stochastic asynchronous scheme. The state of the cell is first updated, then any input is applied to the cell according to the particular input mode. The rule of the cell is then updated. The rule can remain fixed or be updated according to the following adaptive scheme. Whenever the state of a cell is updated, a comparison is made between the response of the cell and that of all other cells possessing the same local neighborhood state configuration. The difference in the number of cells disagreeing and agreeing is calculated and the cell modifies its transition table entry to the opposite value if this difference exceeds a predetermined, individualized, fixed threshold. The cycle is then repeated.

Each input to the host automaton was a complex spatiotemporal pattern derived from the output of a second, input automaton having an identical lattice structure as the host. A fixed correspondence was established between the input and host cells. This provided a mapping between the output pattern of the input automaton and the cells of the host. The input pattern thus consisted of an array of state values, indexed by cell and by time. At time n for the host automaton, the row of the pattern corresponding to time n was sampled cell by cell and applied, according to the input mode, to the corresponding cell of the host automaton. The input automaton was chosen at random using varying combinations: homogeneous/inhomogeneous, linear/complex/chaotic rules, synchronous/asynchronous (random 20%), fixed/adaptive.

Three input modes were studied. In the discrimination mode, a single output from the input automaton was sampled at a 20% rate for a finite duration and this derived pattern provided the input pattern. This provided a single random, but correlated, stimulus. This was then applied to the host automaton as follows: the pattern was sampled in its entirety and the state of the corresponding host cell was changed to match the pattern value only if the pattern value was 1. Thus each presentation of the stimulus remained invariant between trials. This mode explored the ability of the automaton to discriminate between distinct samples.

In the inferential mode, a single complete output from the input automaton was stored as the pattern. This was then applied to the host automaton as follows: the pattern was sampled but this time the cells of the pattern which were to be input to the host automaton were chosen randomly, in this case at a 3% rate. Given a chosen pattern cell, the corresponding cell in the host had its state changed to match the pattern cell only if the pattern value was 1. Each presentation of the stimulus thus varied between trials, constituting a distinct random sampling of the original input pattern. A matching of responses under these conditions reflected the ability of the automaton to respond to the overall structure of the pattern which must be inferred from the random samples presented. In the recognition input mode, a single complete output from the input automaton was stored as the pattern. This was applied to the host as follows: the pattern was sampled as described above. The cells of the pattern which were to be applied to the host were again sampled randomly, this time at a 10% rate. This time, given a chosen pattern cell, the corresponding host cell had its state changed to match that of the pattern cell, regardless of value. Each presentation of the stimulus again varied between trials, constituting a random sampling of the original input. The automaton was studied for its ability to match its response to the pattern. Thus its capacity for pattern completion was studied. This provides one means of pattern recognition, hence the choice of the descriptor: recognition mode.

Figures 1 and 2 display typical runs using a 2-state, 3-neighbor automaton having 75 lattice cells with toroidal boundary conditions. The automaton was simulated under input for 250 time steps. The lattice is displayed horizontally and time vertically. From left to right are the input pattern, the input pattern sites sampled by the automaton during run A, the output pattern generated during run A, the discordance between the output patterns of runs A and B, and the discordance between the input pattern and the output pattern of run A.

In Table 1 are displayed the results of four trials, each consisting of 100 presentations of a stimulus to a single inhomogeneous, asynchronous, adaptive cocktail party automaton using recognition mode. The stimulus varied between the trials. Displayed are the mean (variance) of the Hamming distances between responses and a fixed response template (S), between responses and input pattern (P), between each individual pair of responses (DP) and between each individual pair of randomly generated surrogate patterns having an identical distribution of pattern norms to that of the original sample (RP).

Ham	TIGoRS			
Dist	Strong	Weak	Pseudo	Absent
S	4 (1.4)	31 (3.7)	18 (2.3)	48 (0.9)
P	3 (1.2)	28 (2.0)	21 (4.2)	44 (4.7)
DP	4 (2.1)	31 (5.5)	16 (3.9)	47 (5.1)
RP	39 (6.0)	49 (5.0)	17 (4.4)	49 (4.9)

Table 1:

Clearly stimuli vary in their capacity to induce TIGoRS. The use of surrogate data criteria was inspired by the work of Theiler[35]. The determination of the distribution of interpair Hamming distances and that for a set of suitably constructed surrogates provides a possible means of detecting transient synchronization in vivo. TIGoRS should show itself by a large peak in the low end of the Hamming distance scale. Multiple peaks might arise in the case of multiple synchronizing basins whereas the randomly generated surrogates would show only a single broad peak at the higher end of the scale.

To be truly useful as a solution to the binding problem, it is necessary to demonstrate that stability of the synchronization does indeed exist over time.

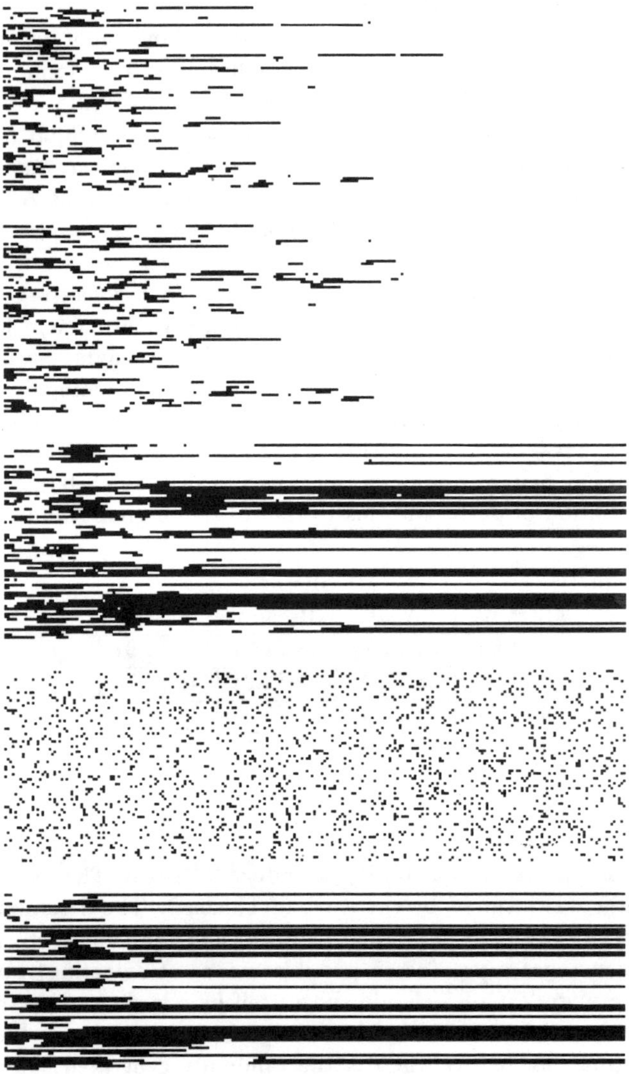

Figure 1

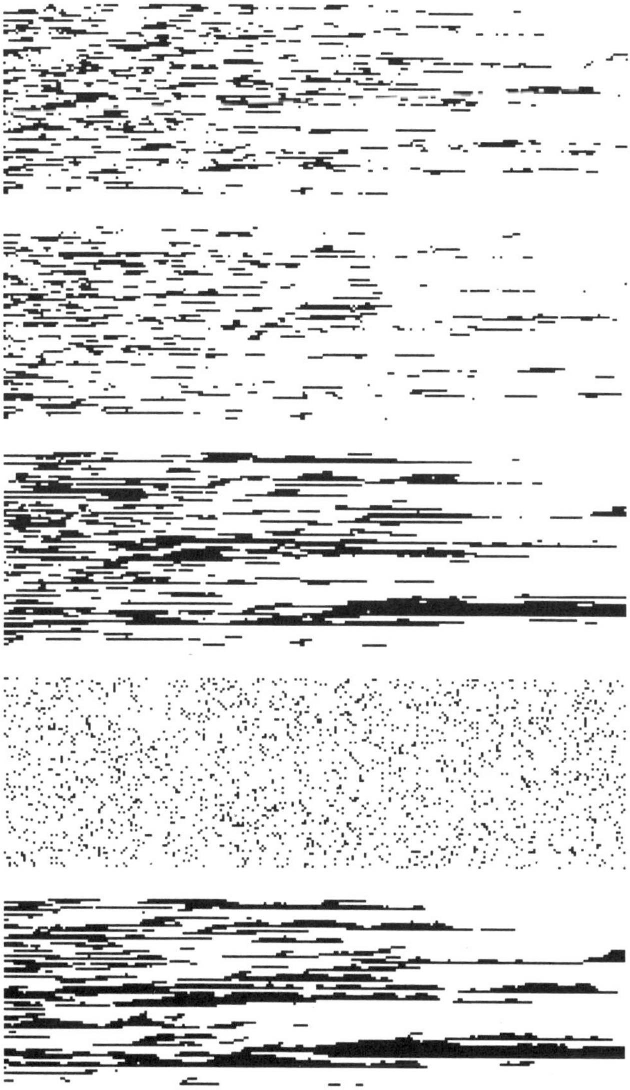

Figure 2

A simulation was thereby carried out in which the automaton was presented with a stimulus followed by a random stimulus of equal duration followed by a second presentation of the stimulus, all via recognition mode. This was repeated for 100 different initial conditions. The mean (variance) Hamming distance between individual states and between states and input pattern respectively were 8 (2.5), 6 (2.5) for the initial stimulus, 48 (1.2), 48 (1.1) for the subsequent random stimulus and 7 (3.2), 5 (3.1) for the repeat presentation of the initial stimulus. The same template was used for all three cases. The strong matching to the template shows that all of the responses to the fixed input stimulus lie within the same neighborhood about the template. Thus transient synchronization occurred on both presentations of the stimulus and the response set was the same in both cases. Thus the response set remains stable over time and bound to the stimulus and therefore can serve as a stable representation of the input stimulus.

It is extremely important to note that the final rule configurations of the automaton following each presentation of the stimulus are different, even if the same initial configuration of rules is used. This is illustrated in Figure 3 which shows the final rule distributions (using the standard Wolfram coding scheme for 2-state, 3-neighbor rules) across the lattice. All four distributions share the same initial rule distribution, input stimulus, degree of asynchrony, adaptation and input mode (recognition). For the last two distributions, the same initial state was used as well. A fixed correspondence between the rule governing a given cell and the stimulus does not exist, in spite of the fact that a high degree of pattern matching is taking place in each case. In other words, the correlation between the stimulus and the system response exists only at the level of the entire system. Any notion of computation can be applied at that level. Indeed the operation of the system at the cell level is stochastic. Invariance of computation is an attribute of the system level, not the cell level. One can have high level computation without low level computation.

The results are quite encouraging. TIGoRS occurred in any of the three input modes. The responses produced varied with the input. An most intriguing finding was that TIGoRS occurred best for inputs from automata which most closely resembled that being stimulated, in this case inhomogeneous and complex, regardless of degree of synchrony and adaptation. This suggests that a network of such automata would show a high degree of TIGoRS in their individual responses, regardless of the specific details of each automaton. This might allow for the emergence of a higher level of pattern structure such as occurs in a swarm or a society.

The degree of pattern completion is remarkable. At times, a mere 2% sample of a pattern is sufficient to enable the automaton to match its response

Figure 3

to the pattern with 92% accuracy. This occurs in the absence of a detailed representation of the pattern in the rules. Indeed the pattern space considered here is vast. Although one cannot give a precise estimate of the number of patterns which can induce TIGoRS, one should note that there are $(2^{100})^{450}$ total patterns (of 450 time step duration). There are some 30^{100} inhomogeneous, complex, synchronous, nonadaptive automata whose patterns the adaptive automata appear capable of recognizing. By way of contrast, a cell assembly version of the system could, at best, encode a number of patterns equal to the total number of subsets of cells which could constitute cell assemblies. Given the present example, this means a maximum of 2^{100} patterns. Adaptation and asynchrony appear to improve the dynamical response range of the automaton without increasing the amount of information which the automaton must carry within it in order to carry out its function. In effect, the information which the automaton needs is stored in its environment.

TIGoRS provides a mechanism through which information is stored, not as an invariant property of the agents which comprise the system, but as a dynamical response, triggered by an appropriate context and determined by a dynamic interaction between that context and the system. There is no engram, nor a fixed encoding. Instead the response is dynamically constructed through the interaction with the environment, being made up on the spot as it were. In effect, the system can be viewed as extracting salient information out of the environment in order to construct its response rather than storing the necessary information within its own structure. Indeed no stable structure exists within these automata. At best, any storage of information can be conceptualized in terms of a probability distribution of responses. This is illustrated in Figure 4 which shows the actual distribution of Hamming distances between states and input pattern for three different asynchrony levels as well as for an autonomous simulation. The response is not a single precise spatiotemporal pattern as would be expected in a temporal coding scheme. Nor is it a coarse rate coding. Instead it is a finely detailed probability distribution within the larger space of response patterns. The support sets for these probability distributions provide a form of dynamic representation.

4.1 Conclusion

It has generally been assumed that the brain of an animal maintains a stable set of behavioral responses to consistent environmental situations by virtue of internally generated and maintained computational processes. These computational processes are themselves, in turn, supported by a low level neural code through which neurons transduce sensory stimuli into a form amenable to

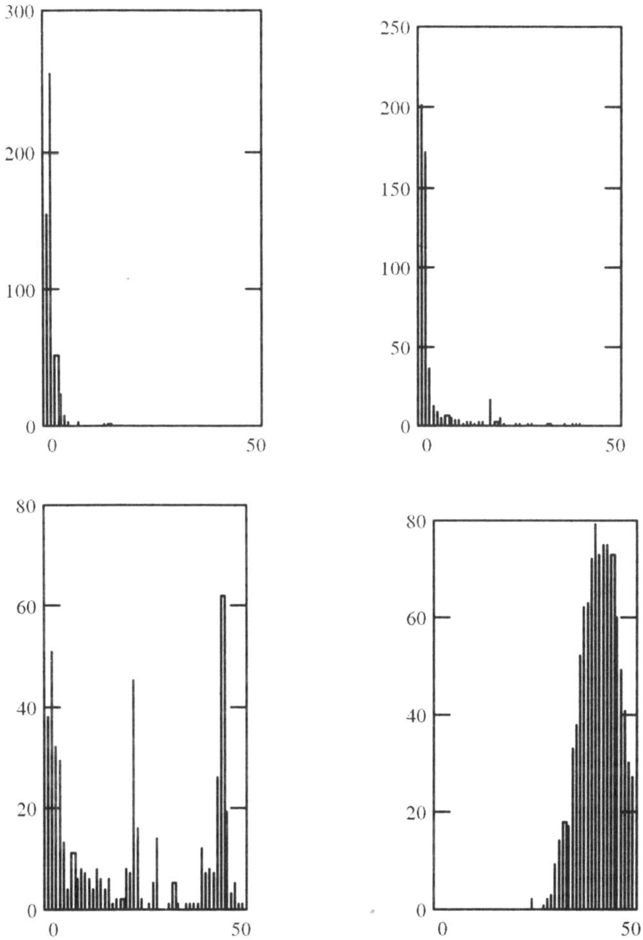

Figure 4

Distribution diagrams of Hamming distances between input and output patterns. The x-axis shows the Hamming distance while the y-axis gives the number of output patterns which occurred at that distance. Clockwise from top left three different rates of synchronization:0%,50%,100% while under a 10% input and an autonomous simulation.

manipulation by these processes. Several different possible encoding schemes have been proposed in the literature. The brief review of the literature presented in this paper suggests that none of the proposed solutions has significant experimental support. Moreover, recent evidence from studies of hippocampal activity in behaving rats suggests that regardless of the neural encoding scheme employed by a neural system, an interpretation of an instance of that scheme may require knowledge of the animal-environment context, both past and present.

This raises concerns about both the existence and utility of a neural encoding scheme. The emphasis upon neural coding derives from an interpretation of the brain as a computational device. However the brain is, first and foremost, a complex adaptive dynamical system. Most salient for the animal is the ability to generate adaptive responses to a changing environment. Computation and coding are just one general mechanism by which this ability might be supported. To search for alternatives, one must first distinguish computational performance from computational competence. This broadens the search to include other dynamical phenomena which may be capable of supporting such adaptive ability.

In this paper one such possibility, TIGoRS, has been presented. A sizable number of complex systems models displays TIGoRS in response to some stimuli. TIGoRS provides a noncomputational mechanism capable of maintaining a high level correspondence between environmental stimuli and system responses without requiring an invariant relationship between the individual components of the system and the particular stimulus. Indeed in the example presented in this paper, that of the cocktail party automaton, the individual units update stochastically, continually modifying their internal rules throughout the presentation of the stimulus. Repeated application of the stimulus results in different rule configurations even though the global response of the system to the stimulus remains stable. Thus it is possible for there to be an invariant response of the system to a stimulus, in this case a probability distribution, even though there is no invariant relationship between the components of the system and the stimulus.

TIGoRS has been put forward as a possible mechanism through which the brain could effect such high level invariance utilizing stochastic components. It is robust to low level noise and to low levels of component failure. As a putative mechanism, it is compatible with the current experimental evidence.

However as defined here it has several shortcomings as a putative mechanism of representation within the nervous system. First of all, the requirement that the binding of response to stimulus be global is not consistent with the fact that most animals do at times display uncertainty and confusion or take inappropriate action even when confronted with a previously familiar experience. Secondly, TIGoRS as defined here applies to the system as a whole whereas within the brain it is presumable that various subsystems might have a quasi independent status relative to one another. It will be necessary to carry out further work with a weaker definition in order to determine whether this criticism holds in general. Thirdly, the dynamics of the cocktail party automaton are wholly unlike that of neurons. Nevertheless TIGoRS has been observed in a simple neural network model, the tempered neural network so there is reason to believe that the phenomenon will occur in biological neural networks even if the particular mechanisms which support it differ from those described here. Fourthly, the response of the system does not show the kind of rate effects observed in systems such as the rat hippocampus. Again it will be necessary to search for this phenomenon in more realistic neural models in order to determine whether TIGoRS can be manifest as a variation around a mean firing rate pattern or only as a Hamming distance variation.

In any case the major goal of this paper has been achieved, namely, to demonstrate that in complex adaptive systems it is possible to demonstrate the existence of invariant relationships between a system and its environment in the absence of such invariance at the level of the individual agents which comprise the system. Computation at the macro-level need not require computation at the micro-level. Moreover micro-level processes may be dependent upon context at the macro-level and may not be fully understandable in the absence of such contextual knowledge. Future theories of neural computation will need to take such high level contextual factors into account. Invariance of function cannot be assume but will require demonstration across a host of differing contexts. As suggested by Cohen and Stewart in *The Collapse of Chaos* , future theories in science will mostly likely need to be *contextual* theories. It is hoped that the present paper will have provided some evidence in favour of such a viewpoint.

References

1. D. Cliff, *Adaptive Behavior* **3 (4)**, 359 (1995).
2. C. Koch and I. Segev in *Methods in neuronal modeling: From synapses to networks*, ed. C. Koch and I. Segev (MIT Press, Cambridge, MA, 1989).
3. C. Langton in *Artificial Life*, ed. C. Langton (Addison-Wesley, Menlo Park, 1989).
4. R. Pfeiffer, *Exp. Brain Res.* **1**, 220 (1966).
5. D.H. Hubel and T.N. Weisel, *J. Physiol. (London)* **195**, 215 (1968).

6. M.N. Shadlen and W.T. Newsome, *Curr. Opin. Neurobiol.* **4**, 569 (1994).
7. W. Softky and C. Koch, *J. Neurosci.* **13**, 334 (1993).
8. Ad. Aertsen and M. Arndt, *Curr. Opin. Neurobiol.* **3**, 586 (1993).
9. R. Eckhorn et. al., *Biol. Cybern.* **60**, 121 (1988).
10. C.M. Gray et. al., *Vis. Neurosci.* **8**, 337 (1992).
11. E. Vaadia et. al., *Nature* **373 (6514)**, 515 (1995).
12. D.E. Cairns et. al., *Neural Comp.* **5(2)**, 260 (1993).
13. G. Gerstein and B. Mandelbrot, *Biophys. J.* **4**, 41 (1964).
14. W. Softky, *Neuroscience* **58**, 13 (1994).
15. Z.F. Mainen and T.J. Sejnowski, *Science* **268**, 1503 (1995).
16. C. F. Stevens and Y. Wang, *Nature* **371 (6499)**, 704 (1994).
17. A. Mason et. al., *J. Neurosci.* **11**, 72 (1991).
18. J. O'Keefe and J. Dostrovsky, *Brain Res.* **34**, 171 (1971).
19. J. O'Keefe and L. Nadel, *The Hippocampus as a Cognitive Map* (Clarendon Press, Oxford, 1978).
20. M.A. Wilson and B.L. McNaughton, *Science* **261**, 1055 (1993).
21. B.L. McNaughton et. al. *J. Exp. Biol.* In press (1996).
22. M.W. Jung and B.L. McNaughton, *Hippocampus* **3**, 165 (1993).
23. G.J. Quirk et. al., *J. Neurosci.* **10**, 2008 (1990).
24. E. Bostock et. al., *Hippocampus* **1**, 193 (1991).
25. T.C. Foster et. al., *Science* **244**, 1580 (1989).
26. U. Neisser, ed., *Remembering Reconsidered* (Cambridge University Press, Cambridge, 1987).
27. W. Sulis in *Proceedings of the World Congress on Neural Networks '93 Vol. IV 449* (Lawrence Erlbaum, New York, 1993).
28. W. Sulis in *Proceedings of the World Congress on Neural Networks '93. Vol. IV 452* (Lawrence Erlbaum, New York, 1993).
29. W. Sulis in *1993 Lectures in Complex Systems,* eds. D. Stein and L. Nadel (Addison-Wesley, New York, 1995).
30. W. Sulis, *Towards a formal theory of collective intelligence*, poster presented at Artificial Life V, Osaka, Japan (1996).
31. W. Sulis, *World Futures.* **39**, 225 (1994).
32. W. Sulis in *Proceedings of the International Joint Conference on Neural Networks '92. Vol. III* (IEEE Press, Baltimore, 1992).
33. W. Sulis, unpublished data.
34. W. Sulis in *Advances in Artifical Life, Lectures Notes in Artificial Intelligence 929* eds. F. Moran et. al. (Springer-Verlag, New York, 1995).
35. J. Theiler, et. al., *Physica D* **58**, 77 (1992).
36. J. Cohen and I. Stewart, *The Collapse of Chaos* (Viking, New York, 1994).

Part I

EMPIRICAL STUDIES

APPROPRIATE ALGORITHMS FOR NONLINEAR TIME SERIES ANALYSIS IN PSYCHOLOGY

CHRISTIAN SCHEIER
AILab, University of Zurich-Irchel
Winterthurerstrasse 190
8057 Zurich
E-mail: scheier@ifi.unizh.ch

WOLFGANG TSCHACHER
University Psychiatric Services Bern: Mid and West Sectors
Laupenstrasse 29
3010 Bern
E-mail: tschacher@spk.unibe.ch

Chaos theory has a strong appeal for psychology because it allows for the investigation of the dynamics and nonlinearity of psychological systems. Consequently, chaos-theoretic concepts and methods have recently gained increasing attention among psychologists and positive claims for chaos have been published in nearly every field of psychology. Less attention, however, has been paid to the appropriateness of chaos-theoretic algorithms for psychological time series. An appropriate algorithm can deal with short, noisy data sets and yields 'objective' results. In the present paper it is argued that most of the classical nonlinear techniques don't satisfy these constraints and thus are not appropriate for psychological data. A methodological approach is introduced that is based on nonlinear forecasting and the method of surrogate data. In artificial data sets and empirical time series we can show that this methodology reliably assesses nonlinearity and chaos in time series even if they are short and contaminated by noise.

1 Introduction

Chaos theory has a strong appeal for psychology for several reasons. First, it provides psychologists with a new vocabulary to conceptualise the complex and often seemingly random behavior of psychological systems. Psychotic episodes, for instance, can now be understood as manifestations of a chaotic system [1]. Second, the mathematical tools of chaos theory allow the formalisation and simulation of psychological theories and hence render the latter more transparent and stringent. Third, nonlinear time series analysis - to a large extent developed by the dynamical systems community - makes it possible to empirically investigate the nonlinearities observed in psychological phenomena. It is therefore not surprising that chaos-theoretic concepts and methods have recently received growing interest among psychologists [2]. In particular, nonlinear modeling techniques and algorithms have been widely used and claims of positive evidence for chaos in psychological data have been published in nearly every field of psychology[1,3,4,5].

However, less attention has been paid to the question of the appropriateness of these algorithms for psychological time series. One rather obvious reason to raise this question is that most of the classical nonlinear techniques have been developed in the physical sciences and hence make strong assumptions concerning the quality and length of the data [6]. More specifically, they often require thousands of data points with high resolution and a small amount of noise. Psychological time series, however, are typically contaminated by a high percentage of noise, are not very fine grained and rather short in length. An other problem with these algorithms is that it is often not possible to determine objectively the significance of the results which obviously renders the interpretation of the latter difficult if not impossible. If we look more closely to the most popular nonlinear method - dimension analysis [7] - even more problems arise. Dimension algorithms seem, for example, to indicate a finite dimension, even when presented with random noise [8]. Another bias is known as the 'edge effect' [9] due to finite data samples in high dimensions. This effect can make even white noise appear to have a finite dimension. Moreover, dimension estimates are sensitive to the amount of noise in the system[10] and require a large amount of data in order to yield reliable results [11]. Hence, it can be concluded that dimension analysis and other classical algorithms of dynamical systems theory such as Wolf's algorithm for calculating the Lyapunov-Exponent [12] are not appropriate for psychological data. Similar conclusions have been drawn in fields such as economics [13] or biology [14,15] that have to deal with related problems, namely the nonlinear analysis of small, noisy data sets.

Consequently, a lot of research has recently been devoted to the development of alternative approaches to nonlinear time series analysis which avoid the mentioned pitfalls of the traditional methods [16]. In this paper, we try to abstract from this very large literature those algorithms that we consider to be most pertinent for the nonlinear analysis of psychological data in that they can deal with short and noisy time series. Moreover, most of these algorithms have been developed outside the physical sciences and have been shown to yield reliable results when applied to short and noisy data. Even more importantly, our emphasis is on using several of these appropriate algorithms since when only one method is used to analyse a time series, the results are expected to be at best incomplete, and at worst wrong and misleading [17]. None of these algorithms have - to our knowledge - been applied to psychological data. Our overall goal is to develop a methodology for the nonlinear analysis of psychological time series which is appropriate in the sense that it satisfies the constraints imposed by psychological systems in general (e.g. very little is known about the dynamical system that generated the data) and psychological data in particular (e.g. short and noisy time series). In essence, two important new methodologies are introduced and applied to artificial and empirical time series: nonlinear forecasting and the method of surrogate data. The nonlinear forecasting of time series data has a relatively recent history [18,19]. However, it constitutes one of the most promising fields of research in nonlinear time series analysis and has recently gained a lot of attention [16]. The even more recently proposed method of surrogate data [20,21] serves the purpose of seeing whether certain classes of models for the data can be rejected. This involves generating many surrogate data sets satisfying the null hypothesis, and computing a statistic (such as the forecast error) for each. It has been shown that this technique is very successful in distinguishing chaos from various random processes even in short and noisy data sets [14]. Moreover, since it is inherently a statistical approach the surrogate data method allows the quantification of the results in a more rigorous manner than was possible with the traditional techniques.

The organisation of this paper is as follows. In Section 2 the nonlinear forecasting methodology and the surrogate data method are introduced. More specifically, a nonparametric simplex algorithm [15], a parametric forecasting algorithm [17], and two surrogate data methods [20,21] are presented. In Section 3 these methods are applied to artificial test data and empirical psychological time series. In Section 4 the results are discussed, the main problems are pointed out and possible further improvements of the proposed methodology are outlined.

2 Methods

2.1 Nonlinear forecasting

One essential property of chaos is its determinism, i.e. chaotic systems obey certain rules. Trajectories of chaotic systems can be predicted for short time scales. However, chaos amplifies noise exponentially and as a result this short-term determinism becomes long-term randomness. The idea behind the forecasting algorithms presented in this section is to use this characteristic to detect chaos in time series. In order to do this, a number of steps have to be performed. The first task is to reconstruct the system phase space from the time series, typically measurements of a single scalar observable at a fixed spatial point. This is of course standard practice in nearly every nonlinear algorithm and we therefore only present the main ideas [22]. From the orginal time series $x(t)$ an embedding space or state space representation of dimension

E is constructed whose points are

$$X(t) = (x(t), x(t-\tau), ..., x(t-(E-1)\tau) \quad (1)$$

$X(t)$ is called a *delay vector* with *embedding dimension* E and *delay time* τ. Takens [23] proved that if $E > 2d$, where d denotes the number of degrees of freedom of the dynamical system, then there is a smooth map $f^T : \Re^E \to \Re$ such that

$$X(t+T) = f^T(X(t)) \quad (2)$$

where the current state is $X(t)$ and $X(t+T)$ is a future state. The problem in nonlinear forecasting is to estimate $X(t+T)$. Since chaotic dynamics does not occur unless f is nonlinear, one has to build nonlinear models to approximate chaotic dynamics. Thus, the task is to find a good representation or approximation for the unknown function f. To avoid additional complexity, we now assume that our time series is embedded in an appropriate state space, and we have determined the embedding parameters E and τ. The problem now is to reconstruct the deterministic rule underlying the data. In the following, we discuss two nonlinear methods of local forecasting: first order and second order approximation.

The simplest (and earliest) forecasting procedure was suggested by Lorenz [24]. Suppose we want to predict the value of $X(k+1)$ knowing a long time series $X(j)$ for $j \leq k$. In first-order approximation [19] - sometimes also called 'method of analogs' - one finds the nearest neighbour to the current value of $X(k)$, say, $X(v)$, and then assumes that $X(v+1)$ is the predicted value for $X(k+1)$. Obviously, this is not really a representation or much of a model and the quality of this prediction can be improved in several ways. One possibility is to take a collection of near neighbours of the point $X(k)$ and take the averaged value of their images as the prediction.

Sugihara and May [15] have introduced a nonparametric simplex method (hereafter abbreviated as SM) that does exactly this. First, the time series is splitted in two parts. The first part is used as a library to predict the points on the second part. Hence, short-term predictions are made that are based on a library of past patterns in a time series as in the example above. Specifically, for each -dimensional point for which one wishes to make a prediction - for each 'predictee' -, all nearby points in the state space are selected. Of these points the simplex with the minimum diameter formed from $E+1$ nearest neighbours is constructed. To estimate the prediction, the evolution of the simplex after T time steps is calculated, weighting the actual distances by original distances from the respective, relevant neighbours. Finally, the correlation between the predicted and the raw data is computed. A decrease in this correlation with increasing prediction time steps suggests chaos whereas a non-decreasing prediction accuracy points to a random or some linear process (see Section 3).

The SM has been applied to biological[25,15], economical[26] and psychopathological [27] time series. In Section 3 it will be shown that the SM is a useful method to distinguish various random and linear processes from chaotic time series even in short and noisy data sets. Let us now turn to a further improvement of the first-order approximation scheme. The main idea is to fit an affine model to approximate the unknown function f of equation (2). This is called *second-order approximation.* The implementation of the 'method of analogs' in second-order approximation can be done in different ways. Polynomials are a good representation because their parameters can be linearly fit to minimise least square deviations (see below). Radial basis functions [17] or neural networks [26] provide other alternatives. One very interesting parametric forecasting method was recently proposed by Casdagli [17]. In this approach, nonlinear models are constructed with a variable smoothing parameter which at one extreme defines a nonlinear deterministic model, and at the other extreme defines a linear stochastic model. Specifically, piecewise-linear approximations to the function f of equation (2) are constructed by using a variable number of k neighbours. In essence, a small value of k corresponds to a deterministic approach to modelling as used in the SM (see above) whereas the largest value of k is equivalent to fitting linear stochastic autoregressive models. Finally, intermediate values of k correspond to fitting nonlinear stochastic models as, for example, proposed by Tong [28]. Since with this approach one compares deterministic with stochastic models, Casdagli calls his method deterministic versus stochastic modelling (DVS).

The DVS algorithm is implemented as follows [17]. First, as in the SM, the time series is divided in two parts: a *fitting set* containing the first N_f data points of the time series, and a *testing set* containing $N_t = N - N_f$ data points, where N is the length of the time series.Again, one chooses an embedding dimension E, a delay time τ, a forecasting step T and a predictee $X(i)$ with $i \geq N_f$ for a T-step-ahead forecasting test. Next, the distances of the predictee from the delay vectors $X(j)$,$1 + (E-1)\tau \leq j \leq N_f - T$, in the fitting set are computed and the k, $2(E+1) \leq k \leq N_f - T - (E-1)\tau$, nearest neighbours of the predictee are found to fit an affine model of the following form:

$$x_{j(l)+T} \approx \alpha_0 + \sum_{n=1}^{E} \alpha_n x_{j(l)-(n-1)\tau}, l = 1, ..., k \tag{3}$$

where the parameters $\alpha_0, ..., \alpha_E$ are computed by ordinary least squares.

Finally, the fitted model (3) is used to estimate a T-step-ahead forecast for the predictee. This procedure is repeated for all data points in the test set. The prediction accuracy can then be computed by using correlation between the predicted and the actual values of the time series. If the correlation increases with the number of E-dimensional nearest neighbours, the model is a linear stochastic process. In contrast if a small number of nearest neighbours shows high predictive power, this points to a nonlinear deterministic process. The DVS algorithm has been successfully applied to short and noisy biological time series [17,29]. In Section 3 the DVS it will be applied to artificial test data and psychological time series.

2.2 Surrogate data

Although the SM and the DVS are superior to traditional approaches in that they can detect chaos in short and noisy data sets they have one major drawback: their results cannot be obtained statistically, i.e. it is still necessary to rely on 'face validity' to be able to conclude if a time series is chaotic or not. In order to overcome this problem, several approaches have been developed, the most promising one being the method of surrogate data (MSD), proposed by Theiler *et al.* [21]. A related algorithm called 'noise versus chaos' (hereafter abbreviated as NVC) was introduced by Kennel and Isabelle [20], the major difference being in the NVC's use of predictor errors from many points in phase space instead of the average prediction error as a distinguishing statistic (see below).

Both algorithms have been successfully applied to short, noisy data from biology [29,21], economics [26] and psychopathology [27]. In the rest of this section, the two methods will be presented. The procedure consists of three parts: First, some linear process is specified as a null hypothesis. Second, suitable random data - the surrogate data - are generated, normalised to the mean and the variance of the original data. Third, a discriminating statistic for the original and for each of the surrogate data sets is computed. The basic idea is that, if the value computed for the original time series is significantly different than the ensemble of values computed for the surrogate data, then the null hypothesis is rejected and nonlinearity is detected. In the following, these three issues will be addressed in more detail.

The *null hypothesis* typically specifies that certain characteristics of the original time series are preserved (e.g. mean and variance) but that there is no further structure in the data. The surrogate data set is then generated to mimic these preserved trivial properties of the raw data but otherwise be random. In this paper, the following null hypotheses were used:

1. *temporally uncorrelated noise*($H_0^{(TUN)}$). The simplest null hypothesis

one can devise about a time series is that there is no evidence for any dynamics at all; that is, that the data are fully described by independent and identically distributed (IID)random variables. To generate the surrogate data, two algorithms can be used. First, if the distribution is assumed to be gaussian, then the surrogate data can be generated from a standard pseudorandom generator. Second, to test the hypothesis of uncorrelated noise with arbitrary amplitude distribution, surrogate data can be generated by shuffling the time-order of the raw data. We will refer to these procedures as *Sur(rand)* and *Sur(shuffle)*, respectively.

2. *linearly autocorrelated gaussian noise*($H_0^{(LAGN)}$).The null hypothesis in this case is that all the structure in the time series is given by the Fourier power spectrum or, equivalently, by the autocorrelation function (ACF). The simplest case is that given by the Ornstein-Uhlenbeck process, which can be produced by

$$x(t) = \phi_0 + \phi_1 x(t-1) + \theta\epsilon(t) \tag{4}$$

were $\epsilon(t)$ is uncorrelated gaussian noise of unit variance. The coefficients ϕ_0, ϕ_1 and θ determine mean, variance, and autocorrelation time of the time series, respectively. The more general case can be implemented by fitting an AR(K)MA(L)-model of the form

$$x(t) = \sum_{k=1}^{K} \phi_k x(t-k) + \sum_{l=1}^{L} \theta_l \epsilon(t-l) \tag{5}$$

where the coefficients $\{\phi_k\}$ and $\{\theta_l\}$ are to be determined by fits to the data, typically using a least squares or an information theoretic criterion, and the $\epsilon(t)$ are some stochastic forcing terms which are specified by the modeler. Surrogate data can then be generated by iterating (4) or (5), where the coefficients have been fit to the original data. We will refer to these procedures as *Sur(Uhl)* and *Sur(ARMA)*, respectively.

An alternative algorithm, based on the phase randomisation of the fast fourier transform, is described in Theiler et. al. [21]. In this case, the surrogate data are generated in two steps:

1. taking the discrete fourier transform

$$\Xi(k) = \sum_{t=0}^{N-1} x(t) e^{2\pi i n k/N} \tag{6}$$

and

2. generating surrogate data sets $s = 1, ..., S$

$$\Xi^s(k) = \Xi(k)[\zeta(k) + i\eta(k)] \tag{7}$$

where ζ and η are independent, real Gaussian random numbers with mean 0 and variance 1/2 and $\eta(k) = -\eta(N-k)$ ensuring $x^s(k)$ to be real. We will refer to this algorithm as $Sur(Fourier)$. There are, of course, many other null hypotheses that can be tested for. For instance, one could test for noisy periodicity as proposed by Kennel and Isabele [20] or the null hypothesis of a static nonlinear filter of linearly correlated noise as used by Theiler *et al.* [21]. In the following, however, we restrict ourselves to the two null hypotheses described above.

Discriminating statistic and technique of comparison. Although, in principle, any discriminating statistic can be used we chose to use a phase-space predictor to obtain the results presented in Section 3. More specifically, the correlation between predicted and raw data was used. Using a predictor has the advantage that it directly investigates one of the essential differences between chaos and randomness, namely determinism. Moreover, it is algorithmic simple and free of any parameters. Theiler et al. also consider correlation dimension as a statistic, but - as mentioned in the introductory section - dimension analysis has several drawbacks in respect to the analysis of short and noisy data. Similarly, as pointed out by Daemmig [30], numerical estimation of even the largest Lyapunov exponent can be problematic in the presence of noise.

Once one has computed the value of the discriminating statistic for the original and the surrogate data, the significance of the difference in the MSD is defined as follows. In our case the statistic of interest is the correlation coefficient ρ. In this case the resulting mean correlation ρ_H is compared to ρ_{raw} and the result is normalised by the standard deviation of the surrogates. Thus the significance with which one can reject the null hypothesis is

$$S \equiv \frac{\rho_H - \rho_{raw}}{\sigma_H} \tag{8}$$

A significant rejection of the null hypothesis occurs when $S \geq 2$ [21]. In the NVC, the set of prediction errors - which are decimated by the factor of the prediction error autocorrelation - on the real data set (A), and the ensemble of prediction errors on surrogate data sets (B) are compared using the Mann-Whitney rank-sum statistic

$$U = \sum_{i=1}^{card(a)} \sum_{j=1}^{card(B)} \Theta(A_i - B_j) \tag{9}$$

Type of process	SM	DVS	MSD	NVC
	$\langle\Omega\rangle$	$\langle\rho_{nl}\rangle$ $\langle\rho_l\rangle$	$\Psi(H_0^{TUN})$ $\Psi(H_0^{LAGN})$	$\langle z\rangle$
non-linear deterministic	>> 0	>>	> 2 > 2	< −3.72
linear stochastic e.g.				
AR(M)	≈ 0	<<	> 2 < 2	> −3.72
MA(L)	>> 0	<<	> 2 < 2	> −3.72
white noise	≈ 0	≈ 0	< 2	>> −3.72

Figure 1: A framework for the nonlinear analysis of psychological time series. $\langle\Omega\rangle$: mean forecasting accuracy one day ahead *minus* mean forecasting five days ahead; $\langle\rho_{nl}\rangle$, $\langle\rho_l\rangle$: mean forecasting accuracy in the nonlinear deterministic end (nl) and the linear stochastic end (l); $\Psi(H_0^{TUN})$, $\Psi(H_0^{LAGN})$: mean MSD significance S for the two null-hypotheses outlined in the text. $\langle z\rangle$: mean z-value. All means are computed using $E = 1...10$; for SM, DVS, and MSD $\tau = 1$, for NVC $\tau = 1...10$

where $\Theta(\cdot)$ is the Heaviside step function: $\Theta(x) = 1$ for $x > 0$ and $\Theta(x) = 0$ for $x \leq 0$. For large sets a standard normal distributed Z-statistic is obtained

$$z = \frac{U - card(A)card(B)/2}{\sqrt{\frac{1}{12}card(A)card(B)(card(A) + card(B) + 1)}} \tag{10}$$

differentiating whether the distribution of the prediction errors of A or B are significantly smaller or larger. With K repetitions for different time delays τ the limits on z

$$\int_{-\infty}^{z} \frac{1}{\sqrt{2\pi e^{-z^2}dz}} = \alpha/K \tag{11}$$

are more trenchant and in our case with $K = 10$ at -3.72 for a one-variable confidence limit$\alpha = 0.01$.We have now reached a point where we can integrate these algorithms into a coherent framework, depicted in Figure 1. In essence, it allows us to distinguish empirically between nonlinear (possibly chaotic) deterministic systems, linear systems such as an ARMA-process and white noise.

Obviously, this is a rather general differentiation. However, it is easily possible to add new features in this scheme. One can, for instance, integrate more elaborated hypotheses such as a static nonlinear filter of linearly correlated noise into the MSD or the NVC. An other alternative would be to use a nonlinear stochastic predictor[28] as a discriminating statistic. In any case we believe that the reliable distinction between the three classes of processes listed in Figure 1 constitutes an important step towards our overall goal, namely the development of a coherent and appropriate methodology for the nonlinear analysis of psychological systems. In Section 4 we will show how this framework can be combined with other time series analysis tools. First, however, let us turn to its application to artificial and empirical time series.

3 Results

3.1 Artificial time series

To test the algorithms, two experiments were conducted. The first experiment served to answer the question whether they can differentiate between white noise, linear and nonlinear (in our case chaotic) processes consisting of very few data points. In the second experiment robustness with respect to noise level was tested. Remember that we analyse the following methods: the simplex method (SM), the 'deterministic versus stochastic'-algorithm (DVS), the method of surrogate data (MSD), and the 'noise versus chaos'-technique (NVC).

Experiment 1: Differentiation. Three categories of processes were generated: (1) random processes, (2) linear processes, and (3) chaotic processes. The main interest was to check whether each of the algorithms was able to differentiate between them. Note that with this length of the data none of the classical algorithms could be used. Figure 2a shows the results obtained by applying the simplex algorithm (SM) to these data. It can been seen that, indeed, the forecasting accuracy - quantified by using the correlation coefficient ρ - decreases with increasing forecasting time step T only for the chaotic time series. This signature of ρ decreasing with T does not arise when the erratic time series is in fact a noisy limit cycle or a first order autoregressive process. Not unexpectedly, no significant forecasting accuracy can be obtained for the IID process. Hence, the SM can discriminate chaotic from linear and random processes. Similar conclusions can be drawn for the DVS algorithm (Fig. 2b). It correctly classifies the chaotic time series in the deterministic end (high correlation with small k values) and the other processes in the stochastic end (high correlation with high k values). However, these conclusions rely on

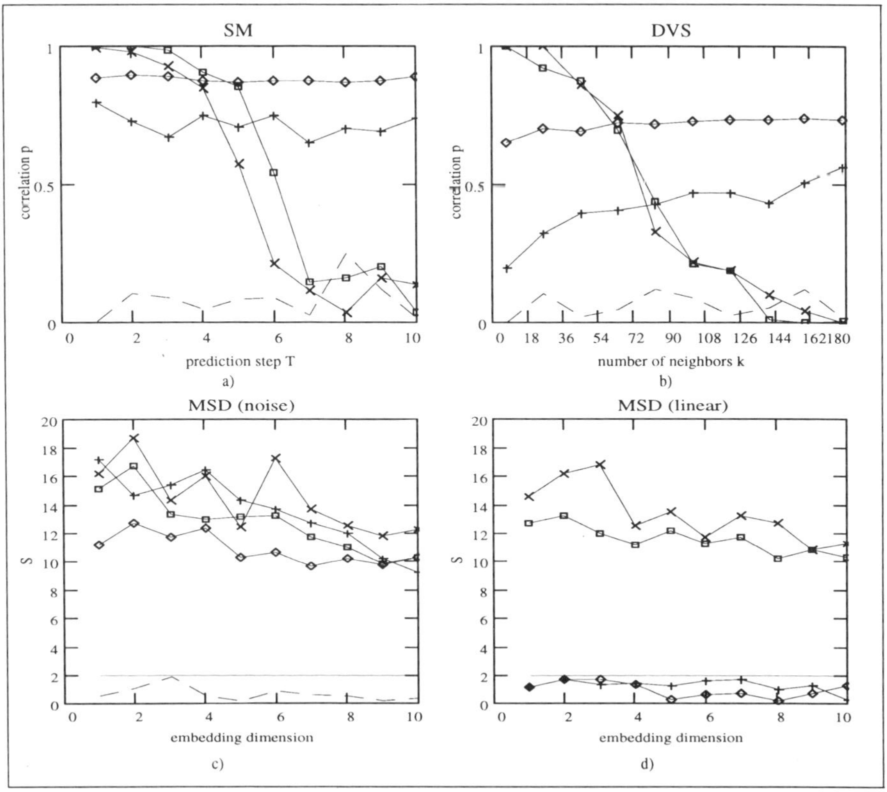

Figure 2: Differentiating chaotic, linear and random processes with the algorithms presented in the text. a) SM, b) DVS, c) and d) MSD. The following processes were used: *crosses*: Henon-Map in the chaotic regime ($x(t+1) = y(t) + 1 - 1.4x^2(t); y(t) = 0.3x(t)$); *squares*: Logistic-map in the chaotic regime ($x(t+1) = 4x(t)(1 - x(t))$); *diamonds*: Sine wave with 50% noise ($x(t) = sin(0.5t) + \epsilon(t), \epsilon \in [-0.5, 0.5]$); *plusses*: first order autoregressive process (AR(1), $x(t+1) = 0.4x(t) + \epsilon(t), \epsilon \in [-1, 1]$); *dashed*: random process generated with a standard pseudorandom number generator. For all calculations $E = 3, \tau = 1$

Type of process	SM	DVS		MSD		NVC
	$\langle \Omega \rangle$	$\langle p_{nl} \rangle$	$\langle p_l \rangle$	$\Psi(H_0^{TUN})$	$\Psi(H_0^{LAGN})$	$\langle z \rangle$
Hénon	0.87	0.21	0.99	17.23	15.89	-16.47
Logistic map	0.85	0.17	0.99	14.77	14.12	-13.38
AR(1)	0.08	0.78	0.34	15.61	1.21	- 0.01
Sinus	0.02	0.81	0.45	12.34	0.98	0.21
Random	0.04	0.12	0.09	0.89	---	2.34

Figure 3: Differentiating chaotic, linear and random processes with the algorithms presented in the text.

the inspection of the plots in Figs. 2a and 2b and hence on 'face validity'.

In order to get quantitative results, the MSD and the NVC algorithm were applied to the test data. The results for the MSD are depicted in Figs. 2c and 2d. First, the null hypothesis of temporally uncorrelated noise was tested and surrogate data were generated with the shuffling algorithm *Sur(shuffle)* described in section 2 (Fig. 2c). The null hypothesis can be rejected for all time series except the IID process. Second, surrogate data were generated using algorithm *Sur(Fourier)*, hence testing for the null hypothesis of linearly autocorrelated gaussian noise (Fig. 2d). In this case, the null hypotheses can be rejected only for the chaotic time series. Thus, the MSD correctly identifies the linear processes and discriminates them from the nonlinear (chaotic) ones. In Figure 3 these results are integrated into the framework presented in the last section. It is easy to verify that we can correctly and reliably distinguish the time series in question. This is an important result since we use very short data sets.

Experiment 2: Discretization In the second experiment, it was tested whether nonlinear determinism can be detected even when the continuous dynamical system which generated the time series is discretized. Discretization of continuous dynamics occurs frequently in psychology where rating-scales are often the only means to obtain relevant data. It is thus important to know if the algorithms presented here are appropriate for discretized dynamical systems. In a first step, the Henon-Map time series used in the first experiment was mapped onto the intervals $[1...I], I \in [5...9]$. Next, the algorithms were applied to these discretized time series. In figure 4 the results of this experi-

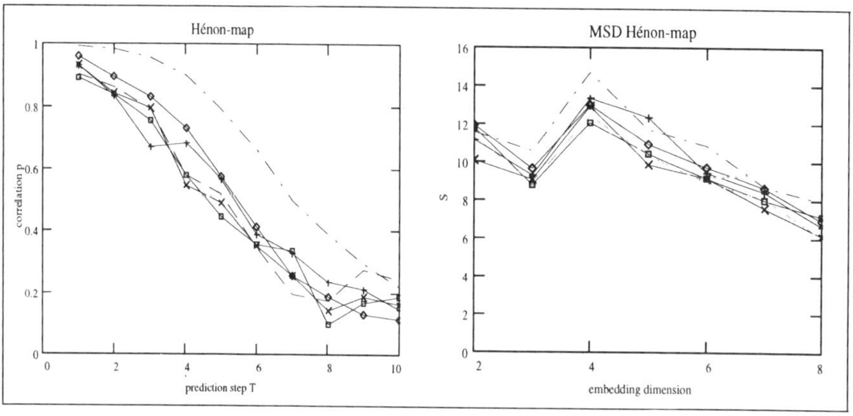

Figure 4: nonlinear forecasting (left) and surrogate statistic (right) for orignial and discretized Henon-Map data. *Dash-dotted*: origninal data; *crosses*: 5-point scale; *boxes*: 7-point scale; *diamonds*: 8-point scale; *dashed*: 9-point scale. $E = 3, N = 400, \tau = 1$. Explanations see text.

ment are shown for the SM and the MSD. Two conclusions can be drawn from this experiment: First, the results obtained for the discretized data are very similar to the ones obtained for the continuous data ($\rho \approx 0.97$). Second, less significant results are obtained for the discretized time series. This is an important outcome since we get more *conservative* results for the discretized data. Hence, results obtained by applying these algorithms to rating-scales data (like the ones presented below) are expected to yield reliable results which tend to be *less significant* compared to results from the underlying continuous system.

3.2 Empirical time series

We present results on clinical data. These results are meant to be illustrative. Our goal is to show that the framework developed in the last sections can indeed yield reliable and meaningful results when applied to psychological time series. We studied 14 patients treated at the 'Soteria Bern'. The Soteria as a small residential clinic specialised for persons experiencing first psychotic manifestations is based on ideas of milieu therapy and affect logic [1,31]. The prerequisite for inclusion in our sample was that the daily manifestations of psychotic symptomatology of a patient could be observed almost completely for a long enough period of time (at least 200 days). The longitudinal course was mapped by the daily rating of a patient's psychoticity by Soteria staff members. A seven-point scale was used as described in Aebi et. al. [31]. At

Patient (sex)	Model	SM	DVS		MSD		NVC
		$\langle\Omega\rangle$	$\langle p_{nl}\rangle$	$\langle p_l\rangle$	$\Psi(H_0^{TUN})$	$\Psi(H_0^{LAGN})$	$\langle z\rangle$
2 (m)	nonlin	0.578	0.53	0.87	9.26	6.66**	-8.32 **
1 (m)	nonlin	0.757	0.32	0.64	4.59	3.42**	-5.12 **
5 (f)	nonlin	0.358	0.43	0.71	3.91	11.34**	-7.12 **
3 (f)	nonlin	0.698	0.55	0.81	15.27	2.18*	-12.55 **
13 (f)	nonlin?	0.671	0.34	0.48	5.13	2.33*	-3.45
9 (m)	nonlin?	0.479	0.44	0.49	11.64	2.28*	-1.88
11 (f)	nonlin?	0.472	0.21	0.20	4.72	2.18*	-3.16
14 (m)	linear	0.790	0.65	0.21	12.22	0.98	-0.23
4 (f)	linear	0.696	0.54	0.11	17.13	1.23	0.47
10 (m)	linear	0.852	0.48	0.01	11.97	0.87	1.09
6 (f)	linear	0.920	0.74	0.32	10.84	1.90	-2.90
8 (f)	linear	0.661	0.44	0.18	15.27	1.72	0.66
12 (m)	noise	0.477	0.12	0.19	1.66	---	2.33
7 (f)	noise	0.174	0.21	0.01	0.80	---	0.77

Figure 5: Results for empirical time series. Explanations see text.

the focus of our interest is the course of psychotic derealisation measured with this scale. We thus restrict ourselves to statements about a specific systems level (namely the level of psychopathological time courses of 200 to 800 days); under these constraints we think it is possible to differentiate types of psychotic dynamics using the nonlinear time series analysis framework presented above.

The identification of different psychotic dynamics will eventually have direct clinical implications. For instance, stochastic systems whose time series do not show serial structure point to a high sensitivity for fluctuating environmental stimuli. These systems are suggested by behavioral theories which take behavior largely under the control of external stimuli. In our study they pose a fundamental null hypothesis since environmental influences on psychoticity are not controlled for in our field data. Nonlinear deterministic (possibly chaotic) dynamics on the other hand point to the existence of an internally controlled, low-dimensional system unfolding relatively autonomously from environmental fluctuations. Moreover, empirical evidence of such dynamics would be a validation of the dynamical disease concept for psychoses[32]. In the context of psychotherapy nonlinear deterministic systems seem compatible with psychoanalytical or system theories.

In figure 5 the results for these data are depicted. The figure shows that in 4 (28%) patients out of 14 all 4 algorithms suggest nonlinear dynamics since we find a rapid decay of the forecasting accuracy $\langle\Omega\rangle$, a higher predictive power for nonlinear models in the DVS, and a significant rejection of the null hypothesis of a linear stochastic process in both the MSD and the NVC. In another 3 patients we can tentatively conclude that their psychotic paths were generated by a nonlinear deterministic system. There is a significant rejection of the 'linear' null hypothesis in the MSD and a steep decay in the forecasting accuracy in the SM. Hence, we find in 7 (50%) out of 14 patients strong evidence for nonlinear determinism. Five time series are best modelled as an autoregressive linear process. Two cases are classified as random. The results of significance tests are summarised under the heading 'model' in Figure 5.

These results suggest that the framework presented in this paper is indeed useful to differentiate several important process models and to reliably assess nonlinear determinism in psychological time series. For further interpretations of these results, see Tschacher et. al. [27].

4 Discussion and conclusions

Our main goal in this paper was to introduce appropriate algorithms for nonlinear time series analysis in psychology. An appropriate algorithm can deal with short, noisy data sets and yields 'objective' results. We presented two new approaches - nonlinear forecasting and the method of surrogate data - and integrated them into a framework that allows for the reliable assessment of nonlinearity and determinism in psychological data. Application of this framework to artificial and empirical data showed that it can differentiate between three important classes of dynamical systems: (1) random walks, (2) linear stochastic systems, and (3) nonlinear deterministic systems. Moreover, the presented framework satisfies constraints imposed by psychological data since it is designed for the analysis of short and noisy time series. The approach outlined in this paper can be improved in several ways. First, it should be complemented with stationarity tests and linear methods. In a first step, one tests whether the time series in question is stationary and otherwise differentiates it. Then one tests for temporal structure in the data. If no structure is detected, the analysis is stopped since the data are white noise. Otherwise, the structure is examined in more detail. It can be linear or nonlinear. If linearity is detected, linear time series methodology is used to further analyse the data. Similarly, if nonlinearity is detected, the nonlinear algorithms presented in this paper can be used to further specify the detected nonlinearity. One can, for instance, test for chaos. It is here where the classical techniques can be helpful.

Another possible improvement of the current framework concerns the amount of variables that can be analysed. Most (if not all) nonlinear algorithms presently allow only univariate analysis of the data. In psychology, however, one often has to use several variables to empirically investigate a given phenomena. In psychotherapy research, for instance, it is obviously important to analyze the interaction between therapeut and client. It is thus necessary to develop multivariate algorithms. This is described in Scheier and Tschacher[33]

All of the presented methods use the deterministic model (2) to analyze the data. The assumption of a purely deterministic process may, however, not be adequate for psychological time series. It is not very likely that all fluctuations in a time series can be explained by chaotic dynamics. Thus, it would be useful to improve the model (2) with respect to dynamical noise. A very interesting first step in this direction has recently been proposed by Nychka et. al. [10].

Finally, it will be important to gain more knowledge about psychological dynamics. Psychological research has to a large extend been tied to cross-sectional studies and thus has little to say about psychological processes[34]. In the future a more elaborated set of process models will have to developed (e.g. different kinds on nonlinearities dependent on the psychological system). We hope to have stimulated such an endeavor with the present paper.

Acknowledgments

We are grateful for the laborious documentations done by Soteria team members. Data acquisition was made possible by the pilot project Soteria (grant of the Swiss National Fund) headed by Luc Ciompi.

References

1. L. Ciompi, System Familie **5**, 133–145 (1992).
2. W. Tschacher*et al.*, *Self-organization in clinical psychology* (Springer, Berlin, 1992).
3. M. Koukkou *et al.*, Biological Psychiatry **33**, 397–404 (1993).
4. R. M. Warner, Behavioral Science **37**, 128-138 (1992).
5. A. Gottschalk, Arch. Gen. Psychiatry **52**, 947–959 (1995).
6. D. Ruelle,*Chance and chaos* (Princeton University Press, New Yersey, 1991)
7. J. Theiler, J. O. Soc. Am. **4**, 1055-1073 (1990)
8. A. R. Osborne *et al.*, Physica D **35**, 357-381 (1989)
9. I. Dvorak and J. Klaschka, Physics Letters A **145**, 225–231 (1990)
10. D. Nychka *et al.*, Journal of the Royal Statistical Society B **54**, 399-426 (1992)
11. D. Ruelle, Proc. Royal Soc.London **427A**, 241–248 (1990)

12. A. Wolf *et al.*, Physica D **16**, 285-299 (1985)
13. C. G. Gilmore, Int. J. of Bifurcation and chaos **3**, 583–589 (1993)
14. F. Mitschke and M. Daemmig, Int. J. of Bifurcation and Chaos **3**, 693–703 (1993)
15. G. Sugihara and R. M. May, Nature **344**, 734-741 (1990)
16. A. Weigend and T. Gershenfield *Predicting the future and understanding the past* (Addison-Wesley,Redwood City,1994)
17. M. Casdagli, Journal of the Royal Statistical Society **54**, 303–328 (1992)
18. J. P. Crutchfield and B.S. McNamara, Complex Systems **1**, 417–452 (1987)
19. J. D. Farmer and J. J. Sidorowich, Physical Review Letters **59**, 845–848 (1987)
20. M. B. Kennel and S. Isabelle, Physical Review A **46**, 234-258 (1993)
21. J. Theiler*et al.* , in *Nonlinear modeling and forecasting*, ed. M. Casdagli and S. Eubank (Addison-Wesley, Redwood City, 1992).
22. , E. Tsonis *Chaos. From theory to applications*, (Plenum Press, New York, 1992).
23. F. Takens in *Dynamical Systems and Turbulence*, ed. D.A. Rand and L.S. Young (Springer, Berlin, 1981).
24. E. N. Lorenz, J. Atmos. Sci. **26**, 636 (1969)
25. M. Dumon *et al.*, in *Proceedings of the first European Conference on Artificial Life ECAL95*314–3241993
26. D. Meier *et al.*, in *Proceedings Proceedings of 1st Int. Conference on Neural Networks in Capital Markets*210–2171993
27. W. Tschacher *et al.*, Biological Psychiatry (, i (n) press).
28. H. Tong *Nonlinear time series: A dynamical systems approach*(Oxford University Press, Oxford, 1990).
29. A. Longtin, Int. J. of Bifurcation and Chaos **3**, 651–663 (1993)
30. M. Daemmig*et al.*, Applied Physics **855**, 121-125 (1992)
31. E. Aebi *et al.*, Zeitschrift für Klinische Psychologie, Psychopathologie und Psychotherapie **41**, 18-30 (1993).
32. L. Glass and M. Mackey *From clocks to chaos. The rhythms of life* (University Press, New York, 1988).
33. C. Scheier and W. Tschacher, *AILab Technical Report AI.96.3*1996.
34. C. Scheier and W. Tschacher System Familie **7**, 11-21 (1994)

LOOKING FOR CHAOS IN TIME SERIES DATA

TIMOTHY L. JOHNSON, KEVIN J. DOOLEY
Department of Mechanical Engineering, University of Minnesota
Minneapolis, MN 55455

Theoretical advances in nonlinear dynamics -- popularized as chaos theory -- have caught the attention of researchers looking for an alternative to the traditional linear modeling paradigm. Concepts from nonlinear systems theory promise more realistic representations of a fundamentally nonlinear world, and offer the potential of detecting deterministic components in what once was explained away as random variation. Much of the theoretical work, however, has focused on known dynamical systems; empirical applications are becoming common, but the use of theoretically derived concepts with limited and noisy data is questionable. This paper offers a review of nonlinear modeling techniques as they apply to empirical time series analysis. Specifically, procedures to estimate dimension and the Lyapunov exponents are described, and other methods of distinguishing between stochastic and deterministic systems are outlined. Techniques of noise reduction, attractor reconstruction, and prediction are also discussed, and an exploratory study of some recent nonlinear statistical methods is presented. Attention is paid to the controversial application of these methods to empirical situations, and a critique of their use is given. It is hoped that this paper will give the reader an integrated overview of this promising modeling paradigm.

1 Introduction

Scientists in the social and physical sciences have come to rely on relatively simple linear mathematical models to describe the behavior of the systems they study. Such models generally rely on two assumptions: first, that the residual error remaining after fitting a model to a set of observed data is due to random variation and, second, that this error can be characterized by a Gaussian (normal) distribution. The traditional principle of parsimony -- using the simplest description possible to characterize the behavior of a system -- is partially responsible for the preponderance of such models. It is also true that linear models are usually the easiest to work with, and that local linear models often suffice to a first degree of approximation for systems at equilibrium.

Recently, however, advances in nonlinear dynamics -- popularized as chaos theory -- have lead researchers to take a critical look at the modeling techniques they employ. Asserting that the world is fundamentally nonlinear, some researchers have begun to argue that linear models present a misleading view of the world. The notion of parsimony, it is argued, should not be used blindly as a guiding principle (Guastello, 1992). Although the limitations of linear models in general, and the popular Gaussian assumption in particular, have been recognized, few alternatives have been readily available.

Developments in nonlinear analysis provide a needed alternative. Concepts from nonlinear systems theory offer the ability to detect low-dimensional, deterministic dynamics (or "chaos") in what was once thought to be unexplainable random variation. The discovery of chaotic dynamics is indicative of the presence of additional factors influencing system behavior, factors dismissed as "noise" in traditional data analysis. These factors, however, may be significant sources of observed behavior, and they interact with nonlinearities in the dynamical system to produce behavior that is apparently unpredictable, yet entirely deterministic. Nonlinear analytical methods offer qualitative understanding into the nature of these factors. Substantial discussions of chaos theory, including empirical modeling techniques, have been published in *The Journal of Applied Econometrics* (1992), *The Journal of the Royal Statistical Society* (*A* , 1990; *B* , 1992), *Statistical Science* (1992), and *Integrated Behavioral and Physical Science* (1994).

Despite theoretical advances in chaotic dynamics, robust empirical modeling techniques have not been developed. A debate currently rages concerning the transfer of modeling techniques developed for known dynamical systems to the more common situation of fitting models to "noisy" empirical data. People on both sides of the debate, however, tend to agree that nonlinear empirical modeling techniques will be developed, providing a valuable supplement to the traditional statistical "tool kit" now used (and sometimes abused) by researchers.

This paper reviews the use of concepts from nonlinear dynamics and deterministic chaos in time series modeling. In particular, methods of analyzing empirical time series data, including the exploration of dimension and the identification of nonlinear deterministic components in system behavior, are described. Results from an exploratory study of some recent nonlinear statistical methods are also presented, and comments on the application of these techniques are offered in the conclusion. It is hoped that this review will enable the reader to gain a integrated understanding of the issues related to the empirical analysis of chaotic systems.

2 Empirical Time Series Modeling

2.1 Introduction

The dynamical behavior of a deterministic system can be described in terms of a set of state variables. These independent variables may be either observable factors or hypothetical constructs, and they comprise the coordinates of what is referred to as the state, or phase, space. Each point in the phase space corresponds to a particular realization of the system, and the time evolution of the dynamical system forms a trajectory through the phase space. The trajectory of a deterministic system converges on a compact set of points known as an attractor; the detection of a low-dimensional attractor is usually taken as evidence of chaotic dynamics. (Note that the attractor in state space does not necessarily reflect temporal evolution of the system; adjacent points on the attractor may be distant in time.)

For a given dynamical system, the values of the d state variables at time t can be represented by the vector $\underline{x}_t = (x_1, x_2, \ldots, x_d)$; a particular time evolution of the system can then be modeled by some function of the past values of $\underline{x}_t$, $f : \mathbf{R}^d \rightarrow \mathbf{R}^d$, such that

$$\underline{x}_{t+1} = f(\underline{x}_t) \quad t = 0, 1, 2, 3, \ldots \tag{1}$$

This difference equation is the trajectory of the dynamical system; in words, it says that the state of the system at time t+1 is a function of the state of the system at time t. Note that this is a discrete characterization of the dynamical system. There is an analogous continuous form, expressed as a differential equation, which states that the rate of change of a dynamical system is equal to some function of its current state (Kaplan, 1993):

$$d\underline{x}/dt = g(\underline{x}) \tag{2}$$

Due to the inherently discrete nature of measurement, however, both linear and nonlinear techniques employ the discrete characterization in empirical analysis.

Unfortunately, it is rarely possible to observe the true trajectory of a dynamical system, and the function f is usually unknown. Often, however, there is an "observer function", $h : \mathbf{R}^d \rightarrow \mathbf{R}^1$, that yields a scalar time series (Dechert and Gencay, 1992),

$$y_t = h(\underline{x}_t) \quad t = 0, 1, 2, 3, \ldots \tag{3}$$

This scalar time series is directly accessible to observation and empirical analysis, and both linear and nonlinear methods focus on the series y_t. One of the fundamental theorems in nonlinear systems analysis states that the observed time series can be used to infer the properties of the true, but unknown, dynamical system.

2.2 ARMA Models

Traditional time series techniques make use of the family of autoregressive moving average (ARMA) models. These models express the current value of a dynamical process as the sum of a series of previous values of the system (the autoregressive portion) and a similar sum of external shocks, or "innovations" (the moving average part). In mathematical notation, a generic ARMA model of order (M,N) takes the form (see Box and Jenkins, 1976):

$$y_t = \sum_{m=1}^{M} a_m y_{t-m} + \sum_{n=0}^{N} b_n e_{t-n} \tag{4}$$

The external shocks, e_t, are often assumed to be uncorrelated white noise. It is also assumed that the system is stationary -- that is, the coefficients in the ARMA model are constant.

Time series modeling is as much of an art as a science, and several heuristic methods have been developed for assessing the order of ARMA(M,N) models (see de Gooijer, et al., 1985; Pukkila, 1990 for reviews of these techniques). The power spectrum of the time series, which identifies the frequency components present in the data, and the autocorrelation structure of the series, which describes the dependence of the current state (y_t) of the process on its past values, are commonly used in the identification process. Particular ARMA models have characteristic frequency and autocorrelation patterns that can aid in their identification. Physical systems thought to have underlying stochastic differential equations are often modeled using the Dynamic Data Systems methodology (Pandit and Wu, 1983) which fits cascaded first (exponential) and second (sinusoidal) order systems.

While ARMA models are relatively basic to use, they are generally unable to distinguish linear stochastic processes from even simple nonlinear deterministic systems (Gershenfeld and Weigend, 1993). Vastly different dynamical systems, for instance, can give rise to nearly identical power spectra. Low-dimension nonlinear dynamics, while entirely deterministic, are ascribed to random variation, ignoring the influence of potentially significant sources of system behavior. The detection of deterministic components in a dynamical system, however, offers improved understanding of system structure and behavior.

2.3 Nonlinear Models and Chaotic Processes

The nature of dynamic processes can be viewed on a continuum, stretching from linear deterministic systems to stochastic ("random") processes. Between these extremes lie low-dimensional (i.e., $d < 10$) nonlinear systems. Time series data from such systems may appear to be stochastic -- successive points in the series are uncorrelated and the power spectrum is flat (i.e., all frequencies are present) -- even though the system generating the data is completely deterministic. For example, the logistics equation with a parameter value of four,

$$y_t = 4y_{t-1}(1-y_{t-1}) \tag{5}$$

is a simple dynamical system that yields a time series that most statistical tests identify as random noise. The behavior of such systems is said to be chaotic.

Several methods have been proposed for distinguishing between stochastic and low-dimension deterministic processes; while both processes produce apparently random behavior (i.e., behavior that appears unpredictable and independent of other factors), the behavior of the latter is completely determined by a small number of factors and the nonlinear nature of the dynamical system itself. Methods of characterizing time series data will be reviewed in subsequent sections of this paper. To date, however, these methods are capable only of detecting nonlinearities in empirical time series data. Techniques that *conclusively* distinguish the time series of stochastic systems from those of deterministic systems of even low dimensions have not been developed *for empirical applications* (Liu, et al., 1992).

Although extensions to empirical characterization, modeling, and prediction have been attempted, nearly all of the theoretical work done in the area of chaotic systems analysis has focused on time series from known dynamical systems. These theoretical applications have been largely successful. The extension of these methods to empirical analysis, however, where the nature of the dynamical system is inferred through the observed time series, is controversial. The causes of this controversy are related primarily to the inherently poor quality of empirical data.

Unlike data generated from a known system, empirical data is limited in quantity and finite in precision, corrupted by noise that (contrary to popular assumptions) is usually not Gaussian or additive, and subject to myriad exogenous factors in the environment and measurement process (Theiler, et al., 1993). The characterization techniques developed for chaotic dynamical systems are not robust to the effects of these factors, and their presence can render calculations meaningless.

As Kaplan (1993) points out, the process of detecting deterministic components in a given time series is poorly defined because *any* series of observations could have been generated from a purely stochastic process. The usual approach to this issue in statistical analysis is to select a particular nonrandom process and determine the likelihood of the observed data arising from the hypothesized process.

While hypothesis tests have been proposed for detecting chaotic dynamics, theoretical work has focused on other methods of characterization. The most common technique has been to estimate the dimension of the attractor in state space; a low estimate of dimension lends support to the presence of a nonlinear deterministic process. Another approach has been to calculate the Lyapunov exponents of the dynamical system. One of the hallmarks of chaotic processes is the rapid divergence of nearby points on a trajectory ("sensitive dependence on initial conditions"); two points that begin arbitrarily close to each other follow independent trajectories over time. The Lyapunov exponents provide a measure of the rate of this divergence and, hence, of the chaotic nature of the dynamical system underlying the data.

The following sections map out the major attempts at estimating both dimension and the Lyapunov exponents. Other methods of characterizing time series and detecting the presence of deterministic chaos in time series data, including hypothesis testing, are also described. In addition, issues related to noise reduction, attractor reconstruction, and prediction are discussed. The relationships between these topics are presented in Figure 1.

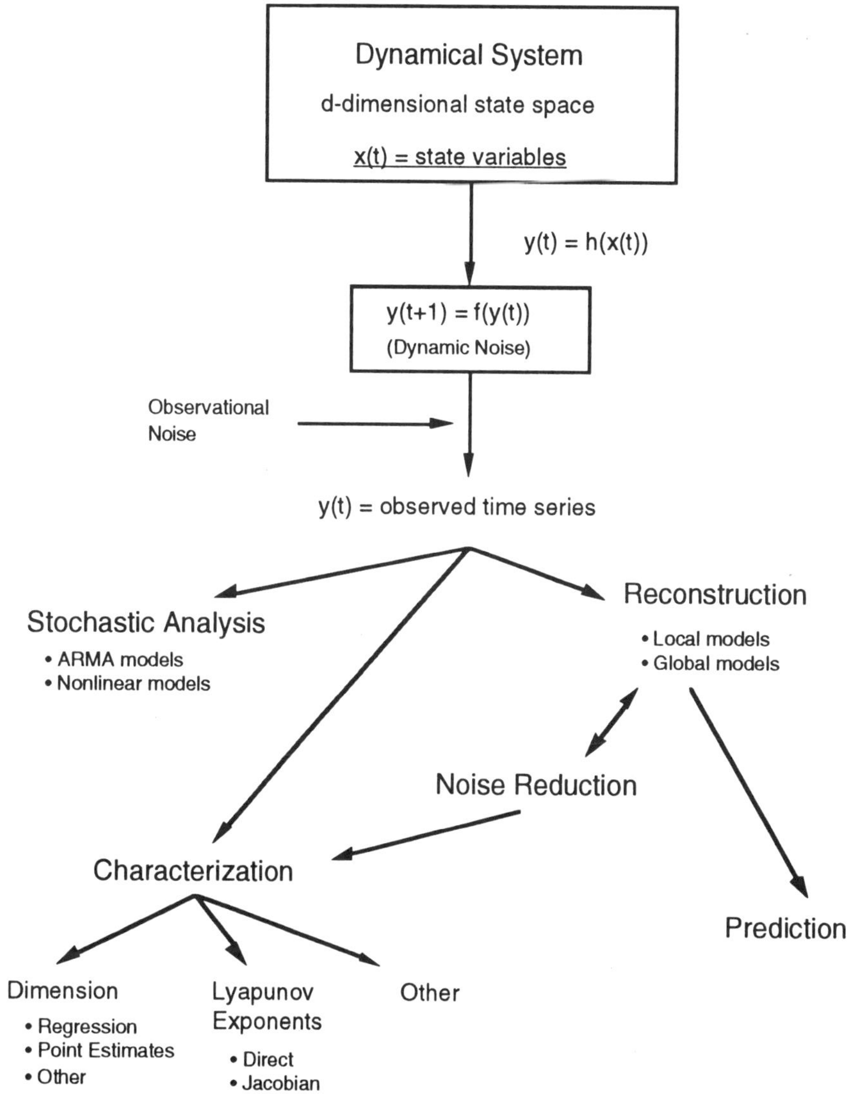

Figure 1: Relationships Between Nonlinear Analytical Methods

3 Characterization Of Dynamical Systems

3.1 Estimation of Dimension

Farmer, Ott, and Yorke (1983) provided one of the first comprehensive reviews of dimension, and made the distinction between metric dimensions and dimensions of natural measure. Metric dimensions, which include the fractal, or box-counting, dimension, attempt to cover the attractor with a mesh of boxes; dimension is estimated assuming a power-law relationship between the number of boxes required for complete coverage of the attractor and the size of the boxes. Dimensions of natural measure, including the information and correlation dimensions, partition the region of the attractor in a similar manner, but estimate the frequency with which the trajectory (the time series on the attractor) visits each region. The different dimensions are generally not equal to each other, although they are related and their numerical ordering is preserved across dynamical systems (Farmer, et al., 1983; Peitgen, et al., 1992).

Estimation of dimension from an empirical time series is based on the fact that the attractor of the system, although inaccessible to direct analysis, may be studied indirectly by an "embedding" of the observed time series. According to Takens' (1981) embedding theorem, provided certain conditions are met, the time-delayed vectors created from the original scalar time series map out a transformation of the true d-dimensional attractor. Such m-dimensional vectors are of the form:

$$(y_t, y_{t-s}, \ldots, y_{t-(m-1)s}) \tag{6}$$

where s, the delay time, is some integer multiple of the sampling time of the data collection process. Note that there are other time scales involved in the embedding process; these will not be discussed here, but see Broomhead and King (1986) for a discussion.

The important implication of Takens' embedding theorem is that the embedded trajectory will preserve the topological properties, including the dimension, of the underlying attractor. It has been shown analytically (see Broomhead and King, 1986; Scargle, 1989) that the dimension of the embedded space, m, will be less than or equal to twice the dimension of the true phase space plus one; that is,

$$m \leq 2d + 1 \tag{7}$$

Low-dimensional dynamics can be detected using this property. The dimension of the embedded space is calculated for successively increasing values of m; evidence for chaotic dynamics is given if at some value of m, m^*, the dimension estimates cease to increase (Smith, 1992). The logic behind this strategy is intuitive: a stochastic system (analogous to the Brownian motion of air molecules in an enclosed container) will fill up whatever space is available; a system that is bound to an attractor will only fill out the space required by the attractor. A two-dimensional circle, for instance, will still require two dimensions when "embedded" in a three dimensional space (Peitgen, et al., 1992).

The Renyi, or generalized, dimensions are the most common type of dimension, and take the generic form (Peitgen, et al., 1992; Pineda and Sommerer, 1993):

$$Dq = (q-1)^{-1} \lim_{e\to 0} \log\Big(\sum_{i=1}^{N(e)} p_i^q \Big) / \log(e) \tag{8}$$

where N(e) is the number of boxes (hypercubes in dimensions greater than three) required to cover the attractor and e is the side length of a box. The natural measure (a measure of the probability mass distribution on the attractor) of the ith box, p_i, is estimated by n_i/n, the number of observed points in box i divided by the total number of points in the time series. Different values of q yield different dimensions.

Dimension estimates typically make use of a power-law relationship between the natural measure and e -- it is assumed that the former scales exponentially with the latter, where the exponent is the dimension. The dimension can be estimated from the slope of the linear region of a log-log plot of the natural measure versus e as e varies over a range of values (Smith, 1992).

The best known members of the Renyi family are the box-counting (or fractal), information, and correlation dimensions; these dimensions correspond to values of q of zero, one, and two, respectively. The box-counting dimension (a metric dimension) is derived from a measure of the average number of boxes covering the attractor, N(e). As the size of the boxes decreases, N(e) increases per a power-law relationship with e:

$$N(e) \approx (\text{constant})\, e^{-D_0} \tag{9}$$

The box-counting dimension is usually estimated from a log-log plot of N(e) versus e; the slope of the linear portion of the graph is approximately D_0. Grassberger (1990) gives an efficient algorithm for box-counting calculations. Hall and Wood (1993) discuss the asymptotic bias and variance of box-counting estimators.

There are, however, two fundamental problems with the box-counting approach: first, the measure does not reflect the distribution of points on the attractor (a distribution that is not uniform); second, N(e) cannot be divorced from n, and the form of this dependence is difficult to determine (Grassberger 1983; Peitgen, et al., 1992). These problems have shifted the focus of research efforts onto the dimensions of natural measure.

The information dimension (see Farmer, et al., 1983; Cutler, 1991), based on Shannon's information theory (Shannon, 1948), is one attempt to overcome the problems associated with the box-counting approach. This dimension can be derived from the generic Renyi formula using l'Hospital's Rule (Peitgen, et al.,

1992). The dimension, however, that has received the most attention both in the literature and in application is the correlation dimension (Smith, 1992).

Rather than using the generic Renyi equation, the correlation dimension is usually derived in terms of the correlation integral (Theiler, 1990; Smith, 1992). For a given time series of length n,

$$C_n(r) = \sum_{i=2}^{n} \sum_{j=1}^{i} I(|y_i - y_j| < r) / (n(n-1)/2) \qquad (10)$$

where I is an indicator function that is equal to one if its argument is true, and is zero otherwise. The correlation integral, C(r), is then defined as the limit of $C_n(r)$ as n tends to infinity. C(r) is a measure of the average number of interpoint distances $(y_i - y_j)$ that are less than the length of a radius r (there are n(n-1)/2 such distances in a time series of n points).

In reality, an infinite number of observations are not available and it is assumed that, for large n (definitions of "large" will be discussed later) and small r, C(r) follows a power-law relationship with r:

$$C(r) \approx (\text{constant})\, r^{v} \qquad (11)$$

where v is the correlation dimension, D_c. Grassberger and Procaccia (1983a, 1983b), and subsequent authors (see Cutler, 1991; Smith, 1992 for references), have proposed regressing log $[C_n(r)]$ on log(r) over a range of values of r to estimate the correlation dimension (the slope of the linear portion of the log-log plot). Note that the correlation integral is related to measures of the Kolmogorov-Sinai entropy of a time series (Pincus, et al., 1991). Variants such as the Apparent Entropy (Pincus, et al., 1991) employ computational algorithms similar to those used in correlation integral calculations. Entropy is a measure of the information content of a time series, and is used to quantify the relative degree of randomness in the observed data.

Several authors, however, have criticized the regression approach to estimation of the correlation dimension. The primary criticisms deal with the fact that sequential values of $C_n(r)$ are not independent, and that the errors in $C_n(r)$ are neither constant nor independent (Theiler, 1990; Cutler, 1991; Smith, 1992). These violations of the fundamental assumptions of least-squares regression render its use questionable. Cutler (1991) also demonstrates how the intercept of the regression can "wander", making the power-law relationship (on which the regression is based) invalid. Cutler (1991), however, does derive a least-squares estimate of the dimension, assuming correlated values of $C_n(r)$. To further circumvent the problem, Cutler (1994) has developed an estimate for correlation dimension which is "independent" of the data, but rather is based on probability distributions. Cutler's claim is that this new definition does not rely on the assumption of an underlying deterministic skeleton, and therefore is valid for a larger class of stationary processes.

To circumvent the problems associated with regression, point estimates of the correlation dimension have been proposed. Smith (1992), assumes that the interpoint distances are independent, and derives a maximum likelihood estimate of the correlation dimension based on the binomial nature of the interpoint distances. If N_j is the number of interpoint distances less than r_j, Smith's (1992) point estimate of the correlation dimension is :

$$D_c = [\log(\sum_{j=0}^{k-1} N_{j+1}) - \log(\sum_{j=0}^{k-1} N_j)] / \log b \qquad (12)$$

where $r_j = ab^j$ ($0 \leq j \leq k$; there are k + 1 values of r used in the calculation: $r_0 > r_1 > \ldots > r_k$), and a and b are constants ($a > 0, 0 < b < 1$). A standard error, necessary for the calculation of confidence intervals, may be derived for this estimate. Liu, et al., (1992) also derive a point estimate for the correlation dimension, but conclude that it performs poorly in the presence of external noise. Noise, which generally exists on a scale larger than that of chaotic dynamics (i.e., when r is less than the standard deviation of the noise, (Smith, 1992)), interferes with all of the dimension calculation methods (Kantz, 1993).

A data set may be nonstationary, in which case the dynamics change over different epochs. The point correlation estimate (Farmer et al., 1983; Mayer-Kress et al., 1988; Mayer-Kress, 1994) allows one to estimate dimension at each instance in time. Skinner, Molnar, and Tomberg (1994) have developed a point dimension estimate which easily discriminates different epochs of dynamical behavior and performs well in lieu of noise.

In addition to noise, a significant problem with dimension estimates, and an issue that has been controversial in the literature, is the number of data points needed for accurate dimension estimates (see Smith, 1988; Ruelle, 1990; Essex and Nerenberg, 1991; Eckmann and Ruelle, 1992). The dimensions are defined in the limit as n tends towards infinity, an obvious impracticality in empirical systems. Although some physical systems may yield time series with thousands of measurements, it is often prohibitively expensive or difficult to collect more than a few hundred observations; in some instances (e.g., macroeconomic data), it is impossible to obtain more than a few hundred points.

Smith (1988) points out that dimension calculations for short data sets tend to be significantly biased, and may converge on estimates of the dimension that are substantially less than d. This biasing effect, which is partially the result of boundary effects on the attractor and issues related to point spacing, is particularly acute for higher-dimensional systems (Ramsey and Yuan, 1989). In addition, the variance of the dimension estimate is dependent on both the number of observations and the embedding dimension; as m increases, n must increase accordingly (see Ramsey and Yuan, 1989; Cutler, 1991). Theiler (1990) also points out that, for a fixed number of observations, the precision of correlation dimension estimates decreases with r since fewer interpoint distances are counted.

When coupled with the presence of noise and the fact that an empirical time series will be attracted primarily to those regions of phase space with the greatest probability density (Smith, 1988), these issues render the usefulness of dimension calculations questionable. Smith (1988) presents a very conservative analysis, and demonstrates that the number of observations required to estimate a given dimension to a specified accuracy can number in the *billions*. Eckmann and Ruelle (1992) are less pessimistic, and report that the number of uncorrelated observations required to determine a given dimension, d, must be greater than $\sqrt{10^d}$; a dimension of six, for instance, requires at least one-thousand observations. Cheng and Tong (1994) argue that the correlation dimension itself, whose estimation requires significant sample sizes, is not the significant parameter--but rather the embedding contains more practical value, and the embedding dimension can be estimated with far fewer points.

While skepticism is prudent, dimension calculations are a valuable means of determining if the continued search for deterministic chaos in a dynamical system is warranted. The dimension estimates are also valuable as a means of comparing processes -- used in relative terms rather than as absolute figures. The need for other methods of characterizing dynamical systems is apparent, however. Nychka, et al., (1992) make the point that, problems with accuracy aside, dimension estimates can only give a static view of dynamical systems -- stochastic or chaotic. A dynamic perspective may be obtained by examining the instability of system behavior as measured by the Lyapunov exponents.

3.2 Analysis of the Lyapunov Exponents

The Lyapunov exponents of a dynamical system measure the rate of divergence of points on arbitrarily close trajectories, providing an estimate of the chaotic phenomenon of sensitive dependence on initial conditions. The exponents also can be used to determine how long sequential data points in an empirical time series remain correlated with each other (Theiler, 1990). The finite precision of any measurement process introduces a certain degree of error in the data; the Lyapunov exponents quantify how fast these errors propagate through the system, rendering the recorded values of correlated observations independent.

The mathematical definition of the Lyapunov exponents is complex, but "intuitive approximations" are given by Berliner (1991) and Cheng and Tong (1992). If dx_t is the separation between two points along one direction in phase space after one time interval, then (Berliner, 1991)

$$dx_t \approx e^{t\mu}\, dx_0 \qquad (13)$$

where dx_0 is the initial separation between the two points, e is the base of the natural logarithm, and μ is the Lyapunov exponent. Note that there is one exponent for each dimension of the phase space, quantifying the rate of divergence (or convergence) of the trajectory in each direction (a positive exponent indicates divergence, a negative exponent convergence).

In mathematical form, the Lyapunov exponents are expressed in the limit as the length of the time series tends to infinity of the natural logarithm of the derivative of *f* (Cheng and Tong, 1992; Wolff 1992):

$$\mu = \lim_{n \to \infty} (1/n) \ln |df^n(x_0)/dx| \qquad (14)$$

Like dimension, the Lyapunov exponents of the true, but unknown, dynamical system can be estimated from an embedded time series. As m, the embedding dimension, approaches 2d+1, the d largest Lyapunov exponents of the embedded space will remain constant while the remaining m-d exponents tend towards negative infinity (Dechert, et al., 1992). Because the embedding preserves the topological properties of the actual dynamical system, the d largest exponents of the embedded time series will estimate the actual Lyapunov exponents.

It has been shown that the Lyapunov exponents of chaotic systems follow certain patterns (Baker and Gollub, 1990). Attention is usually focused on the largest, or dominant, Lyapunov exponent, which is always positive for chaotic systems, and never positive for nonchaotic systems (Dechert, et al., 1992); a positive Lyapunov exponent is generally taken as evidence of a deterministic chaotic process. It has also been shown that the sum of the d Lyapunov exponents must always be negative for dynamically stable systems -- in other words, systems that are bounded (Nychka, et al., 1992).

Unfortunately, like estimates of dimension, estimates of the Lyapunov exponents are acutely sensitive to noise and require an equivalently large number of observations (Eckmann and Ruelle (1992) estimate that the number of uncorrelated observations required for Lyapunov calculations is actually the square of that required for the correlation dimension). In addition, some of the current estimation methods require the determination of the underlying dynamical function, *f*.

Two basic strategies have evolved for estimating the Lyapunov exponents (Nychka, et al., 1992). The first method is the most direct and examines the rate at which neighboring trajectories diverge. Wolf, et al., (1985) presented one of the first algorithms of this type. As different trajectories will be influenced by different random perturbations, however, this approach to exponent estimation is quite sensitive to the effects of noise. Rosenstein et al. (1993) present an "improved" algorithm which appears to perform well with small sample sizes and in lieu of noise.

The second strategy used to estimate the Lyapunov exponents employs the Jacobian matrix of partial derivatives of the function *f*, which must be approximated from the embedded time series (Gershenfeld and Weigend, 1993). Nychka, et al., (1992) use two different methods of function approximation, thin plate splines (nonlinear regression) and neural networks, from which the Jacobian matrix is found. Dechert, et al., (1992) propose a similar method using feed-forward neural networks. Regression methods have also been used to estimate the Jacobian matrix directly (Eckmann, et al., 1986)

If J_t is the m x m matrix of partial derivatives of *f*, $T_t = J_t J_{t-1} \ldots J_1$, and v(t) is the largest eigenvalue of $(T_t^T T_t)$, then (Nychka, et al., 1992)

$$\mu = \lim_{t \to \infty} (1/2t) \log |v(t)| \qquad (15)$$

is the dominant Lyapunov exponent (this assumes that the limit will converge; see Nychka, et al., (1992) for the necessary conditions). A positive estimate of this exponent lends credence to the presence of a low-dimensional chaotic attractor. The remaining exponents may be determined in a similar manner (Kostelich and Lathrop, 1993). The Lyapunov exponents are sometimes expressed in terms of the Lyapunov numbers: the Lyapunov exponents are the natural logarithms of the Lyapunov numbers.

Finally, a dimension can be derived from the Lyapunov exponents. If k is defined such that $\sum_{i=1}^{k} \mu_i \geq 0$, then the Lyapunov dimension is (Baker and Gollub, 1990):

$$D_L = k + |\mu_{k+1}|^{-1} \sum_{i=1}^{k} \mu_i \qquad (16)$$

The usefulness of the Lyapunov dimension becomes apparent for deterministic systems with large dimensions (i.e., $d \geq 10$). Dimension estimation becomes intractable due to the extremely large number of observations required and the dubious nature of the large scaling ratios employed (Peitgen, et al., 1992). The Kaplan-Yorke conjecture (Frederickson, et al., 1983) equates the Lyapunov dimension with the information dimension; this conjecture, however, has been verified for a limited number of systems.

3.3 Other Methods of Characterization

A variety of alternative approaches to identifying nonlinearities in dynamic behavior has been proposed in the literature, and alternate attempts have been made at distinguishing between stochastic and deterministic processes. A few of the more prominent efforts will be discussed here; note that there is overlap with the methods discussed previously.

Theiler, et al. (1992) and Theiler, et al. (1993) discuss the detection of nonlinear dynamics in a time series using surrogate data -- randomly generated data that emulates the autocorrelation structure of the observed time series. If chaotic dynamics are detected in both data sets, then the evidence for chaos is likely to be either an artifact of the autocorrelation or a spurious result of the analytical technique. Theiler and Eubank (1993) suggest using a statistic of the form:

$$|Q_{surr} - Q_{orig}| / \text{StdDev}_{surr} \qquad (17)$$

where Q is a statistic, such as a dimension estimate, that can distinguish between linear and nonlinear systems (Q_{surr} is actually the average of such a statistic, aggregated over several randomly generated surrogate data sets). A large value of this statistic is evidence of nonlinear dynamics.

One of the few hypothesis tests proposed in the literature is due to Brock, Dechert, and Scheinkman (see Brock, et al., 1991). The BDS test takes the form:

$$S(m,r) = \sqrt{n}\,\{C_m(r) - [C_1(r)]^m\} \tag{18}$$

where $C_m(r)$ is the correlation integral (note the change in subscript notation: m is the embedding dimension). S(m,r) is distributed as a normal N(0,q) variable, where q is a function of r, n, and m. Note that $C_m(r) \rightarrow C_1(r)^m$ as $n \rightarrow \infty$ (Theiler and Eubank, 1993).

The BDS null hypothesis is that the data is independently and identically distributed (iid). Liu, et al., (1992), however, point out that the alternative hypothesis is not "chaos", per se. If the linear components in the time series have been removed (e.g., by using the residuals of an ARMA model), rejection of the null is indicative of the presence of nonlinearities in the data, but does not necessarily support the notion that a low-dimensional deterministic process is at work. Theiler and Eubank (1993) also report that the BDS test loses power when used with residual, or "bleached", data.

Three other methods of characterization will be briefly discussed. The first of these is due to Casdagli (Casdagli, 1992; Casdagli and Weigend, 1993). Casdagli's approach fits a series of local models, ranging from deterministic to stochastic, to the time series data. Short-term predictions are made and compared to a portion of the data not used in the model fitting. The prediction errors are then plotted against the size of the neighborhoods used in the models: if the smaller neighborhoods prove better at prediction, support is given to the presence of a low-dimensional chaotic process; if larger neighborhoods yield smaller prediction errors, a stochastic model is favored. This is due to the fact that local models attempt to fit the stochastic signal, degrading predictive ability.

Kaplan and Glass (1992) examine the trajectories of the data points from a time series in small regions of phase space. If a deterministic process has generated the observed data, the individual trajectories will follow similar paths within the region; if the data have come from a stochastic process, the orientation of the trajectories will be random. Kaplan and Glass (1992, 1993) derive a statistic that quantifies this idea and give the limiting cases of the statistic for deterministic and stochastic systems. The authors (1993) also discuss confidence intervals for a similar statistic.

Sugihara and May (1990) also look at the accuracy of forecasts based on models of varying complexity and, like Casdagli, provide a qualitative means of distinguishing between stochastic and dynamic processes. Sugihara and May predict the evolution of a trajectory by examining the dynamics of an observation relative to its neighbors in phase space. The prediction accuracy is then compared to the number of time steps into the future for which a forecast is desired. A stochastic process will produce a relatively flat plot of prediction accuracy versus time, while a chaotic process will yield an inverse relationship between accuracy and the prediction interval. Yao and Tong (1994) and Sugihara (1994) both argue that such modeling approaches are necessary in order to deal with the "ubiquitous" case stochastic chaos.

4 Noise Reduction

In addition to the finite nature of experimental data, the presence of noise has presented the largest obstacle in the application of theoretical concepts to empirical analysis. Noise, which is inherent in empirical data due to the finite precision of the measurement process, will overwhelm deterministic dynamics which exist on a scale smaller than that the noise (measured, for instance, by its standard deviation; see Smith, 1992; Kantz, 1993). Noise is sometimes ignored and attempts have been made to increase the robustness of the various dimension calculations; another approach has been to reduce the noise in the data prior to its analysis, and a few of these methods will be outlined here.

Two types of noise have been identified in the literature: observational, or additive, noise, and dynamic, or system noise (Hammel, 1990; Smith, 1992). As its name implies, observational noise affects individual data points, and includes error due to the measurement process and exogenous forces in the environment:

$$\text{Observational Error} = y_t - p_t \tag{19}$$

where y_t is the observed (measured) value and p_t is the "true" value of the process at time t. Dynamic noise includes errors that affect the time evolution of the system itself (Hammel, 1990):

$$\text{Dynamic Error} = y_t - f(y_{t-1}) \tag{20}$$

Noise reduction methods have focused primarily on observational error; dynamic noise has received less attention.

Traditional statistical methods of noise reduction (e.g., Fourier analysis which is employed to remove high frequency variation from the data) cannot be used with chaotic processes since the frequency spectrum of such dynamical systems is wide-band and flat: high frequency stochastic noise cannot be distinguished from high frequency deterministic dynamics (Kantz, 1993). Some common filtering techniques can also distort the form of the attractor underlying the dynamics, increasing the dimension and adding spurious Lyapunov exponents (Kostelich and Yorke, 1990). In addition, replacement of the original observations by their residuals (the differences between the observed data points and fitted values derived from, say, an ARMA model which has removed the linear components from the data), can distort the underlying attractor, rendering computations meaningless (Theiler and Eubank, 1993). Rosenstein and Collins (1994) show quite clearly (visually) how the Lorenz attractor is distorted when various low, high, and bandpass filters are applied.

Hammel (1990) and Kostelich and Yorke (1990) present iterative noise reduction techniques that successively perturb the observed system to improve its fit to the true dynamical system. Hammel's technique assumes that the functional form of the dynamic system (f) is known. Kostelich and Yorke, on the other hand, make use of the embedding theorem to create local linear models of the uncorrupted trajectory over small regions of the attractor. The improved trajectory from either technique can be used in subsequent analyses.

Hammel's approach is a version of a more general "shadowing" technique presented independently by Farmer and Sidorowich (1991). A shadow trajectory is a completely deterministic orbit that mimics the noisy observed trajectory. Farmer and Sidorowich's method seeks to reduce the difference between the shadow and observed trajectories subject to the constraint:

$$z_{t+1} - f(z_t) = 0 \tag{21}$$

where z_t is the shadow orbit; in other words, there is no error in the iteration of z_t -- it is completely deterministic. A set of nonlinear equations is derived, and their solution yields z_t.

The shadow approach works for both observational and dynamical noise, but assumes that f is known. Farmer and Sidorowich (1991) describe an iterative strategy of function approximation and noise reduction when f is unknown a priori. The authors report that their shadowing approach provides a better reconstruction than the methods of Hammel (1990) and Kostelich and Yorke (1990).

Taking a different approach, Kantz (1993) describes a noise reduction technique that seeks to remove high-dimension noise from low-dimension deterministic dynamics. A geometrical view of the noise is given: the true deterministic dynamics of the embedded time series will exist in an m-dimensional space; the measured data, however, will exist in a "cloud" of points above this space. To reduce the noise, and uncover the true dynamics, Kantz's (1993) method attempts to determine the orientation of the noise, and project this cloud of points back onto the original space of the embedded series. The actual surface of points in the embedded space must first be determined.

Finally, Smith (1992) derives the functional form of the correlation integral assuming additive noise, and outlines the derivation for dynamical noise. Smith's approach avoids estimation of f; this need, however, remains an obstacle to other methods of noise reduction.

5 Attractor Reconstruction And Prediction

Reconstruction of a dynamical system and prediction are related topics: the functional form of the process must be known to produce a forecast. These topics, however, are inherently related to the other analytical methods discussed in this paper. The alternative characterization techniques of Sugihara and May (1990) and Casdagli (1992), for instance, compare the prediction error in stochastic and deterministic models, models which are reconstructed at local levels; the noise reduction methods outlined in the last section function similarly. These topics are discussed separately for convenience only.

There is a bias-variance tradeoff in all modeling approaches (Casdagli and Weigend, 1993). "Weak" models, such as neural networks, tend to include many parameters, fitting both signal and noise. Such models characteristically have low bias, but their precision is poor due to the number of parameters that must be estimated. "Strong" models make use of fewer parameters, but tend to underfit the data. These models include many common statistical techniques and tend to have low variance, but function poorly at prediction due to their higher bias (Gershenfeld and Weigend, 1993).

Due to their sensitivity to initial conditions, chaotic dynamical systems are notoriously difficult to predict: the exponential propagation of small errors renders forecasts inaccurate after a relatively short time interval. This situation even applies to known dynamical systems; whenever the initial conditions cannot be specified exactly, the trajectory of a chaotic system will diverge unpredictably due to the finite precision of the data ("round-off error"). Kostelich and Lathrop (1993), however, speculate that the Lyapunov exponents may not be constant over an entire attractor. If this is the case, there may be regions on the attractor where prediction is feasible.

The remainder of this section sketches out several reconstruction and prediction techniques. Approaches to attractor reconstruction fall into two general categories: those that attempt to create a geometric representation of the dynamics -- using local models, and those that attempt to approximate the functional form of the process -- a global approach (Casdagli and Weigend, 1993).

Broomhead and King (1986) present a geometric approach that uses Takens' theorem to derive a basis for the embedding space. The "singular system approach", uses the autocorrelation matrix of the embedded time series to find the basis that makes most efficient use of the information in the observed data. The attractor is reconstructed by examining how the embedded time series "explores" the space spanned by the derived basis: the principal direction of the observed data is analyzed, and the time series is projected onto it. Noise reduction is an added benefit of this approach.

When relatively long, noise-free time series are available, the local linear method of Kostelich and Yorke (1990) can be used. This approach first covers the phase space of the embedded time series with a grid. A local, linear approximation to the dynamics is constructed at each vertex of the grid using least-squares regression techniques and the points immediately surrounding the vertex. This yields a piece-wise approximation to the underlying attractor. Note that a similar method is employed in Sugihara and May's (1990) noise-reduction strategy.

Sauer (1993) presents a third local linear approach to attractor reconstruction in the runner-up entry to the Sante Fe Institute's *Time Series Prediction and Analysis Competition* (see Weigend and Gershenfeld, 1993). The idea behind Sauer's approach is that vectors similar to a particular embedding vector should share similar dynamics. The nearest neighbors to the embedded vector at the end of the observed time series are first determined, and the best low dimensional space passing through the center of mass of these points is found. The nearest neighbor points are projected onto this space and a regression model is fit to the projections. A regression model is then used to create forecasts.

While Sauer's (1993) geometric approach was found to provide the best long-term forecast in the Sante Fe competition, Wan's (1993) global approach had the lowest short-term error. Wan's method, which won the competition, employed a feedforward, time-delay neural network fit to the entire time series. Such networks, which are capable of approximating any continuous function (see Wan, 1993, for references), are used to determine the functional form of the embedded time series and make predictions.

Finally, Berliner (1991) illustrates the use of statistical techniques in forecasting; these methods assume that the dynamical system is known. Employing Bayesian and maximum likelihood methods, Berliner compares predictions for the logistic equation. Although sensitivity to initial conditions renders quantitative long-term forecasts impossible, the maximum likelihood approach was found to yield good qualitative predictions. Bayesian techniques gave reasonable short-term, quantitative forecasts, but quickly tended towards a constant value as the forecast interval increased.

6 Exploring An Exponential Time Series Approach

We have primarily discussed how to detect the presence of nonlinearity, and potentially chaotic dynamics, in an observed time series. We believe that the key to significant future advancement in this area will come from the theoretical marriage of statistics and dynamics. Sam Wu (e.g., see Pandit and Wu, 1983) realized this, and made great strides forward in the area of machine design and control by explictly equating linear time series models with underlying stochastic differential equations. It would be valuable to extract dynamical parameter estimates, such as dimension, from conventional statistical methods. In this section we wish to begin to study a suggestion by Guastello (1993) in which estimates from a nonlinear (exponential) time series model are directly used to estimate attractor dimension. We have chosen the Lorenz, Rossler, and Henon systems, as they have been studied in detail and have known dimension.

Guastello was one of the first to introduce the idea of using nonlinear statistical (specifically, polynomial regression) techniques to capture nonlinear dynamical information (specifically, catastrophe models; e.g., see Guastello, 1982; 1992). In using a wavecrest polynomial approximation of the logistics equation, he has derived a series of exponential autoregressive models, which we will refer to as EXPO:

$$x(t) = b_1 + b_2 \cdot Q \cdot x(t-1) \cdot \exp(q_1 \cdot x(t-1)) \qquad (22)$$

where b_1, b_2, and q_1 are unknown parameters and Q is a known or unknown bifurcation variable; if Q is unknown it is consumed into b_2. Sometimes a better fit is found by the simpler form:

$$x(t) = b_1 + \exp(q_1 \cdot x(t-1)) \qquad (23)$$

to which additional lags can be added:

$$x(t) = b_1 + \exp(q_1 \cdot x(t-1)) + \exp(q_2 \cdot x(t-2)) + \ldots \qquad (24)$$

The form of this model is attractive in that it can be directly used to obtain an estimate of dimension. Let D be dimension where (Wiggins, 1988):

$$D = \lim (t\text{-->} \infty) \parallel \exp(-at) \parallel ^{1/t} = e^{-a} \qquad (25)$$

Guastello shows (1993) that relative to equations [22] and [23], q_1 is the Lyapunov exponent, and $D = \exp(q_1)$. If multiple lags are involved, then

$$D = \Sigma \{\exp(q_1) + \exp(q_2) + \ldots \exp(q_k)\} - k + 1 \qquad (26)$$

If D<1, the process converges to a fixed point; if D>1, chaotic dynamics may be present. Note that for a linear process, q_1 will tend towards zero and D tends towards one. Note that this is a different interpretation than usual. Typically a random sequence would be considered to have infinite dimension. Because this method works underneath the noise, it finds a single point attractor (the statistical mean), and thus dimension of one.

Guastello has used such models to discover chaotic dynamics in work force populations (1992b), stock exchange prices (1993b), and group behavior (1994). In all of Guastello's models one must normalize the data with respect to scale and location. Location be taken as arithmetic mean, or minimum value in data set, depending on whether there is some contextual meaning to the zero value on the scale. Scale can be corrected for via standard deviation, unless the data are multi-modal, in which case a modal standard deviation should be used (Guastello, 1993).

Guastello differentiates these models from other potential time series forms in that they reflect back upon a solid theoretical construct (catastrophe theory as developed by Thom (1975))--thus they represent a "model" rather than a "curve fit", and therefore will more likely generate fundamental knowledge about the phenomenon in question (Guastello, 1992). Guastello uses comparable linear time series models for benchmarking, using r^2 values as a test of goodness-of-fit.

An autoregressive model (AR) can be used for linear time series fit:

$$x(t) = q_1 \cdot x(t-1) \tag{27}$$

Higher order lags can be added. There are two related models which we would like to study; a quadratic autoregressive (QAR; Tong, 1990):

$$x(t) = q_1 \cdot x(t-1) + q_2 \cdot x^2(t-1) \tag{28}$$

and the EXPAR models (Tong, 1990):

$$x(t) = x(t-1) \cdot \{a_1 + b_1 \exp(-d \cdot x^2(t-1)\} \tag{29}$$

The QAR is attractive in that it is the simplest nonlinear time series form possible; it is a member of the Volterra equations (Tong, 1990). The EXPAR form was motivated by modeling of a physical system in which there is a stable oscillatory mode which leads to a limit cycle, and where the origin acts as a repellor or attractor depending on the current distance away from the origin (Ozaki and Oda, 1978). Note that EXPAR is roughly equivalent to EXPO (equation [22]), except the term in the exponent is squared.

Now let us take a look at the 3 systems being studied. The Lorenz equations are given by (Peitgen et al., 1992):

$$\begin{aligned} dx/dt &= -10x + 10y \\ dy/dt &= 28x - y - xz \\ dz/dt &= -2.67z + xy \end{aligned} \tag{30}$$

1000 points were simulated, using first order Euler and a Δt of 0.01. The attractor is shown in Figure 2. The best fit came from an EXPAR model of order 3; an r^2 value of 0.996 was obtained. A residual plot shows there is a systematic error that in fact resembles the shape of the original attractor. The best EXPO model was of order 1, with an r^2 of 0.766. Finally, an AR model of order one was fit, with an r^2 of 0.885. Predictions tend to lag the actual behavior, so the model is simply tracking the signal. A scatter plot shows the embedded attractor in an even more pronounced manner.

The Rossler equations are given by (Peitgen et al., 1992):

$$\begin{aligned} dx/dt &= -y - z \\ dy/dt &= x + 0.2 \cdot y \\ dz/dt &= 0.2 + xz - 5.7 \cdot z \end{aligned} \tag{30}$$

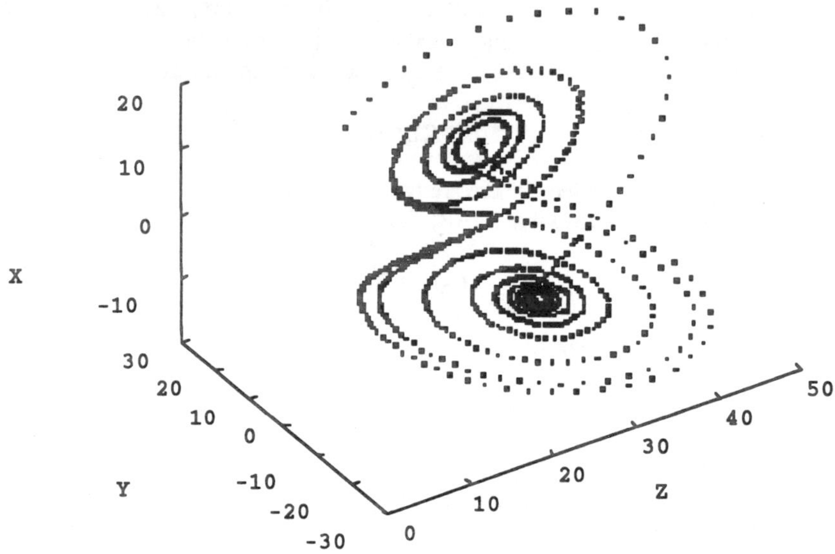

Figure 2: The Lorenz Attractor

The attractor is shown in Figure 3. An EXPAR model of order 2 reveals a "best" fit with r^2 value of 0.969; the fit shows only small systemic error . A EXPO of order 3 was fit with an r^2 of 0.925. The Rossler y-data was first order differenced in order to see if it would improve estimates. A EXPO order one model had an r^2 of 0.741.

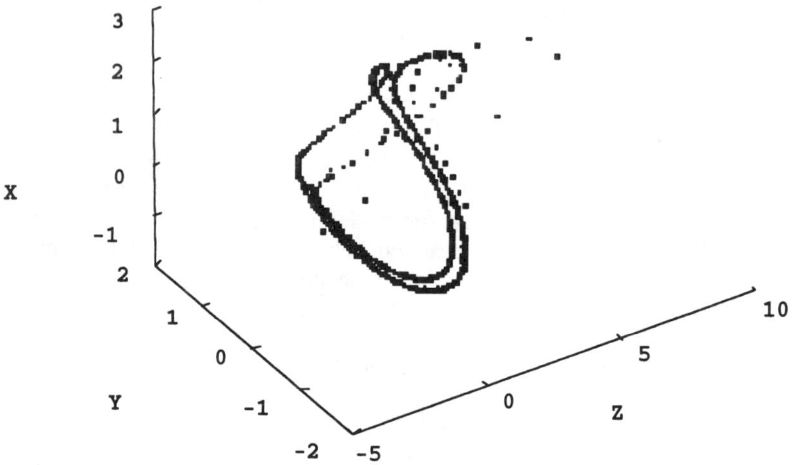

Figure 3: The Rossler Attractor

Finally, the Henon equations are given by (Peitgen at al., 1992):

$$x(k) = y(k) + 1 - 1.4 \bullet x^2(k-1)$$
$$y(k) = 0.3 \bullet x(k-1) \qquad (31)$$

The attractor is shown in Figure 4. The Henon data series were not modeled well at all by either EXPAR, EXPO, or AR models; all had r^2 values below 0.10. The shape of the attractor looks much like a parabola, therefore the QAR model was used; an r^2 value of 0.919 was obtained.

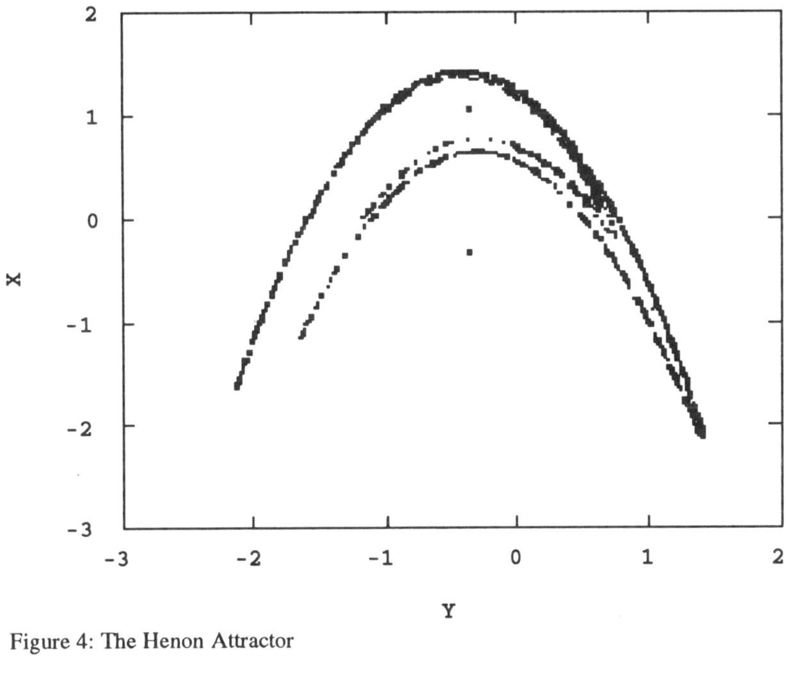

Figure 4: The Henon Attractor

One can see the model fit is quite good. In general, it appears as though various nonlinear time series models are capable of tracking, if not also characterizing the time series output from a chaotic system--more extensive research must be attempted before this statement can be made with any rigor. We found that the estimates were extremely sensitive to embedding lag--this backs up others findings (e.g. Yee et al., 1991). For example, the Lorenz data was generated using a Δt of 0.01. We found that if we observed every single point most models came out to be random walks; estimates became more significant when every fifth point was observed. The Rossler data was much more difficult to model. Various embedding lags and differencing were tried before satisfactory results were found. The problem exists in that in reality, we don't *know* what the correct answer is, so searching for the correct modeling parameters is much more blind. Rapp (1994) summarizes a variety of methods that can be used to identify these lags (tau).

Finally, we investigated the use of EXPO models to estimate the dimension of the attractor. The actual dimension of the Lorenz system is 2.073 (Peitgen, 1992); we found a value of 1.94 using the x-values and 1.84 using the y-values. The actual dimension of the Rossler attractor is 2.00+e (Peitgen et al. 1992). We obtained estimates of 1.68 using the x-values and 1.92 using the y-values.

In order to evaluate the characteristic behavior of the EXPO method a series of experiments were performed. First, 350 z-values was generated from the Lorenz attractor. Figure 5 shows that as the time lag (tau) increases, the estimated dimension D reaches a low of 2.06 (close to the real value of 2.07), and then rises again. A similar experiment with y-values from the Rossler attractor is shown in Figure 6. Notice the dimension steadily decreases until lag 10, in which cases it hits zero and then becomes quite large again; the cycle is repeated in later lags. At the point of collapse D is estimated as 1.83 and 1.97, compared to the true value of just over 2.0.

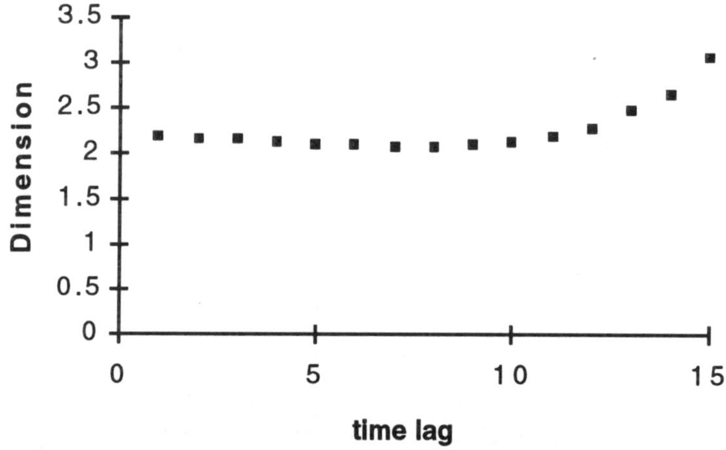

Figure 5: Dimension of the Lorenz System

Next, tests were run to see if EXPO was good at dimension estimation with few samples. Tests failed. Further analysis showed however that it was not the sample size per se but rather the adequate coverage of expanding and contracting attractor epochs which was at issue. The same 350 points which lead to an estimate of D=2.06 for the Lorenz attractor were used. The series consisted of one contraction (48 points, set A), and five expansion loops (B-F, of size 64 ,64, 64, 64, and 46).

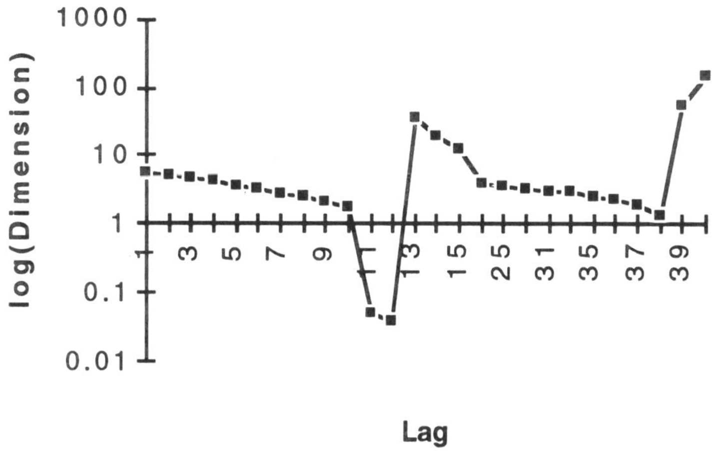

Figure 6: Dimension of the Rossler System

One expansion loop was removed at a time, leading to sample sizes of (304, 240, 176, and 112); D was estimated as (2.10, 2.04, 1.93, 1.76). One can see that when the series only consists of one expansion and one contraction epoch (parts A and B), D is estimated more poorly. When only the contraction epoch is available, D is estimated as 1.41. In summary, sample size is an issue of attractor coverage here.

Finally, noise was added in various amounts to the data set. With the addition of (10, 25, 50, 75, 100, and 150) percent noise, D was estimated as (2.04, 2.00, 1.94, 1.90, 1.87, 1.84). One can see that EXPO's estimate of dimension is quite robust to noise, and would be expected from a statistical approach.

7 Conclusion -- Some Thoughts On The Search For Chaos

The relationships between the analytical approaches used in the study of empirical time series are presented in Figure 1. While characterization and prediction may appear to be final goals, a complete analysis is an iterative process; the means employed depend on the ends, and the ends, of course, follow from the specific application.

The development of techniques within these analytical categories has been both an incremental and a divergent process; the analysis of nonlinear dynamics in empirical situations is an inchoate endeavor, and agreement on a "tool kit" of techniques has not been reached. Differences of opinion abound, an example of which is the differing views given by Smith (1988) and Eckmann and Ruelle (1992) concerning data requirements for dimension estimation. Smith (1988) is quite conservative, reporting data requirements that reach into the billions of observations; Eckmann and Ruelle (1992) are more optimistic, and show that the number of observations required for meaningful analysis is substantial, but modest in comparison to Smith.

A debate also rages between theoreticians and empiricists concerning the application of techniques like dimension estimation and calculation of the Lyapunov exponents to experimental situations. The former tend to be conservative in their assumptions and interpretations, while the latter tend to be less concerned with the exactness of their methods. At this time, both the theoreticians' caution and the empiricists' willingness to experiment are needed.

Aside from problems with the quality and quantity of data, the analytical methods proposed in the literature are computationally intensive. Unless one has access to high-powered computers, the computational requirements can preclude exploratory analysis. In addition, many of the noise reduction and reconstruction techniques are difficult for mainstream researchers to translate from the literature into practical use (Dooley, 1994). Commercial software is available, but often lags the literature in sophistication. Some researchers have made their code available for others to use, but distribution has been limited.

The analytical characterization techniques presented here are best used for comparative purposes only; due to computational difficulties, the literal meanings of dimension and Lyapunov estimates are questionable. These estimates, however, can be used to compare processes, and to study the evolution of a single process over time -- changes in dimension, for instance, can yield significant insight into the structure and behavior of a dynamical system, changes that would not be detected using traditional time series analysis.

The lack of an accepted "tool kit" of techniques poses problems for reporting the results of empirical studies. One of the tenets of the scientific method holds that experiments should be repeatable. Understanding and replication are impossible, however, unless one's methods are known; different approaches to dimension calculation, for instance, may yield significantly different results. Researchers need to clearly document their calculations, describing all algorithms in detail.

Finally, one needs to ask whether efforts at prediction are worthwhile when the propagation of round-off error and other sources of noise restrict forecasts to relatively short time intervals. It may be possible that traditional exponential smoothing gives forecasts that are as good as those derived using nonlinear methods, but are much simpler to compute. While long-term quantitative prediction of chaotic systems is not possible, however, nonlinear prediction techniques may still yield better *qualitative* understanding about the process than traditional methods which average over differences and smooth out the dynamics.

Despite the controversies surrounding their application, nonlinear time series techniques are the focus of an unusual amount of cross-disciplinary research, and they have been employed in a variety of experimental situations. A sample of these applications includes: physiological research (Glenny, et al., 1991), studies of the circulatory process (Skinner, et al., 1991), characterization of the aging process (Lipsitz and Goldberger, 1992), control theory (Vassiliadis, 1994), statistical analysis (Chatterjee and Yilmaz, 1992), and a host of economic problems (Baumol and Benhabib, 1989).

The nonlinear modeling paradigm promises increased insight into the nature of dynamical systems encountered in research, and the ability to detect low-dimensional deterministic dynamics in time series data offers the potential of stronger models. Several specific problems related to the transfer of theoretical methods to empirical situations must first be resolved, however, before the use of these techniques can become wide-spread. These issues have been emphasized in the paper and will be summarized here in conclusion.

First of all is the issue of noise -- both observational and dynamic. Methods of filtering out extraneous noise from the data prior to analysis, without altering the structure of the dynamical system, need to be improved and simplified. In addition, the analytical techniques themselves need to be made more robust to the effects of noise.

Second, issues pertaining to sample size need to be resolved. Many applications preclude the use of nonlinear techniques because of the limited data they generate. To this end, Judd's (1994) recent work is promising. Judd outlines a probabilistic approach to dimension calculation that employs an empirical distribution of the interpoint distances. A maximum likelihood approach is used to estimate dimension. Judd's methods are nonasymptotic; dimension estimates are only applicable to a limited range of scales. This restriction, however, reduces the need for large sample sizes. To date, Judd's method has been verified only for non-fractal systems.

A third issue concerns determination of the statistical properties of the dimension estimates. The work of Cutler (1991), Theiler (1990), and others has provided a good start. Estimates of statistical error are necessary for confidence intervals and hypothesis tests. Related to the latter, tests with "chaos" as the alternative hypothesis need to be developed. The ability to distinguish between stochastic and low-dimension dynamics using empirical data will represent a substantial advance in time series analysis.

Finally, we have presented some experiments which demonstrate that a more direct link between "classical" statistical methods and dynamical estimation may be feasible. Nonlinear time series methods must be studied in greater detail to see how they can be used to uncover the nonlinear dynamics present in observed systems.

Acknowledgements

This research has been supported by 3M Engineering Systems and Technology Laboratory and Honeywell Solid State Electronics Center. Thanks to Steve Guastello for helpful comments.

References

Baker, G. L., & Gollub, J. P. (1990). *Chaotic dynamics: an introduction.* Cambridge, England: Cambridge University Press.

Baumol, W. J., & Benhabib, J. (1989). Chaos: significance, mechanism, and economic applications. *Journal of Economic Perspectives, 3(1),* 77-105.

Berliner, I. M. (1991). Likelihood and Bayesian prediction of chaotic systems. *Journal of the American Statistical Association, 86(416),* 938-952.

Berliner, I. M. (1992). Statistics, probability, and chaos. *Statistical Science, 7(1),* 69-90.

Brock, W. A., Hsieh, D. A., & Le Baron, B. (1991). *Nonlinear dynamics, chaos, and instability: statistical theory and economic evidence.* Cambridge, MA: MIT Press.

Broomhead, D. S., & King, G. P. (1986). Extracting qualitative dynamics from experimental data. *Physica D, 20,* 217-236.

Box, G. E. P., & Jenkins, F. M. (1976). *Time series analysis: forecasting and control (2nd edition).* Oakland, CA: Holden-Day.

Casdagli, M. (1992). Chaos and deterministic versus stochastic non-linear modeling. *Journal of the Royal Statistical Society B, 54(2),* 303-328.

Casdagli, M. C., & Weigend, A. S. (1993). Exploring the continuum between deterministic and stochastic modeling. In A. S. Weigend & N. A. Gershenfeld (Eds.), *Time series prediction: forecasting the future and understanding the past* (pp. 347-366). Reading, MA: Addison-Wesley Publishing Co.

Chatterjee, S., & Yilmaz, M. R. (1992). Chaos, fractals, and statistics. *Statistical Science, 7(1),* 49-68.

Cheng, B., & Tong, H. (1992). On consistent nonparametric order determination and chaos. *Journal of the Royal Statistical Society B, 54(2),* 427-449.

Cheng, B., & Tong, H. (1994). Orthogonal projection, embedding dimension and sample size in chaotic time series from a statistical perspective.*Philosophical Transactions of Royal Society of London A, 348,* 325-341.

Cutler, C. D. (1991). Some results on the behavior and estimation of the fractal dimensions of distribution on attractors. *Journal of Statistical Physics, 62(3/4),* 651-708.

Cutler, C.D. (1994). A theory of correlation dimension for stationary time series. *Philosophical Transactions of Royal Society of London A, 348,* 343-355.

Dechert, W. D., & Gencay, R. (1992). Lyapunov exponents as a nonparametric diagnostic for stability analysis. *Journal of Applied Econometrics, 7,* 541-560.

de Gooijer, J. G., Abraham, B., Gould, A., & Robinson, L. (1985). Methods for determining the order of an autoregressive-moving average process: a survey. *International Statistical Review, 53(3),* 301-329.

Devaney, R. (1989). *An introduction to chaotic dynamical systems (2nd edition).* Redwood City, CA: Addison-Wesley.

Dooley, K. (1994). Complexity in time series modeling. *Society for Chaos Theory in Psychology and the Life Sciences,* October.

Eckmann, J. -P., Kamphorst, S. O., Ruelle, D., & Ciliberto, S. (1986). Lyapunov exponents from a time series. *Physical Review A, 34,* 4971-4979.

Eckmann, J. -P., & Ruelle, D. (1992). Fundamental limitations on estimating dimension and Liapunov exponents in dynamical systems. *Physica D, 56,* 185-187.

Essex, C., & Nerenberg, M. A. H. (1991). Comments on 'deterministic chaos: the science and the fiction' by D. Ruelle. *Proceedings of the Royal Society of London A, 435,* 287-292.

Farmer, J. D., Ott, E., & Yorke, J A. (1983). The dimension of chaotic attractors. *Physica D, 7,* 153-180.

Farmer, J. D., & Sidorowich, J. J. (1991). Optimal shadowing and noise reduction. *Physica D, 47,* 373-392.

Frederickson, P., Kaplan, J. L., Yorke, S. D., & Yorke, J A. (1983). The Liapunov dimension of strange attractors. *Journal of Differential Equations, 49,* 185-207.

Gershenfeld, N. A., & Weigend, A. S. (1993). The future of time series: learning and understanding. In A. S. Weigend & N. A. Gershenfeld (Eds.), *Time series prediction: forecasting the future and understanding the past* (pp. 1-70). Reading, MA: Addison-Wesley Publishing Co.

Glenny, R. W., Robertson, H. T., Yamashiro, S., & Bassingthwaighte, J. B. (1991). Applications of fractal analysis to physiology. *Journal of Applied Physiology, 70(6),* 2351-2367.

Grassberger, P. (1983) On the fractal dimension of the Henon attractor. *Physics Letters A, 97,* 224-226.

Grassberger, P. (1990). An optimized box-assisted algorithm for fractal dimension. *Physics Letters A, 148(1,2),* 63-68.

Grassberger, P., & Procaccia, I. (1983a). Characterization of strange attractors. *Physical Review Letters, 50,* 346-349.

Grassberger, P., & Procaccia, I. (1983b). Measuring the strangeness of strange attractors. *Physica D, 9,*189-208.

Guastello, S. J. (1982). Moderator regression and the cusp catastrophe: Application of two-stage personnel selection, training, therapy, and policy evaluation. *Behavioral Science 27,* 292-272.

Guastello, S. J. (1992). Clash of the paradigms: a critique of an examination of the polynomial regression technique for evaluating catastrophe theory hypotheses. *Psychological Bulletin, 111(2),* 375-379.

Guastello, S. J. (1992b). Population dynamics and work force productivity. *Proceedings of the Second Annual Chaos Network Conference* , 120-127, Urbana, IL: People Technologies.

Guastello, S. J. (1993). Metaphors, Easter bunnies, and empirical verification of chaos theory applications. In F. Abraham, *Chaos theory: secret sect for mathematical mystic versus popular multidisciplinary metamodeling paradigm,* Debate program presented to the American Psychological Society, Chicago, 28 June, 1993.

Guastello, S. J. (1993b). Catastrophes and chaos in NYSE prices: the Crash of 1987. *Proceedings of the Third Annual Chaos Network Conference,* Urbana, IL: People Technologies.

Guastello, S. J. (1994). Does the Chaos Exercise create chaotic behavior? *Chaos Network 6(1),* 7-10.

Hall, P., & Wood, A. (1993). On the peformance of box-counting estimators of fractal dimension. *Biometrika 80(1),* 246-52.

Hammel, S. M. (1990). A noise reduction method for chaotic systems. *Physics Letters A, 148(8/9),* 421-428.

Judd, K. (1994). Estimating dimension from small samples. *Physica D, 71,* 421-429.

Kantz, H. (1993). Noise reduction by local reconstruction of the dynamics. In A. S. Weigend and N. A. Gershenfeld (Eds.), *Time series prediction: forecasting the future and understanding the past* (475-490). Reading, MA: Addison-Wesley Publishing Co.

Kaplan, D. T. (1993). A geometrical statistic for detecting deterministic dynamics. In A. S. Weigend & N. A. Gershenfeld (Eds.), *Time series prediction: forecasting the future and understanding the past* (pp. 415-428). Reading, MA: Addison-Wesley Publishing Co.

Kaplan, D. T., & Glass, L. (1992). A direct test for determinism in a time series. *Physical Review Letters, 68(4),* 427-430.

Kaplan, D. T., & Glass, L. (1993). Coarse-grained embeddings of time series: random walks, Gaussian random processes, and deterministic chaos. *Physica D, 64,* 431-454.

Kostelich, E. J., & Lathrop, D. P. (1993). Time series prediction by using the method of analogs. In A. S. Weigend & N. A. Gershenfeld (Eds.), *Time series prediction: forecasting the future and understanding the past* (pp. 283-295). Reading, MA: Addison-Wesley Publishing Co.

Kostelich, E. J., and Yorke, J. A. (1990). Noise reduction: finding the simplest dynamical system consistent with the data. *Physica D, 41,* 183-196.

Lipsitz, L. A., & Goldberger, A. L. (1992). Loss of 'complexity' and aging: potential applications of fractals and chaos theory to senescence. *Journal of the American Medical Association, 276(13),* 1806-1809.

Liu, T., Granger, G. W. J., & Heller, W. P. (1992). Using the correlation exponent to decide whether an economic series is chaotic. *Journal of Applied Econometrics, 7,* 525-539.

Mayer-Kress, G., Yates, F.E., Benton, L., & Moss, A.J. (1988). Dimensional analysis of nonlinear oscillations in brain, heart, and muscle. *Mathematical Biosciences,* 90, 155-182.

Mayer-Kress, G. (1994). Local measures for nonstationary time-series of phsiological data. *Integrative Physiological and Behavioral Science, July-September, 29 (3),* 205-210.

Nychka, D., Ellerer, S., Gallant, A. R., & McCafferty, D. (1992). Finding chaos in noisy systems. *Journal of the Royal Statistical Society B, 54(2),* 399-426.

Ozaki, T., & Oda, H. (1978). Non-linear time series model identification by Akaike's information criterion. *Proc. IFAC Workshop on Information and Systems,* Compeign, France, October 1977.

Pandit, S. M., & Wu, S. M. (1983). *Time series and system analysis with applications*. New York: John Wiley and Sons.

Peitgen, H., -O, Jurgens, H., & Saupe, D. (1992). *Chaos and fractals: new frontiers of science*. NY: Springer-Verlag.

Pincus, S. M., Gladstone, I. M., & Ehrenkranz, R. A. (1991). A regularity statistic for medical data analysis. *Journal of Clinical Monitoring, 7,* 335-345.

Pineda, F. J., & Somerer, J. C. (1993). Estimating generalized dimensions and choosing time delays: a fast algorithm. In A. S. Weigend and N. A. Gershenfeld (Eds.), *Time series prediction: forecasting the future and understanding the past* (pp. 367-385). Reading, MA: Addison-Wesley Publishing Co.

Pukkila, T., Koreisha, S., & Kallinen, A. (1990). The identification of ARMA models. *Biometrika, 77(3),* 537- 548.

Ramsey, J. B., & Yuan, H., -J. (1989). Bias and error bars in dimension calculations and their evaluation in some simple models. *Physics Letters A, 134(5),* 287-297.

Rapp, P. (1994). A guide to dynamical analysis. *Integrative Physiological and Behavioral Science, July-September, 29 (3),* 311-27.

Rosenstein, M.T., Collins, J.J., & De Luca, C.J. (1993). A practical method for calculating largest Lyapunov exponents from small data sets. *Physica D 65,* 117-134.

Rosenstein, M.T., & Collins, J.J. (1994). Visualizing the effects of filtering chaotic signals. *Computer & Graphics 18(4),* 587-92.

Ruelle, D. (1990). Deterministic chaos: the science and the fiction. *Proceedings of the Royal Society of London A, 427,* 241-248.

Sauer, T. (1993). Time series prediction by using delay coordinate embedding. In A. S. Weigend & N. A. Gershenfeld (Eds.), *Time series prediction: forecasting the future and understanding the past* (pp. 175-193). Reading, MA: Addison-Wesley Publishing Co.

Scargle, J. D. (1989). An introduction to chaotic and random time series analysis. *International Journal of Imaging Systems and Technology, 1,* 243-253.

Shannon, C. (1948). A mathematical theory of information. *Bell System Technical Journal, 27,* 379-423, 623-656.

Skinner, J. E., Carpeggiana, C., Landisman, C. E., & Fulton, K. W. (1991). Correlation dimension of heartbeat intervals is reduced in conscious pigs by myocardial ischemia. *Circulation Research, 68,* 966-976.

Skinner, J.E., Molnar, M., & Tomberg, C. (1994). The point correlation dimension: performance with nonstationary surrogate data and noise. *Integrative Physiological and Behavioral Science, July-September, 29 (3),* 217-234.

Smith, L. A. (1988). Intrinsic limits on dimension calculations. *Physics Letters A, 133,* 283-288.

Smith, R. L. (1992). Estimating dimension in noisy chaotic time series. *Journal of the Royal Statistical Society B, 54(2),* 329-351.

Sugihara, G., & May, R. M. (1990). Nonlinear forecasting as a way of distinguishing chaos from measurement error in a time series. *Nature, 344,* 734-741.

Sugihara, G. (1994). Nonlinear forecasting for the classification of natural time series. *Philosophical Transactions of Royal Society of London A, 348,* 477-495.

Takens, F. (1981). Detecting strange attractors in turbulence. In D. A. Rand and L. -S. Young (Eds.), *Dynamical systems and turbulence* (pp. 336-381) Berlin: Springer-Verlag.

Theiler, J. (1990). Statistical precision of dimension estimators. *Physical Review A, 41(6),* 3038-3051.

Theiler, J., & Eubank, S. (1993). Don't bleach chaotic data. *Chaos, 3(4),* 771-782.

Theiler, J., Eubank, S., Longtin, A., Galdrikian, B., & Farmer, J. D. (1992). Testing for nonlinearity in time series: the method of surrogate data. *Physica D, 58,* 77-94.

Theiler, J., Linsay, P. S., & Rubin, D. M. (1993). Detecting nonlinearity in data with long coherence times. In A. S. Weigend & N. A. Gershenfeld (Eds.), *Time series prediction: forecasting the future and understanding the past* (pp. 429-455). Reading, MA: Addison-Wesley Publishing Co.

Thom, R. (1975). *Structural stability and morphegenesis.* New York: Addison-Wesley.

Tong, H. (1990). *Non-linear time series a dynamical systems approach.* Oxford, England: Oxford University Press.

Vassiliadis, D. (1994). Parametric adaptive control and parameter estimation of low-dimensional chaotic systems. *Physica D, 71,* 319-341.

Wan, E. (1993). Time series prediction by using a connectionist network with internal delay lines. In A. S. Weigend & N. A. Gershenfeld (Eds.), *Time series prediction: forecasting the future and understanding the past* (pp. 195-217). Reading, MA: Addison-Wesley Publishing Co.

Weigend, A. S., & Gershenfeld, N. A. (Eds.). (1993). *Time series prediction: forecasting the future and understanding the past.* Reading, MA: Addison-Wesley Publishing Co.

Wiggins, S. (1988). *Global bifurcations and chaos: analytic methods.* New York: Springer-Verlag.

Wolf, A., Swift, J. B., Swinney, H. L., & Vastano, J. A. (1985). Determining Lyapunov exponents from a time series. *Physica D, 16,* 285-317.

Wolff, R. C. (1992). Local Lyapunov exponents: looking closely at chaos. *Journal of the Royal Statistical Society B, 54(2),* 353-371.

Yao, Q., & Tong, H. (1994). On prediction and chaos in stochastic systems. *Philosophical Transactions of Royal Society of London A, 348,* 357-369.

Yee, H. C., Sweby, P. K., Griffiths, D. F. (1991). Dynamical approach study of spurious steady-state numerical solutions of nonlinear differential equations I: the dynamics of time discretization and its implications for algorithmic development in computational fluid mechanics. *Journal of computational physics 97,* 249-310.

CHAOS IN PSYCHOPHYSICS? HYPOTHESIS TESTING AND NONLINEAR FORECASTING APPROACHES

LAWRENCE M. WARD
Department of Psychology, University of British Columbia
2136 West Mall, Vancouver, B.C. V6T 1Z4 Canada

This paper describes the linear and nonlinear analysis of time series of time estimations. In four experiments subjects produced estimates of 1 sec, 3 sec, random intervals, and "natural" intervals. ARIMA models fitted the time series data well but left variance unexplained. BDS statistics were all significant, rejecting the null hypothesis of no nonlinear structure. Nonlinear forecasting analysis revealed evidence of nonlinear deterministic structure (chaos) in the time series but the data were very noisy, making it difficult to conclude unequivocally that chaos had been discovered in these data. The methods used should be applicable to other data sets.

1 Introduction

In spite of some serious work in psychophysics and related traditions[1,2,3,4] and exhortations to use chaos as a metaphor for psychological processes in general,[5,6] it has not yet been possible to definitively demonstrate the applicability of chaos theory to psychological data. Ideally, we need a success like that of physics in the case of turbulence, both to demonstrate possibility and to delineate where chaos theory might be useful. In this paper, I employ two techniques not often used outside of physics or economics, hypothesis testing and nonlinear forecasting, to explore psychophysical data for nonlinear structure. This type of exploration is intended to supplement more typical studies that use estimates of Lyapunov exponents, correlation or other dimension, frequency analyses, or plots of phase portraits to characterize time series in ways relevant to chaos theory. A similar approach has been applied to economic time series.[7]

2 The Experiments: Constrained, natural, and random rhythms

It has been demonstrated that rhythmical limb movements are usefully described by nonlinear differential equations, such as those that describe the Van der Pol or Raleigh oscillators, operating in parameter ranges that support limit-cycles.[2] Such equations also have chaotic regimes. In such experiments the presence of higher order cycles, or even chaotic noise, from more central sources is often conjectured. It has been demonstrated that the brain, which is inherently a many-dimensional system, in principle could function as a low-dimensional deterministic system under some conditions.[8] Such conditions could be expected to arise when the brain is attempting to control some internal or external process and as such must bring into coherence several neuronal subsystems. Just as in turbulant flow near the transition to chaos the many dimensions of oscillation of the flowing particles come into low-dimensional coherence, so the many dimensions of the brain "particles" (neurons) could come into coherence and be in principle operating according to some low-dimensional, deterministic rule. Crick and Koch[9] have conjectured that the neuronal correlate of consciousness is such a coherent oscillation of brain subsystems. Thus, it is reasonable to expect that instructing people to behave rhythmically is a good way to constrain the brain to produce behavior that reveals such low-dimensional coherence.

I conducted four experiments to this end. In each experiment 6 human subjects produced a single 1000-point time series consisting of the intervals between 1001 successive presses of a computer keyboard spacebar. In the first experiment (1 Second) the subjects were instructed to press the spacebar at 1-second intervals. Two of the subjects from that experiment and four new subjects participated in all of the other three experiments. In the first of these, subjects were instructed to press the spacebar at 3-second intervals (3 Second), in the second according to their own "natural" rhythm (Natural Intervals), and in the third so as to produce a "random" sequence of intervals (Random Intervals). The subjects in the latter three experiments were run in various orders of the three instruction sets. There were no clocks, watches, or other aids to time estimation present for any of the experiments, and it is assumed that subjects produced their intervals according to an internal standard. Thus, the time series analysed were endogenous ones. A simple model of subjects' behavior would be that some cyclic internal process, perhaps different in each condition, was activated and generated a barpress every time it cycled through a canonical state. The question is: can the time series generated by those processes be characterized by low-dimensional, nonlinear deterministic structure, implying that the processes themselves are nonlinear oscillators, which would be evidence of low-dimensional coherence in brain activity?

Tables 1 and 2 show summaries of the data analysed by traditional linear methods. Several aspects of this analysis are notable. First, subjects are fairly good at estimating clock time intervals, and some are better than others at it. Most subjects over-estimated the 1-sec and 3-sec intervals (Table 1), consistent with earlier findings.

Table 1. Pre-chaos data summary for 1 second and 3 second experiments (times in msec).

Subject	Mean	SD	Range Min/Max	ARIMA MA Parameter	95% Confid Interval
1 Second Intervals					
1	1043	114	108/1490	.794	.748/.840
2	1568	187	343/2276	.592	.564/.620
3	1031	207	589/1917	.667	.619/.715
4	1533	221	890/2767	.883	.854/.913
5	1152	264	28/3151	.946	.923/.969
6	1050	66	836/1279	.879	.840/.918
Mean	1230	176	466/2147	.789	
3 Second Intervals					
1	3051	450	18/4589	.712	.662/.762
2	3589	62	1191/5922	.813	.722/.854
3	3018	308	751/4165	.550	.488/.611
				.229	.161/.296
4	2263	673	16/7445	.597	.520/.674
5	4312	918	658/8222	.794	.749/.838
6	4521	1269	16/8362	.504	.433/.575
Mean	3458	706	608/6451	.600	

Second, natural interval estimates (Table 2) were the shortest and their average is about the same duration that humans typically estimate without error (700 msec). Random interval estimates (Table 2) were somewhat longer than the 1-sec estimates and had the largest ranges and standard deviations. Clearly subjects performed differently in the different experimental tasks.

Tables 1 and 2 also summarize an ARIMA analysis of the data of the four experiments. In this analysis, autocorrelation and partial autocorrelation functions were examined to determine which type of model would fit best. None of the time series were stationary (a chronic problem with psychological time series) and so all were differenced once to render them so. The resulting time series were analysed by fitting the most appropriate model, AR or MA,. As can be seen in Tables 1 and 2,

Table 2. Pre-chaos data summary for random interval and natural interval experiments (times in msec).

Subject	Mean	SD	Range Min/Max	ARIMA MA Parameter	95% Confid Interval
Random Intervals					
1	1515	1166	137/6609	1.327	1.268/1.386
				-.349	-.410/-.288
2	860	790	14/8892	*null autocorrelation func.*	
3	1972	1634	10/24573	*null autocorrelation func.*	
4	1440	1086	11/9186	. 711	.651/.770
				.262	.202/.322
5	2159	1580	34/9939	.832	.766/.898
				.140	.075/.206
6	1892	1891	101/10852	*null autocorrelation func.*	
Mean	1640	1358	51/11675	.487	
Natural Intervals					
1	987	173	494/1467	.576	.527/.626
2	1000	444	241/4093	.567	.516/.619
3	319	98	139/737	.820	.788/.852
4	682	151	13/1254	-.355(AR)	-.450/-.259
				.579	.474/.683
5	246	96	48/904	.720	.657/.783
6	808	361	335/1745	.369	.314/.423
Mean	674	221	212/1700	.469	

the best model was generally ARIMA (0,1,1) (e.g., one MA parameter), in which random shocks persist for only one observation. Occasionally an additional MA parameter was needed and in one case an AR parameter (negative) was required. All fits were very good and the ACFs and PACFs of the residuals of the fits for all series were uniformly null. The different instruction sets were also differentiated by this analysis, in that MA parameter size was largest for the 1 second estimates, next largest for the 3 second estimates, and smallest for the natural intervals. Moreover, the random instructions produced some dramatic behavior: three subjects had null autocorrelation functions (expected for a random series) and another required an additional negative MA parameter. The other two needed additional MA parameters to describe their data, in contrast to their behavior in the 3 second and natural interval conditions. Thus, all subjects behaved differently in the random condition, although not all behaved randomly, consistent with earlier results. Although this is the first time such data have been analysed in this way, and useful linear theories could be developed to explain these results, I now turn to analysis of these data for nonlinear deterministic structure.

3 Hypothesis testing with the BDS statistic

This approach has been taken by a group of economists[10,11] based on work done by physicists on what is called the *correlation integral*, $C_m(\epsilon,T)$, where T is the length of a time series and ϵ the size of a neighborhood in phase space.[12] The correlation integral measures the extent to which the different 'm-histories' of a time series clump together (relative to the neighborhood size ϵ) in phase space. An m-history is a set of successive terms in a time series where m is the number of terms considered (the embedding dimension). For example, in the time series {3, 5, 8, 2, 5} the m-histories of length 2 are {3,5}, {5,8}, {8,2}, and {2,5}. Each m-history is taken as a set of Cartesian coordinates in phase space. *M*-histories from random time series fill up phase spaces of any dimensionality uniformly, so that for a random process $C_m = (C_1)^m$, while those from time series generated by chaotic processes tend to clump together in a subspace in the strange attractor and thus C_m does not change with m when m is larger than the dimension of the attractor. Brock and his asscoiates[13,14] have developed a family of statistics based on the quantity $D_m = C_m - (C_1)^m$, that measures the extent to which the m-histories of a particular time series diverge from those that characterize a series derived from IID (Independent Identically Distributed) noise. I have used the statistic below in the present work:

$$\mathrm{BDS/SD} = (\mathrm{T-m+1})^{1/2}\, \mathrm{D_m} \,/\, \mathrm{SD} \tag{1}$$

where *SD* is the standard deviation of this statistic.[14] *BDS/SD* is distributed N(0,1) under the null hypothesis of an IID time series. Thus, this statistic can be treated as a standard z-test. A sufficiently large z-score indicates rejection of the null hypothesis of IID for a particular time series with probability of Type I error equal to α. Unfortunately, departures from linearity other than those indicative of nonlinear deterministic structure, particularly conditional heteroscedasticity, can be detected by the *BDS/SD* statistic.[7] Also, some nonstationary linear processes and even some stationary nonlinear processes of less interest here, such as nonlinear stochastic processes, can generate sufficiently large values of the statistic to reach statistical significance. Thus, a large value of *BDS/SD* is not sufficient evidence that the underlying process is chaotic. However, it should be applied to short time series, like the present ones, to obtain statistically convincing evidence for nonlinear structure.[15] Measuring *BDS/SD* for some psychophysical data is thus a reasonable first step in obtaining evidence for chaos in psychophysics.

Table 3. Summary of the BDS/SD values.

Subject	1 Second	3 Second	Random	Natural
1	6.65	16.54	3.44	15.94
2	10.05	17.24	5.18	22.03
3	8.43	11.08	6.08	33.56
4	14.06	15.69	4.01	14.46
5	15.00	14.70	9.63	8.15
6	5.97	6.77	8.24	7.59
Mean	10.02	13.67	6.10	19.96

BDS/SD statistics were calculated for the normalized residuals from the ARIMA analysis described above. Since the differenced series were stationary, and the ARIMA residuals had all detectable linear structure removed, it is unlikely that any linear process would give rise to large values of the statistic. This is the recommended procedure for using the *BDS/SD* statistic.[14] Table 3 summarizes the results of this analysis (ϵ varied for the 1 Second experiment results but was set at 0.0625 for the others; $m = 5$ for all experiments). Not shown in Table 3 are the results of calculating *BDS/SD* for 10 different random shuffles of each series in the 1 Second experiment. These all gave mean *BDS/SD* near zero, as they should for random series. Thus, the large values of *BDS/SD* shown are not artifacts of the particular numbers produced by the removal of linear structure. Table 3 shows that there is considerable evidence for nonlinear structure in these data. All *BDS/SD* statistics are very large and all are significant at the $\alpha = .01$ level ($z = 2.33$). Thus, it is possible to reject the hypothesis that the residuals of the ARIMA analysis are IID for all subjects in all of the experiments. Interestingly, the average value of the statistic varies across the experiments, being smallest for the random intervals, intermediate for the time estimation conditions, and largest for the natural intervals. Although it is not really appropriate to use *BDS/SD* as a measure of *amount* of nonlinearity, these differences are suggestive that the evidence for nonlinear structure is greatest in the natural intervals series.

4 Nonlinear forecasting with the Casdagli algorithm

Out of sample forecasting is a relatively new approach to discovering deterministic structure in time series. The crucial result is that of Takens,[16] who proved that the delay vectors $\mathbf{x_i} = x_i , x_{i-\tau} , ..., x_{i-(m-1)\tau}$ of embedding dimension $m > 2d$ and delay time τ, selected from an observed scalar time series generated by a D-dimensional attractor of a deterministic dynamical system with d degrees of freedom, fill out a "reconstructed" attractor that is diffeomorphic to the original attractor. Thus, inferences about the system generating the time series can be based on results with the time series. Farmer and Sidorowich[17] showed that for such time series excellent short-term out-of-sample forecasting was possible with a *local approximation* to the time evolution, i.e. using only nearby states in the phase space. Casdagli[15] refined their and his own previous work and presented an explicit algorithm for doing such forecasting that contrasted the results for low-dimensional deterministic processes with those for linear and nonlinear stochastic processes. He used this technique to confirm the presence of low-dimensional chaos in some physical systems, such as fluid flow near near the transition to turbulence and two coupled diodes. He also demonstrated that some time series thought to be characterized by low-dimensional chaos, like measles outbreaks and EEG data, failed to provide significant evidence of it.

Casdagli's[15] algorithm is fairly simple to describe although somewhat complex to implement in a computationally efficient way. Given a time series, $x_1 , x_2 , ...,$ x_N , divide it into two parts, a fitting set and a test set. Choose embedding dimension m, delay time τ, and forecasting time T. From the test set choose a delay vector $\mathbf{x_i}$ for a T-step ahead forecast. Compute the distances of $\mathbf{x_i}$ from all delay vectors $\mathbf{x_j}$ in the fitting set and choose the k nearest neighbors of $\mathbf{x_i}$. Fit a linear prediction equation ($m+1$ terms) to the delay vectors in the fitting set by ordinary least squares and use the obtained coefficients β_j to make the T-step ahead forecast for $\mathbf{x_i}$. Calculate the absolute error for the prediction. Repeat for all delay vectors in the test set to obtain the overall root mean square error (RMS error) for this value of k and m. Repeat for values of k from $2(m+1)$ to the number of delay vectors of size m in the fitting set.

Vary the embedding dimension m and study the curves of RMS error versus k for different m.

I implemented this algorithm on a HP Vectra 486/66U computer (runs at 66 MHz) with 24 Mb of RAM and a 250 Mb hard disk system. The program is written in C and I used singular value decomposition to estimate the linear model for each

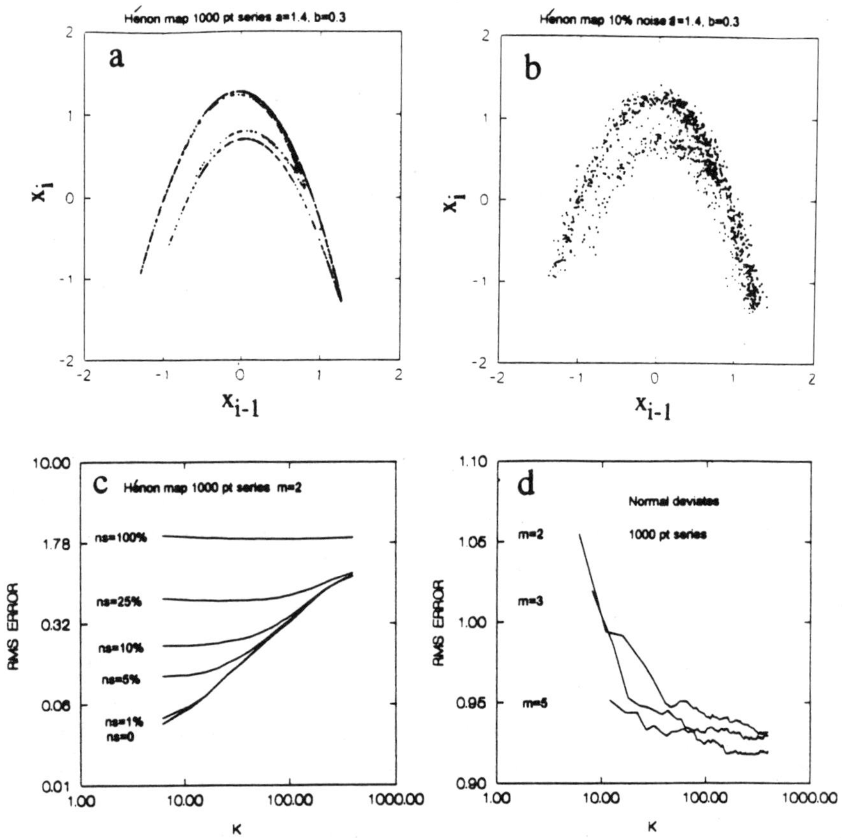

Figure 1. (a) the Henon attractor, (b) Henon with 10% noise; (c) forecasting result for the Henon map (0% through 100% noise); (d) forecasting result for 1000 normal deviates.

forecast. In all implementations, I used $\tau = 1$ and $T = 1$. My program is limited by the largest memory model available to the compiler so that I could only vary k up to 400 for 1000-point time series with a fitting set of 800 points and a test set of 200 points. This limitation should not be serious since low-dimensional deterministic dynamics yield very small error for very low k, and very large error for large k, so that degradation of forecasting from small to large k is dramatic, usually several hundred percent. Figure 1c (ns=0; the lowest curve) displays such a result for the Henon map:

$$x_i = 1 - 1.4\, x^2_{i-1} + 0.3\, x_{i-2} \tag{2}$$

The strange attractor of the map is shown in Figure 1a. Observational noise has serious effects on this forecasting curve. Figure 1b shows the Henon attractor with 10% observational noise added. Figure 1c shows the effects of these and other noise amounts on the forecasting curve. The noise decreases forecastability at low k and decreases the degradation of predictability with increasing k that is diagnostic of low-dimensional deterministic structure. When the added noise dominates, forecastibility will be poor and will be greatest at largest k, as in Figure 1d, which shows the results for a 1000 point series of normal deviates. For intermediate levels of noise, the forecastability curve has minimum error at moderate k (near or greater than k = 100) and not much difference between the minimum and maximum errors (a few %). Such a pattern is also expected for a nonlinear stochastic process, so such results are inconclusive.

I applied this algorithm to the time series described above. All time series were first differenced to render them stationary since the algorithm fails for nonstationary series. A range of m values was explored to find the minimum error curves for each series. Typical results for one subject who participated in all four experiments are shown in Figure 2a-d. As shown in Figure 2d, the strongest evidence for low-dimensional nonlinear deterministic structure comes from the natural interval series. Here the curves for 5 of the 6 subjects were similar to the one shown. Minimum error is clearly reached at low k (6,8,11,13,33, and 48 for the six subjects). The other conditions also show some evidence for predictable nonlinear structure, although not as clearly as for the natural intervals. Unfortunately, degradation of predictability at high k is not great. For the natural interval experiment it ranges from 13% to 24%. Even extrapolating the curves to k = 800 only adds about 5 percentage points, yielding 18% to 29%. The other experiments give less, as low as 5% in some cases. So these results are in the ambiguous range. They are either very noisy or nonlinear stochastic.

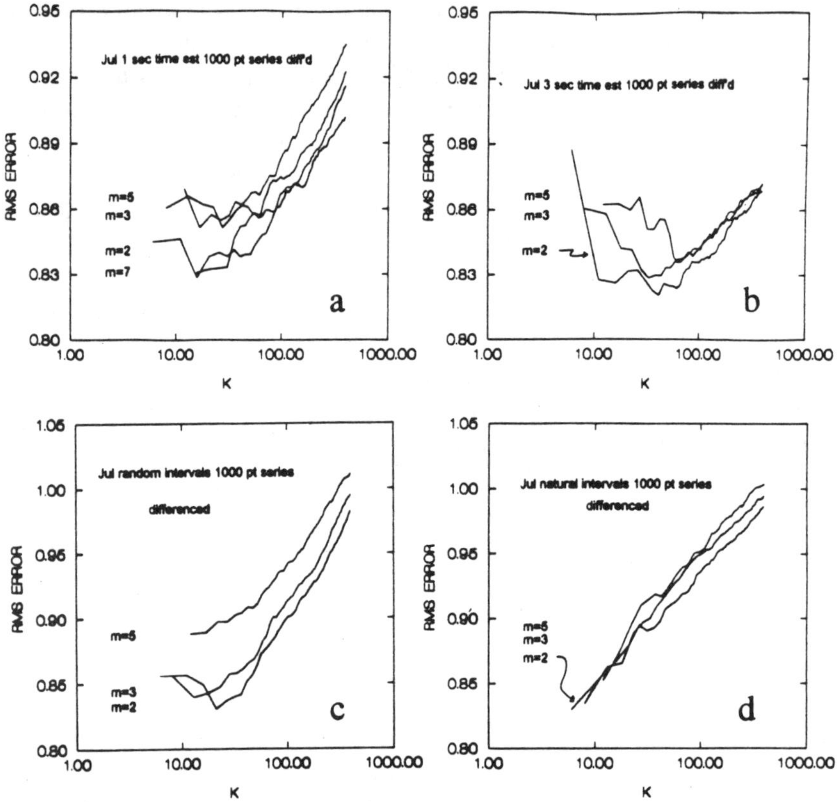

Figure 2. Forecasting results for a representative subject.

One encouraging sign is that not a single one of the curves for the 24 available time series resemble those of Figure 1d. Thus, there is no evidence whatsoever in these data for a linear stochastic model. Another encouraging sign comes from an experiment reported by Ward and West (1994). In that experiment subjects attempted to iterate the following version of the logistic equation:

$$x_i = 3.66 (x_{i-1}) (1- x_{i-1}) \tag{3}$$

Figure 3 shows the results of applying the forecasting algorithm to both the series generated by iterating Equation 3 and to the series generated by one typical subject. The curve for the logistic is typical, with excellent predictability at low k decreasing with increasing k. The curve for the subject's data is similar but at a higher level of error. It does not show the bowing that observational noise produces in such curves, rather it is parallel to the logistic curve. This implies that the subject's data contain low-dimensional nonlinear structure, although we also showed that it can't be the same as the logistic, since the subject's phase plot is not the same as that of Equation 3. Since some of the curves from the present experiments resemble the subject's in Figure 3, albeit with less prediction improvement for small k, it seems reasonable to assert the presence of nonlinear structure there also.

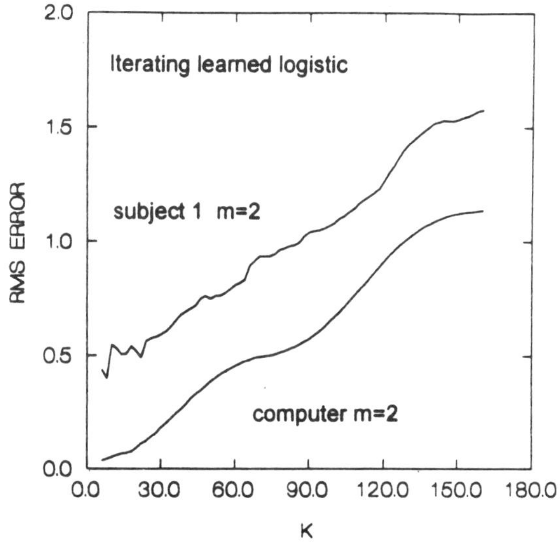

Figure 3. Forecasting curves for the logistic and a subject attempting to iterate it

5 Conclusions

The analyses described above reveal the presence of nonlinear structure in psychophysical data. A conservative conclusion would be that this nonlinear structure is not low-dimensional chaos since forecasting is not very good, and improvement not very great, for low k values. It is possible that nonlinear stochastic models would have some success in accounting for this structure, but it is also possible that what we are seeing is a very noisy manifestation of low-dimensional chaos. This noise problem will persist, since psychophysical data are among the least noisy in psychology and even here the noise is substantial. Techniques for either removing the noise or modeling noisy chaos are needed. We are experimenting with the effects of intrinsic noise on forecastability in nonlinear time series on the assumption that it is most characteristic of the noisy coherence we are expecting in the brain.

Acknowledgements

This research was supported by a grant from the Natural Sciences and Engineering Research Council. I thank Narly Golestani for research assistance.

References

1. R.A. Gregson, *Nonlinear psychophysical dynamics.*(Lawrence Erlbaum Associates, Hillsdale, NJ, 1988).
2. B.A. Kay, *et al*, *J. Exptl. Psych.: Hum. Percep. Perfor.* **17**, 183-197 (1991).
3. A. Neuringer and C. Voss, *Psych. Sci.* **4**, 113-119 (1993).
4. L.M. Ward in *Fechner Day 91* ed. G.R. Lockhead, (International Society for Psychophysics, Durham, NC, 1991).
5. S. Barton,. *Amer. Psych.*, **49**, 5-14 (1994).
6. P.R. Killeen in *The computer and the brain: Perspectives on human and artificial intelligence*, ed. J.R. Brink and C. R. Haden (North-Holland Elsevier Science Publishers, Amsterdam 1989).
7. D. Hsieh, *J. Finance,*1839-1877 (1991).
8. C.A. Skarda and W.J. Freeman, *Behav. Brain Sci.*, **10**, 161-195 (1987).
9. F.H.C. Crick and C. Koch, *Sem. Neurosci.*, **2**, 263-275 (1990).
10. W.A. Brock in *Beyond Belief: Randomness, Prediction and Explanation in Science*, ed. J. Casti and A. Karlqvist (CRC Press, Boca Raton FL, 1991)
11. W.A. Brock and A. Malliaris, *Differential Equations, Stability and Chaos in Dynamic Economics.* (North Holland Press, Amsterdam, 1989). .
12. P. Grassberger and I. Procaccia, *Phys. Rev. Let.*, **50**, 346 (1983)..
13. W.A. Brock, W.D. Dechert and J. Scheinkman, *A test for independence based on the correlation dimension.* (Department of Economics, University of Wisconsin, Madison, University of Houston, and University of Chicago, 1986)..

14. W.A. Brock, D. Hsieh and B. LeBaron, *A Test for Nonlinear Dynamics.* MIT Press, Cambridge, MA, 1990). .
15. M. Casdagli, *J. Royal Stat. Soc.*, **54**, 303-328 (1992).
16. F. Takens, in *Dynamic Systems and Turbulence*, ed. D. Rand and L.-S. Young (Springer Verlag, New York, 1981).
17. J.D. Farmer and J.J. Sidorowich, *Phys. Rev. Let.*, **59**, 845-848 (1987).
18. L.M. Ward and R.L. West, *Psych. Sci.*, **5**, 232-236 (1994).

INTER- AND INTRA-TRIAL DYNAMICS IN MEMORY AND CHOICE

KEITH CLAYTON & BARBARA FREY
Department of Psychology
Vanderbilt Universirt
Nashville, TN, USA

1 Introduction

Models of memory and choice typically focus either on the dynamics within the experimental trials or between them. Within-trial (intra-trial) dynamics involve processes initiated by the cue and ended at the subject's response. The central interest of contemporary information processing models of memory and choice are intra-trial dynamics (for example, Gillund & Shiffrin[1]; Hintzman[2]). In these models, the dynamics to be captured include the processes of cue encoding, retrieval, and decision. Between-trial (inter-trial) dynamics, on the other hand, characterize the impact that the actions and outcomes of the current trial have on performance during the next. Phenomena observed in studies of inter-trial dynamics include learning, forgetting, sequential effects in choice response time (see for example, Link[3]; Luce[4]), and priming effects in tests of memory (see for example, McNamara[5]).

Results recently reported by Cooney and Troyer[6] suggest the fruitfulness of examining inter-trial dynamics in a task, the Sternberg task, typically used to study intra-trial processes. The Sternberg task involves presenting subjects with a short list of items, called the memory set, and testing with recognition tests in which a single item is presented. The subject judges whether the test item is or is not a member of the memory set, and the principal dependent measure is the duration between the time the test item is presented and a recognition response is executed. The task was originally designed by Sternberg[7] to investigate processes in recognition tests of short-term memory for small lists, that is, intra-trial events, and involved manipulating the size of the memory set. Two procedures have been used, one involving a new list for each trial, and a "fixed-set" procedure, in which the subject is repeatedly tested on the same memory set.

Cooney and Troyer's innovation was to treat the successive response times in the Sternberg task as a time series. Being interested in the dynamics involved when the same memory is tested over and over, Cooney and Troyer employed the fixed-set procedure. Five letters (C, J, M, P and X) constituted the memory set for each of a total of 250 test trials for 5 adult subjects and 5 children. Half the trials were positive (the test item is a member of the memory set and the correct response is "Yes"), and half were negative (a foil is presented and the correct response is "No").

Cooney & Troyer's principal analysis was to compute correlation integrals as suggested by Grassberger and Procaccia[8]. This is an auto-correlation technique, to be described more fully later, that provides an estimate of the correlation dimension. From a theory-construction perspective, the advantages of the technique are two-fold. First, it is designed to help evaluate whether the time series comes from a deterministic or a random process. Seconds, if the data are the result of a deterministic process, the size of the correlation dimension instructs us about the number of variables needed to account for the data.

Cooney and Troyer's estimates for the correlation dimension, for all five of their adult subjects, and four out of five of their younger subjects, were consistent with a deterministic system with dimension three or less.

We saw three important implications from these results. First, the possibility that reaction-time measures are the result of a deterministic process stands in sharp contrast to the traditional view of reaction time as a random variable, or as the result of a probabilistic process. Second, constructing a deterministic model of the data is greatly facilitated by the finding of dimensionality of three or less. Finally, we felt the Cooney and Troyer results encourage the development of a general dynamic memory model that links the goals of accounting for the dynamics within and between trials. As we see it, such a model would specify the dynamics from cue to response in such a way that the initial conditions for a trial would depend on the trial-ending states of the preceding trial. That is, the model would incorporate a loop between intra- and inter-trial events clearly linking the dynamics between and within trials.

We report two experiments, both designed to yield time-series data in choice response-time tasks. The first experiment replicates Cooney and Troyer's experiment. The second was motivated by the goal of applying dynamic analyses to response-time tasks more generally. For this purpose, response-time tasks are treated as on a continuum, ordered on the basis of their memory requirement. At one end would be the **simple response-time** task, that is, simply responding to the onset of a signal. Next, would be a simple **two-choice classification** task in which the only memory requirement is to remember the response rule. (For example, if X, press key "1"; if O, press "2"). Next, is a two-choice, same-different, task in which the response rule is to make one response if the stimulus is the same as last presented (previous trial), otherwise make the other response (**same-different lag-one** task). Deepening the memory involvement, the next task requires a same-different judgment, "same" if the test stimulus is the same as the stimulus presented two trials back, etc. (**same-different lag-two** task). In the second experiment, subjects were given the two-choice classification task, the same-different one-lag task, and the same-different two-lag task.

2 Experimental Methods

We present the methods for both experiments, and then present the data.

2.1 Experiment 1: Cooney & Troyer Replication

Procedures

The procedures were designed to follow closely those used by Cooney and Troyer. Each trial began with five asterisks presented in the center of the computer monitor. Subjects initiated the trial by depressing the 2 button on the numeric key pad. They were instructed to keep the 2 button depressed until they saw the test letter. The asterisks were erased at the button press, replaced two seconds later by the memory set (C,J,M,P,X), displayed for two seconds. The memory set was then erased, followed by a two-second delay, then the presentation of the test letter. On half of the trials, the subject would see a test letter that matched the memory set (e.g., M), and the test letter for the other half of the trials was selected randomly from the letters not in the memory set. The subjects were instructed to maintain high accuracy and to respond as quickly as possible by pressing either the "yes" or "no" key (i.e., the 1 and 2 keys, respectively). Response latency was measured from the time the test letter was presented to the time a response key was pressed.

Subjects BP and JB participated for 250 trials. Subject BF participated in the same Sternberg paradigm on two occasions. The two sessions differed from each other and from the sessions for the other two subjects by the number of trials. Session one consisted of 210 trials and session two consisted of 500 trials. The purpose of session two was to examine the effect of trial number on the estimate of the correlation dimension.

Subject KC participated for 256 trials under slightly different procedures. Specifically, a response was made to the test item (2 key pressed) and then a classification response was made (1 key if "yes", 3 key if "no"). Response time was measured from the first to the second response. These procedures are very close to those used by Cooney and Troyer, but as seen below, produced results indistinguishable from those from the other subjects.

Subjects

There were four subjects, one female and three males. Two subjects are undergraduate students, one is a graduate student, and one is a faculty member at Vanderbilt University.

Apparatus

All data were collected using IBM-compatible PCs. Response latencies, display timing, and display synchronization were controlled through either Microcomputer Experimental Laboratory (MEL) software or True Basic software utilizing an internal clock board.

2.2 Experiment 2: Two-choice Classification vs. Lag-one vs. Lag- two

Procedures

Subjects initiated each trial by pressing a key. After initiating the trial, a fixation point was displayed for 400 milliseconds. The fixation point was then erased, followed by presentation of the test letter. In all conditions the test letter was randomly selected from two possibilities: X or O. Subjects were instructed to maintain high accuracy and to respond as quickly as possible by pressing either the 1 or 2 key. Three conditions were used. In the **classification** task, the subject's task was to press the 1 key when the letter O was displayed and to press the 2 key when the letter X was displayed. In both same-different tasks, the subject was to make a same-different judgment between the current test letter and a previously presented test letter. If the current test letter matched the previous test letter, the subject was instructed to press the 1 key for "same". Otherwise, the subject was to press the 2 key for "different". In the **same-different lag-one** condition, the match was between the current test letter and the letter presented on the immediately preceding trial. In the **same-different lag-two** condition, the match was between the current letter and the test letter presented *two* trials earlier.

Each test condition was given in a session consisting of 100 practice trials followed by 500 experimental trials. Response latency was measured from the time the test letter was presented to the time a response key was pressed. In addition, since interesting dynamics may involve the decision to initiate a trial, the time from the last trial to initiating the next was also recorded.

Subjects

One male and one female participated, both graduate students at Vanderbilt University. The female had participated in Experiment 1, the male had not.

Apparatus

The same apparatus, using the MEL software, was used as in the experiment above.

3 Results

3.1 Estimating the correlation dimension

In this report, we focus on the application of the correlation integral analysis. We do this because it was the primary choice of Cooney and Troyer, and because it has become a widely-used tool for dynamic analyses. It has received an enormous amount of critical attention, with concern about the effects of data length, delay length, etc., (for recent reviews see Abarbanel, Brown, Sidorowich, and Tsimring [9], and Pritchard and Duke [10]). And alternative dimension estimators have recently been suggested (e. g., Judd [11]).

The correlation integral analysis, a form of auto-correlation, is briefly described here, so that we can highlight some issues, but Grassberger and Procaccia [8] should be consulted for details. The time series values are first "embedded" in successive N-tuples of values. For example, if $N = 3$, the time series values are treated as successive triplets. Each N-tuple is then treated as a point in N-dimensional space (the "embedding dimension") and the distances among all pairs of points in that space are calculated. For each dimension a function, $C_r(I)$, is computed relating the proportion of distances in that space that are less than a given distance, or radius, r. The linear portion of the function relating $log[C_r(I)]$ to $log(r)$ is used to estimate the correlation dimension for that embedding dimension. The estimated slope of the linear region is the estimate of the correlation dimension.

The "dimensionality" of the time series, in the sense used by Cooney and Troyer, is then evaluated by determining whether and where the slopes of the functions stabilize, or "saturate", as the embedding dimension is increased. If the data were produced by successive independent samples from some probability distribution, that is, are simply "stochastic noise", then the slopes would not saturate, they would continue to equal the size of the embedding dimension. If the dimension of the data were as low as three, say, then the slopes should stabilize at embedding dimension three.

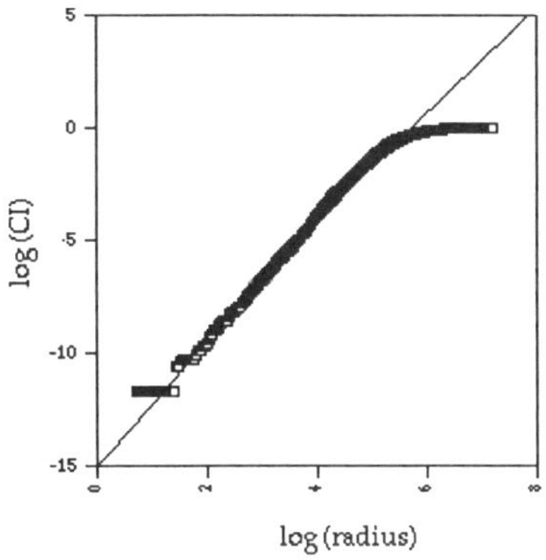

Figure 1: A plot of $log[C_r(I)]$ versus $log(r)$ for one subject (with embedding dimension equal to 3), showing the sigmoidal shape typically obtained.

Although the computational steps for calculated $C_r(I)$ are unambiguous, determining the linear portion of the slopes is not. Consider, for example, the next figure, plotting $C_r(I)$ from one of our subject's time series data, for an embedding dimension of size three.

Figure 1 illustrates the function typically observed when $log[C_r(I)]$ is plotted against $log(r)$. As has frequently been noted, the function seems to have three parts: an initial irregularly rising portion, an intermediate linear part, and a final curve leveling off to $log(0)$. It is the middle linear portion, the "scaling" region, that is used to estimate the correlation dimension. What is needed is a rule for selecting the linear portion, to allow the calculation of a best fitting straight line and its slope. Several algorithms have been suggested. The program Chaotic Data Analyzer, uses "the middle one-quarter of the vertical scale" (Sprout & Rowlands[12]).

One point deserves emphasis: any algorithm that uses a portion of the function extending in either direction beyond the linear portion will result in an underestimate of the slope. This point occurred to us after initial attempts to compute the slopes. Our first efforts involved plotting the $log[C_r(I)]$ functions and locating the linear portion by visual inspection. However, when we compared our results with those produced by using the Chaotic Data Analyzer program, we met consistent disagreement. Our slope estimates were virtually always higher. Since we were confident that our estimates used only the linear portion of the function, and most of the computations performed by the Chaotic Data Analyzer are opaque to the user, we decided to forego its use.

Nevertheless, given the extensive time needed to locate the linear portion by visual inspection, an objective computer algorithm is preferred. After some practice comparing possible algorithms with the results obtained from visual inspection with these data, we developed such an algorithm. The algorithm we settled on is described next.

Algorithm for determining slope

First, the values of $log[C_r(I)]$ for each r are calculated following Grassberger and Procaccia, yielding the function for $log[C_r(I)]$ against $log(r)$. Then, following Albano, Mees, de Guzman, & Rapp [13], the first and second derivatives of this function were calculated. The first derivative was calculated using

$$d_i' = \frac{C_i(i+5) - C_i(i)}{r_{i+5} - r_i} \tag{1}$$

where the index i moves over the values of r. The second derivative was calculated using

$$d_i'' = \frac{d_{i+1}' - d_i'}{r_{i+1} - r_i} \tag{2}$$

The lower boundary for the scaling region was defined as the first point at which a zero first derivative was preceded by ten non-zero first derivatives. The transition from the middle linear portion of the correlation integral to the upper region was estimated using the second derivative. As the points approach the upper region, the second derivative decreases. The upper boundary was calculated by testing the variance of the second derivative for the point at which the variance of eight consecutive points was less than a fixed value of four. Once the middle linear region was so identified, slopes and r^2 indices of fit were computed.

3.2 Experiment 1: Cooney & Troyer Replication

Figure 2 shows the estimates of slopes for each embedding dimension, treating each subject's data separately. The estimates from the two sessions of subjects BF are also plotted separately. Except possibly for the long session of subject BF, none of the functions appear to be leveling off by embedding dimension 10. Certainly in no case is there evidence for a dimensionality of 3 or less.

For all subjects, the estimated slopes begin to fall short of the embedding dimension for dimension larger than five or six. It is conceivable that the slopes might eventually saturate somewhere beyond ten. On the other hand, slopes from pure noise also produce values less than the embedding dimension when the data set is finite (see Rapp [14]).

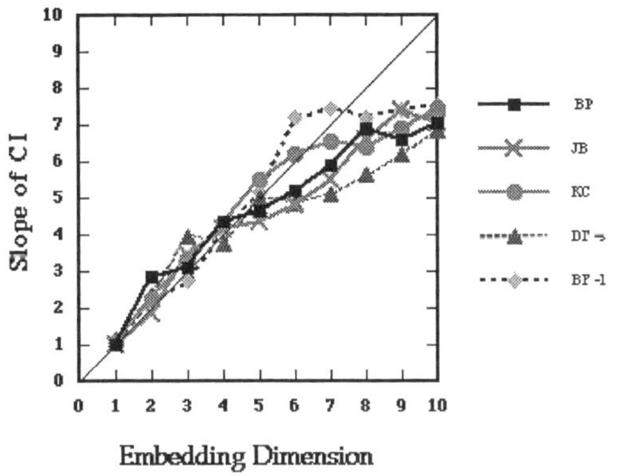

Figure 2: Slopes as a function of embedding dimension for all four subjects. Two plots are presented for subject BF, one for the shorter 210-trial session, labeled 's', the other for the longer, 500 trial- session, labeled 'l'.

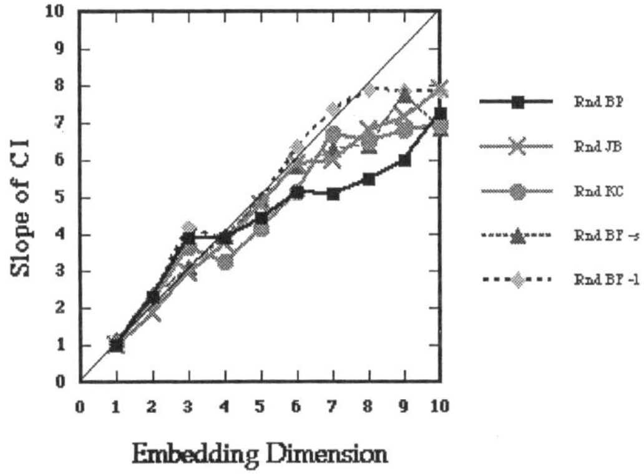

Figure 3: Slopes as a function of embedding dimension for all four subjects. Scrambled data trials.

The slope estimates were then repeated after the time series were scrambled. Scrambling the data acts as a control because it preserves the distribution of the reaction time measures across trials but destroys any autocorrelation. The informative comparison, then, is between the dimension estimates for the unscrambled and their scrambled control.

The results from scrambling the data are shown in Figure 3. The slope estimates from the scrambled data show the same form as for the unscrambled data. There is no evidence here for a dimensionality less than 10.

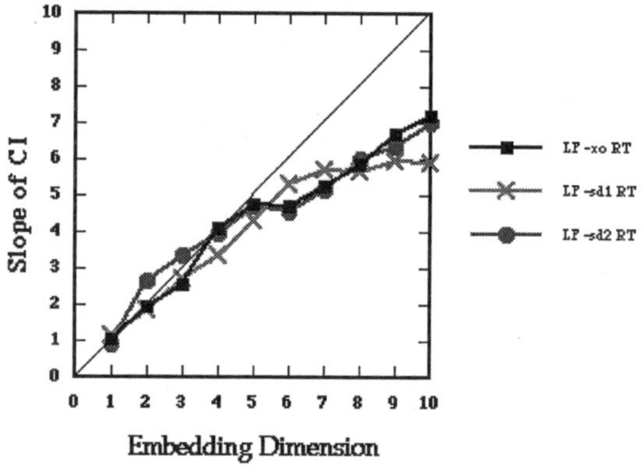

Figure 4: Slopes as a function of embedding dimension for one subject, all three tasks. RT data.

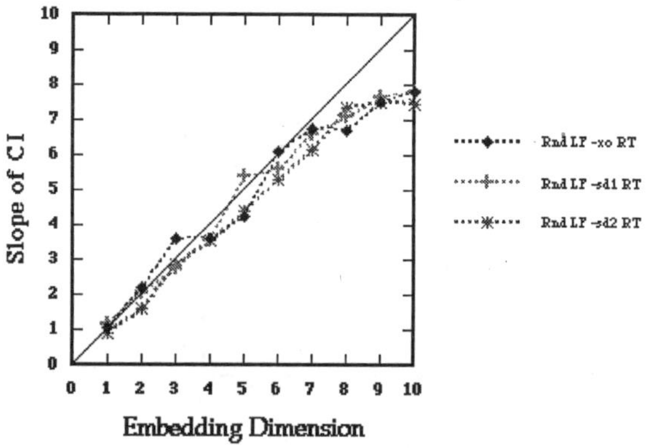

Figure 5: Slopes as a function of embedding dimension for one subject, all three tasks. RT data. Scrambled data trials.

3.3 Experiment 2: Two-choice Classification vs. Lag-one vs. Lag- two

Slope estimates were obtained for each of the three tasks for both subjects. The results for both were similar, so only the data for one will be shown. Figure 4 shows the estimates from the reaction time data unscrambled. The estimates from the scrambled data are in Figure 5.

Overall, the data for all three tasks resemble those from Experiment 1. No evidence is obtained for dimensionality less than 10, and the scrambled data is similar to the unscrambled data.

Slope estimates from the trial-initialization latency data are presented in Figure 6 (unscrambled) and 7 (scrambled).

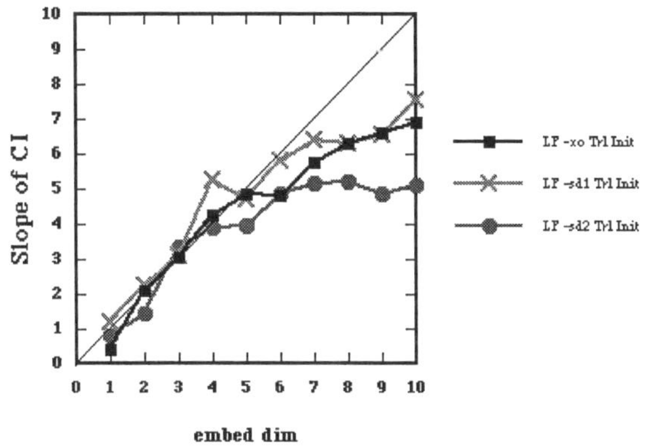

Figure 6: Slopes as a function of embedding dimension for one subject, all three tasks. Trial-initiation latency data.

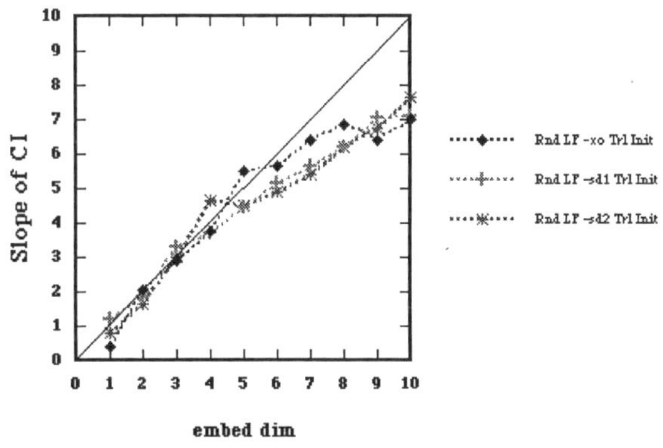

Figure 7: Slopes as a function of embedding dimension for one subject, all three tasks. Trial-initiation latency data. Scrambled data trials.

Again, the slopes fail to level off before dimension 10. There is a hint in the lag-two same-different task that the slopes level off near a value of five (Figure 6), but this trend is not apparent in the data from the other subject.

3.4 Discussion

To summarize the main findings: in four different two-choice reaction-time tasks, ranging from a simple classification task to the Sternberg recognition task, no evidence is found for dimensionality less than ten, using the Grassberger-Procaccia algorithm.

Why are our Sternberg results different from those of Cooney & Troyer? Slight differences in experimental procedure remain, but they seem minor. In any case, we have not managed to generate an interesting theoretical reason for investigating procedural differences, so this research direction remains unlikely for us. Our current guess is that Cooney and Troyer's algorithm for locating the scaling region may have allowed for an underestimate of the slopes.

Although the results fail to replicate Cooney and Troyer's low-dimensionality finding, they do not deflect us away from the theoretical enterprise that their results stimulated, namely, evaluating the application of nonlinear dynamic models to the phenomenon in the empirical literature of memory and choice.

4 Linking Inter- and Intra-trial Dynamics

The model sketched here is developed with the goal of accounting for dynamics occurring both between and within trials of choice and memory tasks. Completely filled in, the model is intended to simulate the dynamics from the presentation of a cue or probe, through the execution of a response, the reception of feedback and other inter-trial activities, back to the presentation of the next cue or probe.

4.1 Dimensions of the model

Figure 8 shows the state variables, or dimensions, of the system. The figure maps the dynamics of six variables among three agents: the environment, working memory, and long-term memory.

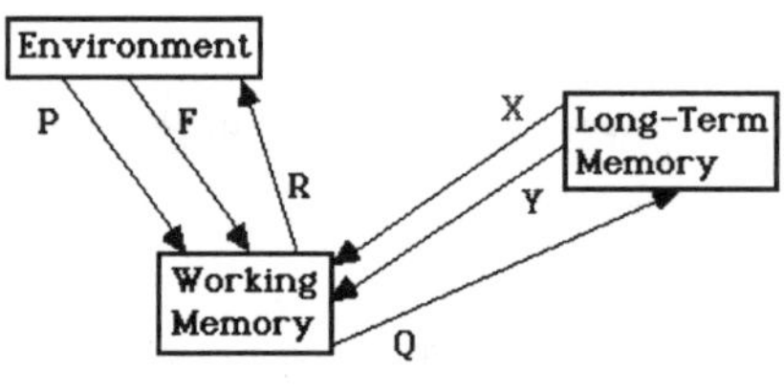

Figure 8: The variables P, F, R, Q, X, and Y are the dimensions of the model.

The effects of processing the **probe**, or cue, from the outside environment is indexed in the variable **P**. Its initial value depends in part on the outcome of the preceding trial, i.e., is a function of feedback, reflected in **F**, and its growth rate and asymptote could index the "quality" ("transfer-appropriateness") of the probe.

As the value of **P** increases in working memory, it allows a **query** of long-term memory, indexed by **Q**. **Q** is positively related to **P**.

The principal variable for within-trial dynamics is the index **X**, which is the measure of memory "strength" associated with eliciting the correct response. The higher the value of **X**, relative to competitive dispositions measured by **Y**, the higher the likelihood of the correct response. **X** is a property of working memory. It may be thought of as a measure of amount of retrieved "information", or as a collection of dispositions that permit the correct response. **X** arises from a query of long-term memory.

A measure of the strength of all response dispositions that compete with the correct response is provided by **Y**. Both **X** and **Y** reflect retrieval from long-term memory.

An index of the **response** made at the trial end is **R**. It is a function of the dispositions to make the correct response or an error, **X** and **Y**, and of some decision criterion, which can be altered by feedback. Response *probability* depends on the relative magnitudes of **X** and **Y**, and response *latency* depends on the time for a response criterion to be reached.

The **feedback** from the environment subsequent to a response is indexed by **F**.

4.2 Three empirical constraints on the model

So far we have outlined a "meta-model", which yields specific models as details are supplied. The details provided here are guided by three constraints demanded by our current understanding of memory performance. First, we assume that working memory has *limited capacity*. To reflect limited capacity, the variables **X** and **Y**, reflecting the "contents" of working memory, are limited in magnitude. The second constraint is *competition*. We assume that inter- and intra-trial dynamics reflect competition between incompatible alternatives. In intra-trial dynamics, **X** and **Y** compete for the finite capacity of working memory. To reflect this, we limit the magnitude of the *sum* of **X** and **Y**. Third, we assume that competition can lead to *temporary dominance*. That is, we consider any model of intra-trial competition incomplete unless it can simulate initial dominance by one response tendency eventually replaced by a different response tendency, even in the absence of feedback.

An example of the kind of temporary dominance that we consider an obligatory capability of an intra-trial memory model is shown in an experiment recently reported by Hintzman and Curran [15]. In this experiment, subjects were presented with a list of study words and then given a recognition test that included some new words similar to the old words. Hintzman and Curran used the response-signal technique, in which the test item is followed at a variable lag by a signal requiring an immediate response. The experimenters used a range of signal lags so that the course of accuracy could be graphed over time. Results for (incorrectly) responding "old" to the new, but similar, words are shown in Figure 9. As can be seen there, the disposition to incorrectly judge these words as old temporarily dominates the correct response to say "new".

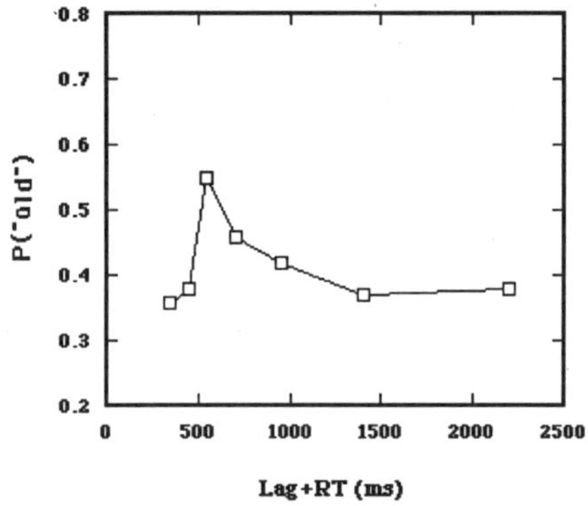

Figure 9: An example of temporary dominance revealed by the response- signal method. The probability of incorrectly responding "old" to similar items rises then falls as the delay of the signal to respond increases. Re- drawn from data plotted in Fig. 4 in *Hintzman&Curran* [15], "Similar 2" condition.

4.3 Dynamic Equations for intra-trial dynamics

This report focuses on intra-trial dynamics, and dynamic equations are developed for **X**, the variable indexing dispositions to respond correctly, and their competing dispositions, reflected in the variable **Y**. Here we place them in lower case and employ subscripts for discrete intervals of time.

To ease into the full dynamics, we start with the logistic map,

$$x_{t+f} = r * x_t \tag{3}$$

which has the growth in x to depend on the rate parameter r. This version of the logistic map, of course, grows without limit. To place a limit, the parameter K is introduced (Beltrami [16])

$$x_{t+f} = x_t + \left[1 - \frac{x_t}{K}\right] * r * x_t \tag{4}$$
$$(1 < r < 3)$$

in such a way that x_t is limited by K. If $x_t = 0$, the term in brackets equals 1 and growth is maximal. As x_t approaches K, growth goes to 0.

In this form, over time x_t becomes stable or chaotic depending on r. For r between 1 and 2, over time x_t grows sigmoidally up to K. The system bifurcates at $r = 2$, and for increasing r we have period-doubling, attaining chaos at about $r = 2.57$.

But we need competition, another grower, y, and its growth rate r_y. That is, we need two growth equations.

$$\begin{aligned} x_{t+f} &= x_t + \left[1 - \tfrac{x_t}{K}\right] * r_x * x_t \\ y_{t+f} &= y_t + \left[1 - \tfrac{y_t}{K}\right] * r_y * y_t \end{aligned} \tag{5}$$

Stated this way, however, x and y are not coupled. There is no competition. How do we introduce competition? It depends on what the two growers are competing for. In the current model, we have them competing for working memory. Thus, it is their sum, $(x_t + y_t)$, that limits their growth, that is limited to the carrying capacity K. All this yields the competing system:

$$\begin{aligned} x_{t+f} &= x_t + \left[1 - \tfrac{x_t+y_t}{K}\right] * r_x * x_t \\ y_{t+f} &= y_t + \left[1 - \tfrac{x_t+y_t}{K}\right] * r_y * y_t \end{aligned} \tag{6}$$

(Those familiar with van Geert's [17] competition model may notice that the present model is slightly different. The reason for the difference is that, unlike van Geert, we have assumed that x and y compete for the same "carrying" capacity K)

Does the model, as assembled so far, allow for temporary dominance? To find out, we simulated it. It turns out, and this result was not foreseen, that any time a chaotic grower ($r > 2.57$) is pitted against a stable exponential grower ($r < 2$), starting at the same value, the chaotic grower dominates temporarily and then loses out. It appears that in this kind of competition, chaos is a pitiful competitor. On reflection, this makes sense. An exponential grower can only increase, but a chaotic grower sometimes decreases. When this happens, the stable grower has to benefit by the chaotic decrease, because they are competing for the same capacity, eventually taking over. Although this outcome surprised us, other students of chaotic competition may have already noticed this effect.

However, in our simulations, if two non-chaotic growers are pitted against one another, they simply divvy up the capacity in direct proportion to their relative growth rates. There is no "take-over" in the sense of the loser decreasing to zero magnitude. Also, temporary dominance is not accompanied by a decline in the loser, unless the loser is chaotic. These results stem from the way competition has been built in. The magnitude of one grower influences the other only indirectly by claiming some capacity. To have the size of one grower directly effect the size of the other, we add another term.

$$\begin{aligned} x_{t+f} &= x_t + \left[1 - \tfrac{x_t + y_t}{K}\right] * r_x * x_t - c * x_t * y_t \\ y_{t+f} &= y_t + \left[1 - \tfrac{x_t + y_t}{K}\right] * r_y * y_t - c * x_t * y_t \end{aligned} \tag{7}$$

Here we add c $(0 < c < 1)$, so that each grower is diminished in direct proportion to the size of the other. Simulations of this version of the revised model produce temporary dominance under some parameter values, as illustrated in the next figure.

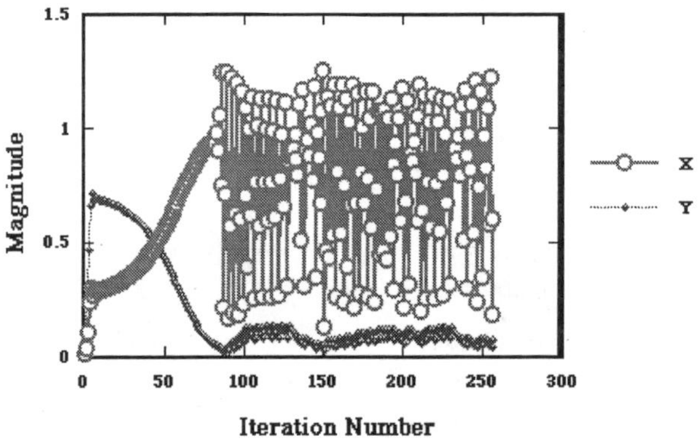

Figure 10: Simulation of the model showing temporary dominance by y. Parameter values for this simulation are $x_0 = .01, y_0 = .15, r_x = 2.99, r_y = 1.0, f = 1, K = 1$.

The figure shows x and y starting at about the same level but y quickly achieving dominance. At about iteration 50, x, which has a higher growth rate, takes over. Also note that the winner becomes chaotic after becoming dominant. This unexpected feature has provocative implications for a model striving to clarify the relationship between intra- and extra-trial dynamics. Depending on the precise rule specifying the decision to execute a response, a trial terminating in a chaotic value for x has to have highly variable consequences for the initial conditions of the next trial.

5 Summary and Conclusions

The present research began with an attempt to replicate Cooney and Troyer's apparent finding of low dimensionality in the successive reaction times of a Sternberg recognition memory task. Using analyses proposed by Grassberger and Procaccia, and procedures similar to Cooncy and Troyer, we did not replicate their finding, neither with the Sternberg task, nor with three simpler choice reaction time tasks. We have focused here on the Grassberger and Procaccia analysis because of its wide-spread use and because it was the major analysis of Cooney and Troyer. In the process, we encountered serious problems involving locating the scaling region of the correlation integral function. Specifically, we have seen that any algorithm that includes more than the scaling region results in an underestimate of the dimension. Although not treated here, we are aware of, and recommend use of, more recently advanced techniques for evaluating time series data, including the use of surrogate data (Theiler, Eubank, Longtin, Galdrikian, & Farmer[18]), and, perhaps more useful for psychological applications with relatively short data sets, Judd's estimator (Judd[11]).

We have also proposed a dynamic model intended to account simultaneously for reaction time and accuracy dynamics across a broad range of memory and choice tasks. The model was formulated within three constraints that we consider imposed by the empirical literature: performance depends on a) competition, b) is a product of a working memory that has limited capacity, and c) occasionally shows temporary dominance. The model is incomplete, needing especially development of details for inter-trial dynamics. So far the model simulates chaotic and non-chaotic dynamics, depending on the values of its control parameters. It remains to be seen whether and how the model has heuristic value. Evaluating this last point is one of our current tasks.

References

1. G. Gillund & R. M. Shiffrin, *Psychological Review* **91**, 1 (1984).
2. D. L. Hintzman, *Psychological Review* **93**, 411 (1986).
3. S. W. Link,*The Wave Theory of Difference and Similarity* (Lawrence Erlbaum Associates, NJ, 1992).
4. R. D. Luce,*Response Times: Their Role in Inferring Elementary Mental Organization* (Oxford University Press, NY, 1986).
5. T. P. McNamara, *Journal of Experimental Psychology: Learning, Memory, and, Cognition* **20**, 507 (1994).
6. J. B. Cooney & R. P. Troyer, *Journal of Experimental Child Psychology* **58**, 200 (1994).

7. S. Sternberg in *Acta Psychologica 30 Attention and Performance II*, eds. R. Robertson & A. Combs (North-Holland, Amsterdam, 1969).
8. P. Grassberger & I. Procaccia, *Physical Review Letters* **50**, 346 (1983). P. Grassberger & I. Procaccia, *Physica* D **9**, 189 (1983).
9. H. D. Abarbanel, R. Brown, J. J. Sidorowich, & L. S. Tsimring, *Reviews of Modern Physics* **65**, 1331 (1993).
10. W. S. Pritchard & D. W. Duke, *International Journal of Neuroscience* **67**, 31 (1992).
11. K. Judd, *Physica* D **56**, 216 (1992). K. Judd, *Physica* D **71**, 421 (1994).
12. J. C. Sprout & G. Rowlands,*Chaos Data Analyzer: Users Manual* (American Institute of Physics, NY, 1992) p. 21.
13. A. M. Albano, A. I. Mees, G. C. de Guzman, & P. E. Rapp in *Chaos in Biological Systems*, eds. A. V. Holden., & L. F. Olsen (Plenum Press, NY, 1987).
14. P. Rapp in*A Chaos Psychology Reader*, eds. R. Robertson & A. Combs (Lawrence Erlbaum, NJ, in press).
15. D. L. Hintzman & T. Curran, *Journal of Memory and Language* **33**, 1 (1994).
16. E. Beltrami,*Mathematics for Dynamic Modeling* (Academic Press, MA, 1987).
17. P. van Geert, *Psychological Review* **98**, 3 (1991).
18. J. Theiler, S. Eubank, A. Longtin, B. Galdrikian, & J. D. Farmer, *Physica* D **58**, 77 (1992).

APPROACHES TO THE NONLINEAR DYNAMICS OF REFLEXIVE EYE MOVEMENTS

M. Shelhamer
Departments of Otolaryngology — Head and Neck Surgery and Biomedical Engineering, The Johns Hopkins University, School of Medicine, Baltimore, MD 21287 USA

A type of reflexive eye movement — optokinetic nystagmus (OKN) — is examined with the tools of nonlinear systems analysis. Time-delay reconstruction of the state-space attractor of OKN reveals interesting dynamics, which are further explored with the correlation dimension. The dimension of OKN decreases from approximately 3.6 to 2.6 during 120 seconds of stimulation, underscoring the nonstationary nature of this physiologic system. Surrogate data sets, created by randomizing various aspects of the OKN time series, produce slightly increased dimensions, indicating that the dynamics are determined more by the population of fast and slow eye movements in OKN than by any specific relationship between movements. Nonlinear prediction of the sequence of fast phases shows a slight deterministic nature in the starting and ending position of fast phases, which can be interpreted physiologically. Other parameters of OKN appear to be more random. Implications of these findings for mathematical modeling and oculomotor physiology are discussed.

1 Introduction

The study of eye movement control is an interesting problem in sensorimotor processing. As pointed out by Robinson[1], several simplifying characteristics of eye movements make them an idealized system for the general study of the neural control of movement: the load (the eyeball) is constant, there is only a single "joint" about which motion takes place, the goal of eye movements are understood, and much of the neural anatomy and signal processing are known. For these reasons, we approach eye movements as a study in sensorimotor processing and brain control of movement. A more thorough report of some of the matters presented here is in preparation[2,3].

2 Background — Eye Movements

There are several types of eye movements. *Saccades* are rapid ballistic movements that bring the eyes to rest on a new part of the visual field, as when reading. *Smooth pursuit* is used to follow objects that move in the visual field. *Vergence* is a disconjugate movement (the eyes move in opposite directions from one another) that is used when shifting gaze between objects at different distances. The vestibulo-ocular reflex (VOR) is a reflexive movement that compensates for motion of the head so as to keep the line of sight stable in space — it is what allows you to read on a jostling subway car, for instance. The VOR is driven by the vestibular system, which is the non-auditory part of the inner ear, sometimes known as the "organ of balance". Finally, related to the VOR, is optokinetic nystagmus (OKN). This report deals with the analysis of OKN.

The term "nystagmus" refers to a pattern of eye movement in which the eyes move slowly (less than about 100 deg/sec) in one direction, and occasionally make rapid saccadic movements in the opposite direction. The alternation of these slow and fast phases resembles a shaky sawtooth waveform, as seen in the figure below.

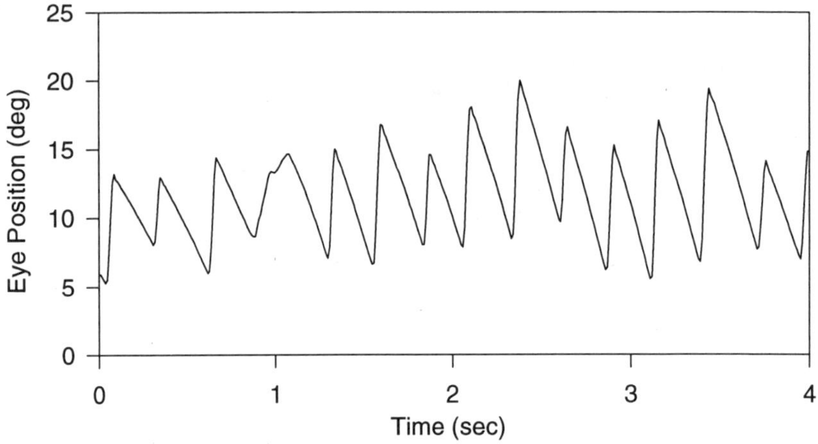

OKN is a type of nystagmus that is evoked by a wide-field moving visual stimulus. A subject is placed inside a large drum that rotates about a vertical axis. The interior of the drum is covered with a random pattern (sometimes a series of vertical stripes is used). When the drum rotates at a constant velocity (60 deg/sec for the data presented here), the eyes reflexively move in an attempt to track the moving visual pattern. The slow phase eye movements track the field with a gain of approximately 0.8 (the eye velocity is about 80% of the pattern velocity). Fast phases intermittently move the eyes in the other direction, to pick up a new piece of the visual field before they reach the limit of their motion. OKN is a reflex — a person presented with such a stimulus can modulate to some extent the gain of his or her OKN, but cannot suppress OKN totally (unless given a fixed visual target).

This explanation of OKN is straightforward and easy to understand. Unfortunately, it is not the whole story. The optokinetic system (the brain mechanisms that drive OKN) is intimately related to two other systems: the smooth pursuit system and the vestibular system. The relation to smooth pursuit is easy to see; both pursuit and OKN involve the generation of smooth tracking eye movements that are needed to follow the motion of an object with the head still. Pursuit can often be a confounding factor when trying to interpret OKN data. The detailed properties of the pursuit and optokinetic systems are, however, slightly different, and many of the brain structures that drive each of them are not identical. Perhaps the most important difference between the two types of eye movements, at least conceptually, is that OKN is reflexive while pursuit is voluntary; you can simply refuse to track a single moving object in your field of view (lack of smooth pursuit), while a wide-field pattern moving at a constant velocity almost certainly will induce OKN regardless of your attempts to suppress it. Thus one way to stimulate OKN with minimal interference from pursuit is to instruct a subject to simply look straight ahead at the moving visual pattern and *not* attempt to track it voluntarily ("stare OKN").

There is a closer relation between OKN and the vestibular system, although it can at first be a bit more difficult to understand. The vestibular system senses motion of the head and creates the appropriate compensatory postural reflexes. People who have a loss in vestibular function, through disease or trauma, suffer acutely from a variety of ailments including difficulty in walking (especially in the dark) and nausea brought on by head movements. An intact and properly functioning vestibular system is needed to generate the proper motor commands to perform these tasks as we have become used to performing them: without conscious thought. One of the important vestibular reflexes is the vestibulo-ocular reflex (VOR). This reflex is easily demonstrated by rotating a person in a dark room and measuring the resulting eye movements. As the head moves to the right, for example, the VOR moves the eyes to the left (relative to the head), so as to keep the line of sight more or less fixed in space. The fact that this occurs in darkness is one piece of evidence that it is a *vestibular* reflex, rather than smooth pursuit or OKN that would be stimulated by relative motion of the visual world in the direction opposite to head motion.

The vestibular system is good at transducing rotations of the head that last for a very short time (less than a second). This is because of the mechanical properties of that part of the vestibular system — the semicircular canals (SCC) — that sense head rotation. The SCC are basically rings of fluid, three on each side of the head, oriented so that head rotations in any direction can be sensed. When the head rotates, the fluid in any given canal will lag behind due to inertia. A small flap called the cupula sits in the lumen of the canal and is deformed by this fluid motion. Deformation of the cupula is transformed into a change in afferent firing rate in the vestibular nerve. This is how information on head rotation makes its way to the brain. The weak point in this system is the mechanical nature of the cupula and the fluid. As the head is rotated at a constant velocity for more than a fraction of a second, the cupula slowly returns to its original shape as the fluid in the canal "catches up" with the canal itself. A microelectrode placed in the vestibular nerve will show that the firing rate increases abruptly when the head begins to rotate, but then decays back to resting level over the next 20 to 30 seconds of maintained rotation. Thus the canals are heavily damped acceleration sensors, or adequate velocity sensors, over the frequency range of typical head movements[4]. The VOR —

the eye movements that compensate for the head rotation — are driven (after a few interneurons) by this vestibular afferent information; if eye movements are recorded during this same procedure they too will decay in intensity (the slow phases of nystagmus will become slower) over 20 to 30 seconds.

When the head moves with a step of velocity, the vestibular system senses the rotation, and the afferent neurons begin firing. Under such constant velocity rotation, the afferent firing rate decays with a time constant of approximately 6 seconds. However, the slow phase velocity of the induced nystagmus (it is the slow phases that compensate for the sensed rotation of the head) decays with a time constant of 20 seconds or more. This perseveration of velocity information in the pathway from vestibular afferents to oculomotor control is termed "velocity storage"[5]. OKN stimulation, even with the head still, excites the central vestibular pathways in the same way; velocity storage is manifest in OKN as a slow increase in the velocity of the slow phases over the first approximately 20 seconds of stimulation. We will see this characteristic again in the subsequent discussion of OKN stationarity.

The point of this discussion of the mechanics of the vestibular system is that another mechanism must aid the VOR during maintained head rotations in the light, in order to correct for the decaying information carried by the vestibular afferents. This is the optokinetic system. The VOR starts big and decays slowly. OKN has complementary dynamics; after an initial transient, the velocity of the slow phases builds up slowly. The time constant of the decay of VOR and that of the increase in OKN are very nearly equal. If the head is rotated in the light, then as the vestibular information (and the VOR) decays, the optokinetic information (and OKN) rises, to keep the slow phase velocity of the compensatory eye movements approximately constant for as long as the head rotation lasts. Indeed there are brain-stem sites, very early in the processing of vestibular afferent flow, where the optokinetic and the vestibular afferent information have combined to produce what is in essence an internal representation of the motion of the head — a representation constructed from the combination of visual and vestibular senses. These long-term (20 to 30 second) dynamics are important for some of the results to be presented later. But they also make it clear how OKN can be used as one measure of the function of the central vestibular system (the neural pathways that process vestibular information).

The *short-term* dynamics of OKN are of more direct concern to us. Inspection of the previous OKN time series plot shows some interesting features. Quite obviously, OKN is not periodic. The slow phases do not drive the eyes to the same point with each slow phase, the fast phases do not bring the eyes back to the same point each time, the fast phases are not periodic, and there is some variation in the velocity of consecutive slow phases. Yet the stimulus — a constant-velocity drum rotation — is in steady-state. This variability is intriguing. Why do the eyes not go to the extreme of their position range with each slow phase and each fast phase? Is there a rule — that so far has escaped us — that determines when fast phases will occur, and how large they will be? This report presents our initial attempts to answer these questions.

3 OKN Dynamics

Previous attempts to model the generation of OKN have relied largely on statistical characterizations[6,7]. More recent work strongly suggests a functional connection between the fast and slow phases of nystagmus[8,9,10]. These studies show that many of the properties of the slow phases of nystagmus (velocity, duration, starting position) may be affected by the behavior of the fast phases. This is in distinction to the prevailing view that the slow phases compensate for (real or perceived) head motion, with the fast phases merely preventing the eyes from reaching the limits of rotation. Rather, it is suggested that there may be a fast phase strategy with the goal of extending the linear range of the VOR; it does so by triggering a fast phase when the eyes move far enough that neural responses on either side of the midline start to approach inhibitory cut-off. This paradigm predicts a deterministic relationship between fast and slow phases, and between successive fast phases.

Observations made in two other studies are suggestive of nonlinear dynamics at play in the generation of nystagmus. First, there is a nonlinearity in the relationship between OKN stimulus velocity and the resulting slow phase eye velocity, and also between retinal slip velocity (the difference between pattern velocity and eye velocity: how quickly images move across the retina and cause degraded vision) and slow phase eye velocity[11]. Eye velocity increases with stimulus velocity up to a point, and then decreases. A similar effect exists between eye velocity and retinal slip velocity. This suggests that there may be a nonlinear structure to slow phase velocity generation. In another study[12], histograms of the interval between OKN fast phases were determined for different stimulus velocities. At lower stimulus velocities (less than 15 deg/sec), there is a type of bifurcation; the histograms change from unimodal to multimodal. The higher modes in the histogram were, moreover, approximately integral multiples of the main mode. This might be interpreted as a form of period-doubling as the stimulus intensity changes. It is not yet clear what relevance, if any, these studies may have. They do, however, show that OKN displays some of the characteristic signatures of nonlinear and chaotic systems.

4 Dynamics — Deterministic, Random, Chaotic

A *dynamic system* is one whose behavior over time can be described by a system of differential equations. These equations describe how the important quantities in the system relate to each other and change over time. These "important quantities" are known as *state variables*; the values of the state variables give complete information about the system at any point in time.

A *deterministic* system is one whose behavior is completely specified for all time. Knowing the set of differential equations, along with a set of initial values for the state variables, would allow someone to calculate the behavior (the trajectories of the state variables) for all time, infinitely far into the future. There is a special case of deterministic systems, however: those that are chaotic. (Actually it turns out that chaotic systems are not very rare at all, but rather seem to be the norm for most nonlinear systems of any complexity.) A chaotic system is deterministic (described by differential equations, not probabilistic descriptions), yet its behavior is unpredictable in the long term. This is because of the system's extreme sensitivity to initial conditions; small (ultimately unmeasurable) differences in the starting conditions lead to exponentially-increasing differences in the system trajectories. Thus, in the long term, for practical purposes, the behavior is unpredictable. Yet the system is deterministic, and reasonable predictions can be made of the behavior in the short term. We will see this property most vividly when we discuss predictability again in a later section.

A random system, on the other hand, is inherently unpredictable because it contains probabilistic or statistical elements. Noise, uncorrelated from moment to moment in time and following no discernible pattern, is random. Some correlation can be imposed on a white (uncorrelated) noise process by filtering it (producing a colored noise). In such a case, there is some predictability in the noise series[13], and it is not always obvious how to distinguish this from the chaotic case.

5 Dynamics in State Space

One way to gain insight into the dynamics of a system is to plot the values of the state variables, with respect to each other, over time. The resulting plot shows trajectories of the state variables in a mathematical abstraction known as *state space*. Each state variable lies along a principal axis of the state space. For example, the position and velocity of a simple harmonic oscillator (such as a frictionless pendulum) can be the state variables of that system. (Other sets of state variables are possible, but these two quantities give us all the information that there is to know about the system at any given time. Together with the differential equations of the system, we can describe the behavior of the system for all time.) A region of state space to which the trajectories of a dynamic system converge is known as an *attractor*. In this case, we can plot the position of the pendulum on the x-axis, and the velocity on the y-axis; the resulting attractor looks like the left plot in the next figure. If we add friction to the pendulum, its oscillations will gradually become smaller and slower, as shown in the next plot to the right in the figure. The Lorenz attractor, at the far right, is a famous example formed by three coupled nonlinear differential equations. A noise sample (not shown) would have no discernible dynamic structure.

Trajectories in state space provide a visual and intuitive way to examine the dynamics of a system. As we will see, they can also form the basis of mathematical approaches to analyzing these dynamics.

6 Attractor Reconstruction

We would like to form the state space attractor of a system, using a signal that is measured from it. There are, however, problems in generating the actual state space and the attractor (see below). We therefore use a common method known as time-delay embedding to reconstruct the attractor from a single measured time series. This is done by choosing time-delayed values of the measured time series (in our case, eye position) $y(i)$. The points $x(i)$ on the reconstructed attractor are:

$x(1)=[\, y(1), y(1+L), \ldots, y(1+(M-1)L)]$

$x(2)=[\, y(1+J), y(1+J+L), \ldots, y(1+J+(M-1)L)]$

$x(N)=[\, y(1+(N-1)J), y(1+(N-1)J+L), \ldots, y(1+(N-1)J+(M-1)L)]$,

where:

$y(i)$ = measured eye position
N = number of points on attractor
M = embedding dimension (number of "state variables")
L = interval between components of each point x
J = interval between first elements of successive points x.

The total time spanned by each vector, $(M-1)L$ (the "window" of the embedding), should be large enough that the state variables are independent, but small enough that they approximate the derivatives of the signal. There have been several different suggestions for choosing the value of the time delay L or the window $(M-1)L$[14,15,16,17]. The value used here is 0.3 seconds, approximately twice the first minimum in the autocorrelation function.

This embedding technique has the advantage that the actual state variables of the system need not be identified and used in further work in the state space. Indeed the actual state variables may not be known or readily available. Furthermore, the state variables of a dynamic system are often related to one another via the mathematical process of differentiation; position, velocity, and acceleration form a common set of state variables for many mechanical systems. Differentiation of signals, such as position to produce velocity, is a noisy process (differentiators are high pass filters and so tend to amplify any high-frequency noise that is present). The time-delay method does not require differentiation, while it approximates in some sense the process of taking a derivative by using values of the time series that are close in time.

Examples of attractors that have been reconstructed using this procedure are shown in the next figure. Time-delayed values of a single variable were used to create each of these plots. The left plot is for the same damped oscillator shown in the previous figure, and the right plot is the Lorenz attractor also shown earlier. Note that, although the precise form the attractor changes when the time-delay method is used, the overall shape (and more precisely, the topological dimension, to be discussed) of the attractor does not change[18,19]. Thus we can feel confident in using the time-delay reconstruction in our analysis of OKN.

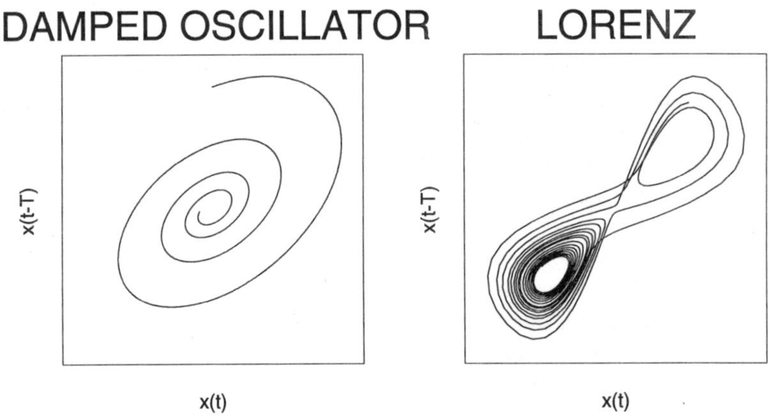

7 OKN in State Space

We obtained a 120-second sample of OKN eye movements from a normal human subject. The subject sat on a stationary chair, within a surrounding drum with a diameter of 1.5 m. The interior of the drum was lighted, and covered with randomly-placed black rectangles on a white background. The drum rotated at a constant velocity of 60 deg/s to the subject's right. Eye movements were recorded with the scleral search coil technique[20]. The subject sat within two orthogonal magnetic fields, each oscillating at a different frequency. A topical anesthetic was applied, and a small coil attached to a scleral contact lens was then placed on the subject's eye. Phase-sensitive amplifiers measured the voltage induced in the eye coil by each magnetic field; this voltage varied according to the orientation of the coil (and therefore the eye) with the field. The system was calibrated by having the subject look alternately at fixed targets at a known angular displacement before the recording session. The overall bandwidth of the coil system and associated instrumentation is more than 40 Hz. The eye position signal was sampled and stored on computer disk.

Three-dimensional views of the attractor from this OKN sample, reconstructed from 2000 points (4 sec) sampled at 500 Hz, are shown in the figure below. The time delay, L, is 0.3 s.

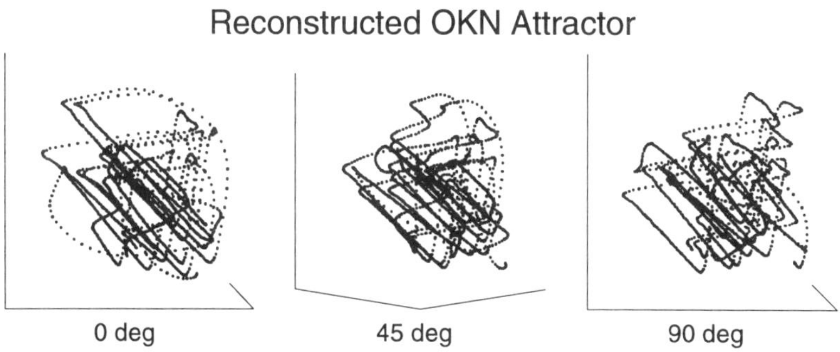

The points on these attractors are evenly spaced in time; one can judge the relative velocities of the trajectories by looking at the distance between consecutive points. The several parallel segments, with points closely spaced, are from the slow phase tracking components of the OKN signal. The fast phases are the wildly scattered semicircular loops that eject from one end of the slow phases and return to the other end. In this view of the attractor, there seem to be two separate "sheets" of slow phases, roughly parallel to one another and oriented from upper left to lower right in the figure. The fast phases take the trajectory from one sheet to another.

8 Background — Dynamics and Dimension

With a reconstructed attractor in hand, we can attempt to measure its fractal or Hausdorf dimension. The dimension is a measure of the attractor's volume, or "bulk", under different scales of measurement. The classic example of dimensional concepts is the measurement of the length of a coastline. The measured length gets larger as smaller measuring sticks are used. There is often a power law that describes how the measured length increases as the unit of measurement decreases in size. The exponent of the power law is one type of dimension. A common dimension estimate that can be readily computed from (discrete) experimental data is the correlation dimension[21,22,23], which is found from the spatial structure of the reconstructed attractor (see below). The computations are straightforward, and many physiologic phenomena have been studied with the correlation dimension[24,25,26].

Refer back to the second figure, with the oscillators and the Lorenz attractor, for some dimension examples. The oscillators, with and without damping, form a single line in the state space, equivalent to a topological dimension of 1. From the point of view of the topologist, they are simply long strings (one-dimensional lines) bent into various shapes. Trajectories in state space cannot cross each other; otherwise the system would not know which trajectory to follow if started from that crossing point, and the system would not be deterministic. Noise has infinite dimension, meaning that is will occupy more and more space — more dimensions in the state space — as it becomes available. Although the Lorenz trajectories appear to cross, this simply means that we are not plotting them in a large enough space — that is, one with enough dimensions. The Lorenz attractor has a dimension of 2.06. Obviously this is a special case — a noninteger dimension. This attractor is an example of a *fractal* object. The attractor occupies a finite volume in the state space, and yet the trajectory is infinitely long — it never repeats itself (it is not a periodic system). This counterintuitive structure requires a novel approach for its accurate description, and the dimension approach to be described provides it. (For a more detailed discussion of dimensions see, for example, the books by Tsonis[27] and Edgar[28].)

9 Correlation Dimension

The correlation dimension is a very common estimate of the fractal dimension of a time series. The Grassberger-Procaccia method[21,22] is in turn a very common way to estimate the correlation dimension. The computation proceeds as follows. Let $C(r)$ be the fraction of attractor points within distance r of a reference point (on the attractor). $C(r)$ is expected to grow as r^d. The exponent d quantifies the attractor's spatial structure; it is known as the correlation dimension d_{corr}. If the dimension is fractional and less then the dimension M of the phase space in which the attractor is reconstructed (see below), then there may exist an underlying nonlinear system, possibly with chaotic dynamics.

To compute the dimension, first create a time-delay reconstruction of the attractor, as described previously. The attractor is embedded in a state space of integer dimension M by using M state variables in the reconstruction. Within this M-dimensional space, a preliminary correlation dimension is found (by a process described below). The embedding space is then increased in dimension (M is incremented by one, decreasing L to keep the window $(M-1)L$ constant), and another correlation dimension found. This process is repeated for a range of M values. These preliminary correlation dimensions will increase as the embedding dimension M increases, up to a point. When the dimension saturates with increasing embedding dimension, the saturation value is the correlation dimension estimate, d_{corr}.

To find the sequence of correlation dimensions, begin by selecting a reference distance r. A count is made of the number of points $x(j)$ on the attractor that are within the reference distance of some reference point $x(i)$. Each point is used in sequence as the reference point. The proportion of those pairs of points meeting the distance inequality $|x(j)-x(i)|<r$ is called the correlation integral, $C(r)$. This integral is found for decreasing values of r. The limiting value of $log(C(r))/log(r)$ is the correlation dimension. In practice, the slope of $log(C(r))$ vs. $log(r)$ is plotted as a function of $log(r)$. The range of r over which the slope is nearly constant is known as the scaling region. In this report, the range of r over which the slope varied less than 10% was found, and a least squares fit of $log(C(r))$ vs. $log(r)$ over this range gave d_{corr}. Values computed in this way were compared across the different embedding dimensions. Those values that did not vary from one another by more than 10% (as long as this occurred for at least four values of embedding dimension) were averaged to find the final value of d_{corr}. Embedding dimensions in this study ranged from 5 to 10.

10 OKN Dimension and Nonstationarity

Our initial computations of the correlation dimension of a small sample of OKN[29] produced a value of approximately 2.8. This noninteger value suggested that that particular OKN attractor may be a fractal object, and that the system that generated it may have chaotic dynamics. There are many caveats to this conclusion, however, and it remains to be verified via other methods.

We next applied the correlation dimension to the analysis of OKN *stationarity*. Consecutive 10-second segments of human OKN were analyzed; the OKN had been evoked by 120 seconds of continuous visual stimulation. The dimension decreased almost monotonically over the course of the recording session, in a total of three subjects tested. The next figure shows this result in one subject.

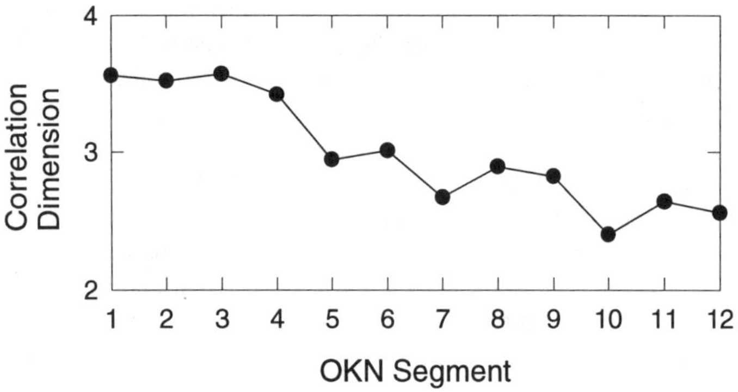

The first and last ten-second OKN segments were examined further, to see how the changes in dimension are manifest in the time domain and in the state space. The following figure shows the OKN time series early (0-10 sec, dimension 3.6, left top) and late (110-120 sec, dimension 2.6, right top) in the test session.

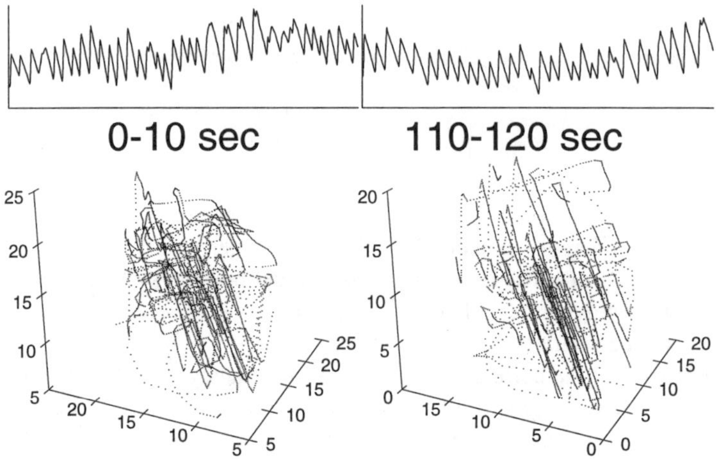

The decrease in dimension is associated with an apparent "simplification" of the OKN signal — it has a slightly more periodic nature in the late segment. Trajectories in the state space (three-dimensional time-delay reconstructions of the same two 10-second segments, bottom plots) line up more closely and are more parallel.

The time course of the decrease in dimension coincides approximately with the time course of "velocity storage" — the perseveration of afferent activity in the vestibular nuclei[5]. The decrease in dimension might also be due to fatigue of the smooth pursuit system, which aids the optokinetic system by means of voluntary tracking of moving visual targets. This result provides a way to quantify stationarity in extended duration OKN. It might be useful as a test for the relative presence or absence of smooth pursuit tracking in OKN; this could be useful clinically in determining the relative functions of the vestibular and the pursuit systems.

We intend to follow up on this result in future work by examining the dimension of vertical and torsional eye movements. (Torsional eye movements are those made about the optical axis of the eye.) The vertical and torsional VOR and OKN have little or no velocity storage[30,31], while there is essentially no torsional smooth pursuit[32,33]. By comparing the dimensionality of these different eye movements, over several minutes, we can gain insight into the physiologic meaning of the dimension of OKN. Specifically, we may be able to determine if the decrease in dimension over time is due to velocity storage or pursuit.

11 OKN Dimension and Surrogates

Further investigations of the dimension of OKN involved the use of *surrogate data* techniques, using a modification of the methodology described in the literature for the detection of nonlinearities in time series[34,35]. In this approach, the original OKN signal is scrambled in some way, to produce a new signal that retains some of the properties of the original OKN but randomizes other properties. One surrogate data set was made by separating the OKN into its component fast phases and slow phases, scrambling these independently, and then reassembling them without deliberately keeping any fast or slow phases adjacent to one another (although of course fast and slow phases were alternated). The dimension of this surrogate was then computed and compared to the dimension of the original OKN. An increase in dimension indicates that some randomness has been introduced by the shuffling process, while no change in dimension indicates that the scrambling has not changed whatever it is that determines the "essential structure" of the OKN. Two other surrogate data sets were also examined — in each case the OKN was broken into pieces; in one case each piece contained a fast phase and the subsequent slow phase, and in the other case each piece contained a slow phase and the subsequent fast phase. These pieces were then randomly scrambled. The dimension results are given in the next table.

Data	Correlation Dimension
Original OKN	2.5
Scramble (Fast-Slow) Pairs	3.1±0.2
Scramble (Slow-Fast) Pairs	3.0±0.2
Scramble Both	3.0±0.2

The dimension increase in each of the surrogates suggests that noise has been introduced in the scrambling process, by destroying deterministic relationships among nearby fast and slow phase components of the OKN. An earlier analysis of OKN[29] had a different result with a different type of surrogate data. The surrogate in that study was made up of slow phases of uniform slope, with second-order statistics of fast-phase size and timing identical to those of the original OKN. In that case no reliable dimension could be found for the surrogate. The results of these two studies together indicate that the unique structure of OKN is not solely due to correlations between adjacent fast and slow phases, or even between adjacent slow phases or adjacent fast phases. Rather, there appears to be a relationship between the population of slow phases and the population of fast phases in OKN.

12 OKN Prediction

Nonlinear forecasting or prediction[36,37,38] begins with a time-delay reconstruction of the attractor in state space. In the simplest implementation, two points, consecutive in time, are found on the attractor. A linear regression is made on these points in space, and a prediction made of the next point (one time step into the future) on the attractor. This next point is already known, and therefore we can determine the error in the prediction. This procedure is carried out over the entire attractor, yielding an average prediction error. Then predictions two steps into the future are made, and so on. The prediction error will increase with the number of steps ahead that the prediction is attempted. The correlation between the predicted data and the actual data is found by calculating a correlation coefficient over all predictions made in a given data set. The behavior of the correlation coefficient as the prediction is carried further into the future, away from the reference data point, indicates the nature of the system under study. A purely random system will have uniformly poor prediction (near zero correlation) for all time steps, a classic deterministic system will have uniformly good prediction, and a chaotic system will have good short-term prediction that rapidly drops off (one of the defining conditions of a chaotic system). The prediction approach to system analysis has the advantage that relatively small data sets can be used, which is good for physiologic systems which are notoriously nonstationary. As pointed out by others[39], it may be more fruitful to ask "how predictable" a system is than if it is random or deterministic *per se*, and our work addresses that question.

Prediction results for some sample systems are shown in the next figure. They were produced by applying the prediction method described above, for 1 to 10 time steps into the future. The y-axis shows the correlation between the predicted values and the actual values; a correlation of 1.0 means perfect prediction, while 0 means that the prediction was no better than guessing. A sine wave provides an example of a simple deterministic system; prediction is good (near 1.0 correlation) for all time steps. Noise provides the other extreme; prediction is uniformly poor for all time steps; even the first step into the future is poorly predicted. A sine wave with additive noise is intermediate; correlation (prediction quality) does not change appreciably with subsequent time steps. The logistic equation, with parameter chosen to produce chaotic dynamics, is different from these. Prediction is good at small time steps, because the system is deterministic. But the predictions break down and become poor at later time steps, reflecting the unpredictable nature of chaotic dynamics. These examples provide a basis on which to judge results from OKN.

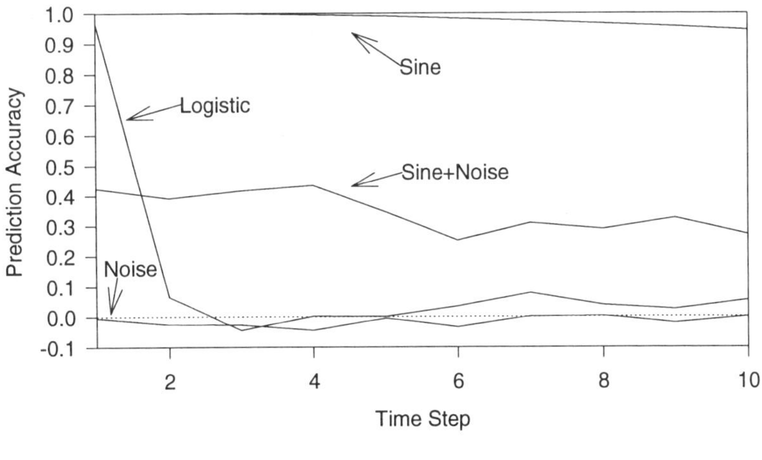

We analyzed several data sets, extracted from OKN, and attempted to predict the future values of each sequence. The sequences were: slow phase velocity, amplitude of each fast phase, time between fast phases, position of the eyes at the start of each fast phase, and position of the eyes at the end of each fast phase. Prediction results are shown in the following figure.

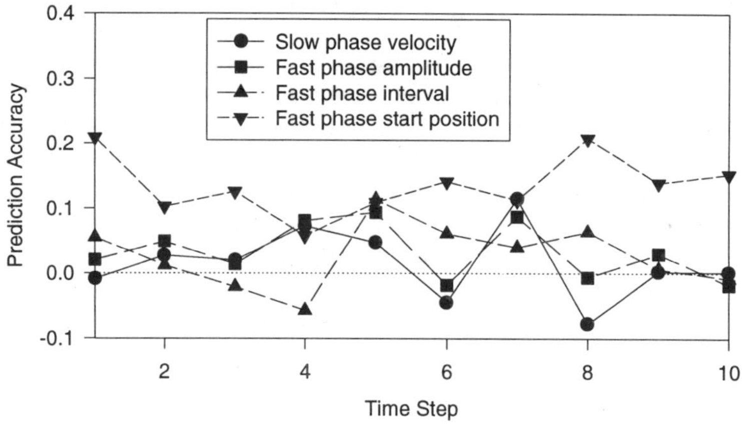

There is little or no predictability in the sequence of successive fast phase intervals or sizes, or slow phase velocities. There is some predictability in the sequence of fast phase starting positions and (not shown) ending positions — there is some pattern to the eye positions at which fast phases begin and end. This has some biological meaning, since the eyes make fast phases in order to pick up and track a portion of the visual field in which it is expected that something interesting will appear[7]. This result lends some support to the idea that the fast phases of OKN are not programmed merely to interrupt the slow phases to keep them from exceeding the limits of eye rotation. Rather the fast phases have an important role of their own — to choose an interesting and important point in the visual field at which to begin the next slow phase tracking.

13 Discussion

Some of the enthusiasm has dissipated from the initial flurry in the study of "chaotic" systems, physiologic systems included[39,40]. "Proving" that a system exhibits chaotic behavior is not as easy as it used to be; even obtaining reliable dimension estimates can be difficult and fraught with pitfalls[41]. Newer techniques, and the realization of the potential shortcomings of some of the early work with laboratory data, have led to more cautious and more realistic assessments of the value of dynamical systems analysis in physiology. In this report we present the initial studies of the dynamics of a neurologic system — control of reflexive eye movements — as a paradigm of sensory processing and motor control. The system is rather straightforward and direct, yet reflexive eye movements have significant variability. It is tempting to speculate that there may be some purpose to the variability. It is even more tempting to speculate that it is deliberately created by the central nervous system. If these things are true, then we should be able, in turn, to quantify the variability, look for correlations with other behavioral data, and model the neural processes involved.

The results presented here do not definitively decide the issue of chaos in the oculomotor system. (In fact, it should be noted that these results are somewhat preliminary, and the study of more subjects and a more thorough application of the computational approaches may lead to some variation from the results reported at this time.) The fact that the dimension algorithms converge to a non-integer value is slightly suggestive of chaotic dynamics. State-space prediction of fast phase and slow phase parameters, however, supports the view that most of the variability in OKN is random. Only the eye positions at the beginning and end of the fast and slow phases appear to have any deterministic pattern to them, and that is small. It remains to reconcile these results. We can only speculate at this time as to whether the variability that we see in OKN eye movements is deliberate, and if so what its purpose might be. Such speculations are, however, important for thinking about how the brain coordinates voluntary and reflexive movements.

Does dimension really tell us anything useful about the system in question? It is not clear that anyone has demonstrated this clearly, at least for physiologic systems. In theory, the dimension should approximate the number of independent variables that would be needed in a mathematical model of the system, although the jump from knowing the dimension to actually creating an appropriate model is a large one. In more general terms, the dimension might be thought of as a measure of the complexity — the richness of the dynamics — of a system. The dimension can then be used in relative terms — one set of stimulus parameters may lead to a dimension that is less than that with another set of values, for instance, and we may be able to draw some conclusions about the relative complexity of the system in these cases. This is the approach that we take here in the study of OKN.

One true test of our understanding of a physiologic process is to make a mathematical model that reproduces the qualities of the real thing. In the vestibular and oculomotor systems especially, modeling plays an ever-increasing role in synthesizing the mass of behavioral and neurophysiologic information and in suggesting new experiments. Several simplifying features of the oculomotor system, as described previously, make it an ideal paradigm for the study of sensorimotor control and neural signal processing, and so our analysis and modeling efforts may have implications beyond the fields of vestibular and oculomotor function. New mathematical and computational techniques for the analysis of dynamic systems make it possible to ask questions about the behavior of physiologic systems that were not until recently practical to address. It is now possible to create mathematical models of some phenomena that were previously considered random, or at least to determine to what extent a given system's behavior is random or deterministic. The prospect of finding new and useful information about how the nervous system processes information, by examination of variability, is appealing.

Recent work has given some credence to the thought that physiologic variability is healthy, and that the body takes deliberate steps to invoke or at least to allow some variability in physiologic processes[42,43,44,45]. Deliberate variation, so the theory goes, confers on a system the flexibility to adapt and respond effectively to environmental changes and to react to new situations[46]. The situation is far from clear, however, even in the relatively straightforward systems involved with oculomotor control. We have, for example, preliminary results from a separate set of experiments that indicate that adaptability and variability may be inversely related. Briefly, we are studying the rate at which subjects can adapt the amplitude of their saccadic eye movements, and relating this to the scatter in their saccades. Evidence so far suggests that those subjects with high variability may be the slowest adapters — opposite to what the foregoing discussion would predict. Thus the role of physiologic variability remains an unanswered question. Hopefully the type of work presented here, with novel analyses of variability and dynamics in a well-defined, well-understood, and thoroughly-studied system, can help us to approach the answer.

Acknowledgments

The author's work is supported by grants from NASA and the Whitaker Foundation.

References

1. D.A. Robinson in *Progress in Brain Research: Vol. 64. The Oculomotor and Skeletal-Motor Systems: Differences and Similarities*, ed. H.J. Freund, U. Büttner, B. Cohen, J. Noth (Elsevier, New York, 1986).
2. M. Shelhamer, On the correlation dimension of optokinetic nystagmus eye movements. I. Computational parameters, filtering, and stationarity. Manuscript submitted for publication.
3. M. Shelhamer, On the correlation dimension of optokinetic nystagmus eye movements. II. Surrogate data. Manuscript submitted for publication.
4. V.J. Wilson and G. Melvill Jones, *Mammalian Vestibular Physiology* (Plenum Press, New York, 1979).
5. T. Raphan *et al*, *Exp. Brain Res.* **35**, 229 (1979).
6. C.D. Balaban and M. Ariel, *Biol. Cybern.* **66**, 203 (1992).
7. K.-S. Chun and D.A. Robinson, *Biol. Cybern.* **28**, 209 (1978).
8. S.C. Cannon and D.A. Robinson, *J. Neurophysiol.* **57**, 1383 (1987).
9. H.L. Galiana, *IEEE Trans. Biomed. Eng.* **38**, 532 (1991).
10. C.G. Rey and H.L. Galiana, *Biol. Cybern.* **69**, 395 (1993).
11. W.A. Fletcher *et al*, *Exp. Brain Res.* **81**, 46 (1990).

12. M. Cheng and J.S. Outerbridge, *Vision Res.* **14**, 1053 (1974).
13. A.R. Osborne and A. Provenzale, *Physica* D **35**, 357 (1989).
14. A.M. Albano *et al*, *Phys. Rev.* A **38**, 3017 (1988).
15. A.M. Albano *et al*, *Physica* D **54**, 85 (1991).
16. Th. Buzug and G. Pfister, *Physica* D **58**, 127 (1992).
17. W. Liebert and H.G. Schuster, *Phys. Lett.* A **142**, 107 (1989).
18. N.H. Packard *et al*, *Phys. Rev. Lett.* **45**, 712 (1980).
19. F. Takens in *Lecture Notes in Mathematics: Vol. 898*, ed. D.A. Rand and L.-S. Young (Springer-Verlag, New York, 1981).
20. D.A. Robinson, *IEEE Trans. Biomed. Electron.* **BME-10** 137 (1963).
21. P. Grassberger and I. Procaccia, *Phys. Rev. Lett.* **50**, 346 (1983).
22. P. Grassberger and I. Procaccia, *Physica* D **9**, 189 (1983).
23. J. Theiler, *J. Opt. Soc. Am.* A **7**, 1055 (1990).
24. A. Babloyantz and A. Destexhe, *Proc. Natl. Acad. Sci.* **83**, 3513 (1986).
25. A. Babloyantz and A. Destexhe, *Biol. Cybern.* **58**, 203 (1986).
26. D.T. Kaplan and R.J. Cohen, *Ann. N. Y. Acad. Sci.* **591**, 367 (1990).
27. A.A. Tsonis, *Chaos: From Theory to Applications* (Plenum Press, New York, 1992).
28. G.A. Edgar, *Measure, Topology, and Fractal Geometry* (Springer-Verlag, New York, 1990).
29. M. Shelhamer, *IEEE Trans. Biomed. Eng.* **39**, 1319 (1992).
30. S.H. Seidman and R.J. Leigh, *Brain Res.* **504**, 264 (1989).
31. S.H. Seidman *et al*, *Vision Res.* **40**, 89 (1995).
32. R.J. Leigh *et al*, *Exp. Brain Res.* **75**, 221 (1989).
33. D. Straumann *et al*, *Vision Res.* **32**, 1067 (1992).
34. S.J. Schiff and T. Chang, *Biol. Cybern.* **67**, 387 (1992).
35. J. Theiler *et al*, *Physica* D **58**, 77 (1992).
36. M. Casdagli, *Physica* D **35**, 335 (1989).
37. J.D. Farmer and J.J. Sidorowich, *Phys. Rev. Lett.* **59**, 845 (1987).
38. G. Sugihara and R.M. May, *Nature (London)* **344**, 734 (1990).
39. P.E. Rapp, *The Biologist* **40**, 89 (1993).
40. D. Ruelle, *Proc. R. Soc. London Ser. A* **427**, 241 (1990).
41. J. Theiler, *Phys. Rev.* A **34**, 2427 (1986).
42. D.T. Kaplan *et al*, *Biophys. J.* **59**, 945 (1991).
43. L.A. Lipsitz and A.L. Goldberger, *JAMA* **267**, 1806 (1992).
44. J.E. Skinner *et al*, *Circ. Res.* **68**, 966 (1991).
45. J.E. Skinner *et al*, *Am. Heart J.* **125**, 731 (1993).
46. C.A. Skarda and W.J. Freeman, *Behav. Brain Sci.* **10**, 161 (1987).

IDENTIFICATION OF CHAOTIC BEHAVIOR IN WARFARE

T. TAGAREV
Mladost 4, bl.443, vh.3, ap.59
Sofia 1715, Bulgaria

D. NICHOLLS
2833 Noble Fir Court
Woodbridge, VA 22192

The purpose of this work was to determine if chaos is present in war. Data sets were drawn from the grand strategic, strategic and operational levels of war. Several qualitative tests as well as one quantitative test indicated the presence of chaos. Therefore, we concluded that war is chaotic. The implications of this finding in the context of war include ways to use computer simulations to find centers of gravity, a need to place emphasis on nonlinear processes, some advantages in creating fractal organizations, and the possibility of multiple attractors. Several potential sources for nonlinearity in warfare were also described.

1. Introduction

For the last thirty years, the study of chaos has intrigued investigators, prompting many to see a great future for the study and application of chaos theory. In science and engineering, chaos theory has significantly improved our understanding of phenomena ranging from turbulence to weather to structural dynamics[1]. Chaos theory has even been used to drastically improve our ability to control some dynamic systems[2]. In the social sciences, there has been considerable interest in whether social phenomena, previously thought to be random, have an underlying chaotic order. Several mathematical tests for chaotic behavior have been applied to historical data from both the stock market and cotton prices. These tests indicate that these economic phenomena are chaotic and so have a deterministic basis (i.e. are governed by rules) as opposed to being random. Naturally, this has gotten some business attention and at least two firms are now using chaos theory to guide their financial advice[3,4].

There is evidence that warfare might also be chaotic. Firstly, strategic decision making, an integral part of war, has been found to be chaotic[5]. Secondly, non-linearity, which is a requirement for chaotic behavior, appears to be a natural result of Clausewitzian friction[6]. Thirdly, some computer combat simulations[7] and arms race simulations[8] have been found to exhibit chaotic behavior. This paper discusses our work showing that chaos is present in warfare and the implications of chaos theory in the context of warfare.

1.1 Historical Data and Levels of War

For identification of chaotic behavior historical data from grand strategic, strategic, and operational level of war were used. Under the impact of the technology and the ever increasing complexity of interactions, the boundaries between the levels in a modern war become fuzzier. Thinking of war as being divided into levels, however, is convenient and practical for its planning, study and execution. For the purposes of our study we used the definitions for the levels of war given by Warden[9].

The *grand strategic* level of war is the place where a country decides whether and how to prepare for war, whether to participate in a war, who its allies will be, and how it envisions the state of peace. The *strategic level* of war concerns the overall conduct of the war, the allocation of forces for various theaters, and includes direct and indirect application of military and other national resources in support of the national policy[10]. The *operational level* is concerned with employing the allocated forces in a theater of operations to achieve the strategic objectives of the war. The lowest level of war is the *tactical level*. This is the level where opposing forces physically meet in battles to achieve specific and unambiguous objectives.

At the grand strategic level of war US government budgetary data from 1940 through 1992[11] were examined. These data reflect decision making at the highest level under the influence of bureaucratic, domestic and international political conditions and aggregate such factors as technological advances, overall condition of the economy at key stages in the business cycle, and long term changes in the structure of the economy. Arms race and arms sales could be other potential sources of data for chaotic behavior analysis at the grand strategic level. In fact, data related to arms sales aggregate the influence of many input factors like the political relations between the exporter and the importer of weapons; the political, economic and military balance in the particular region; the competition for regional dominance; the political and economic relations among exporters of weapons themselves, and so forth.

At the strategic level, daily or weekly US fixed-wing aircraft losses during the entire Vietnam war [12] were used to determine the presence of chaos in war. These data absorb a lot of factors which not only had great impact on the outcome of the war, but are inextricably linked with each other. We consider that these data reflect the strategic level because they aggregate information about the activity of different services in interaction with an enemy during different campaigns and operations at different places for the whole period of the military conflict.

The number of US casualties during the advance of 12th Army Group through Western Europe in World War II were used for analyzing the presence of chaos at the operational level. This is considered to be the operational level because the data comes from a single campaign limited geographically and with a single set of objectives.

1.2 Methods

The analysis of the results from application of qualitative and quantitative tests for identification of chaotic behavior in historical data from grand strategic, strategic, and operational level of war support the thesis that chaos is present in war. The appearance of time series and phase portraits were examined for signs of chaos. The phase space was reconstructed using Takens' time delay technique[13]. The correlation dimension was defined using the Grassberger-Procaccia[14] algorithm. The slope of the curve **C(r)** versus **r** was estimated using a least square fit for a region, limited by the value of the root mean square error for the linear approximation of the curve. Finally, two-dimensional Poincare maps were examined for patterns, space-filling and fractal characteristics.

2. IDENTIFICATION OF CHAOTIC BEHAVIOR

2.1 Grand Strategic Level

As an example of historical data on grand strategic level we used the US defense spending for the last 53 years. Figure 1 represents the time dependence of the defense spending as percentage of the Gross National Product (GNP) and as percentage of the federal outlays. The curves have ragged appearance and periodic cycles are not distinguishable. Figure 2 represents the phase space trajectory for the first data set.

Figure 1: Time dependence of US defense expenditures (FY87$)as a percentage of both GNP and federal outlays

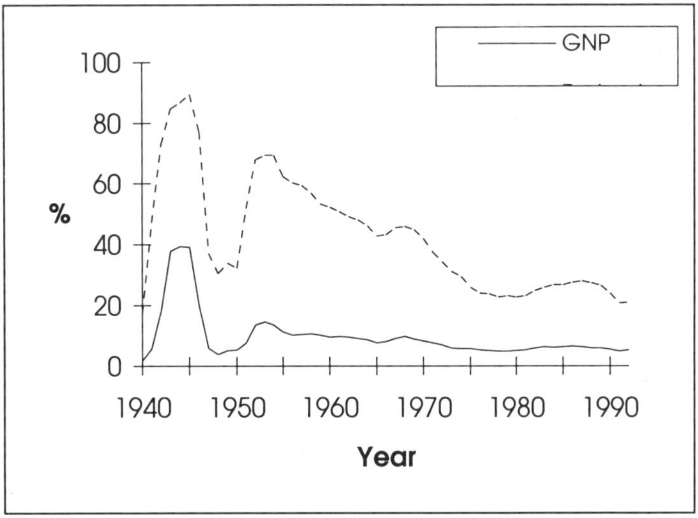

Figure 2: Phase space trajectory for defense spending as a percentage of GNP.

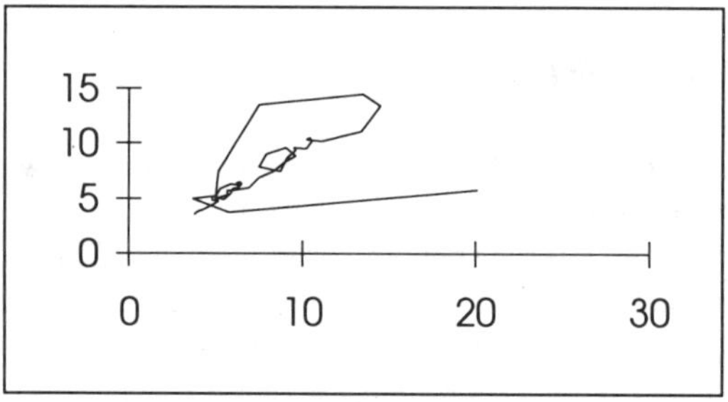

The data for y-axis is constructed from the original data (x-axis) with a time delay of one year.Though this is hardly an evidence for presence of patterns and attractors, the trajectory is confined in a region of the phase space, which may be interpreted as a sign of chaotic behavior. Applying the Grassberger-Procaccia algorithm for the first set of data (defense spending relative to GNP), we estimated a dimension of 1.21 and for the second (defense spending relative to federal outlays), 1.07. The dimensions are low and fractal, which is a clear sign of chaos.

The fact that chaos is present on grand strategic level is supported by the conclusions about the character of the strategic decision making[5] and the examination of a simple model of the arms race[8].

2.2 *Strategic Level*

Figure 3 represents the time dependence of US fixed-wing aircraft losses in Southeast Asia during the Vietnam war. The graph has a ragged appearance, exhibits recurrent nature at irregular intervals, and the region of values is limited, which is a qualitative sign of chaos.

The presence of chaos on the strategic level is further confirmed by the estimation of the correlation dimension for the same time series. Figure 4 presents the convergence of the estimated dimension for all aircraft losses (2.9) and for the losses in the air (3.2).

Figure 3: Time dependence of the US aircraft losses in Vietnam.

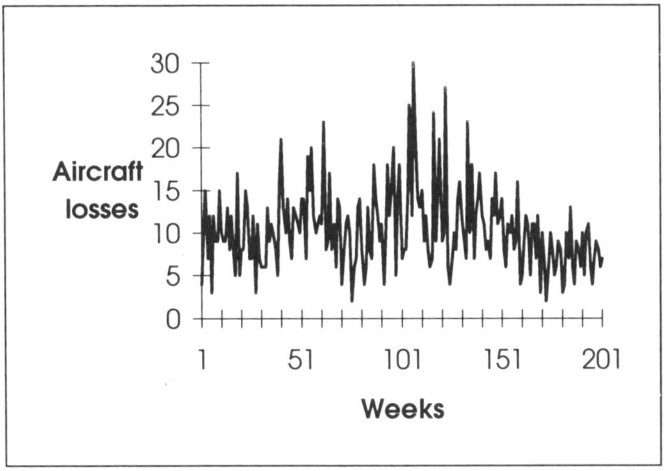

Figure 4: Estimated fractal dimension for weekly US aircraft losses

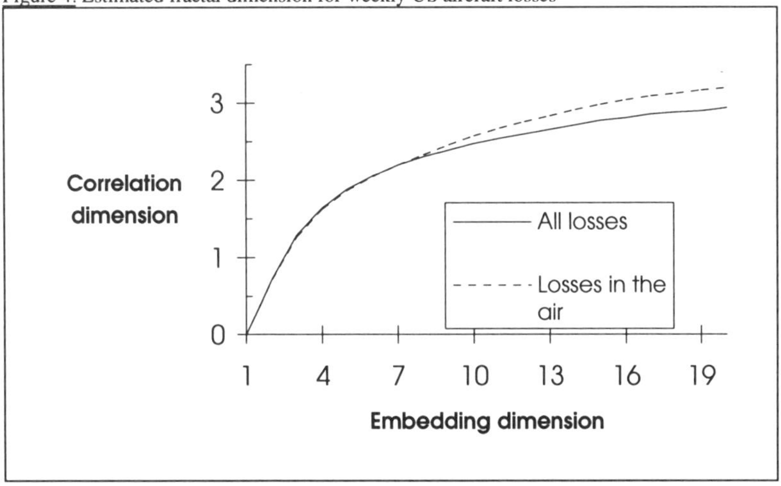

Figure 5 shows the convergence of the estimated fractal dimension for weekly and monthly data. Since the aggregation of the weekly data on a monthly basis is approximately equivalent to the linear operation of integration, the closeness of the estimates of the fractal dimension is another confirmation of chaotic behavior and reliability of the tests used.

Figure 5: Comparison of estimated fractal dimensions for weekly and monthly aircraft losses.

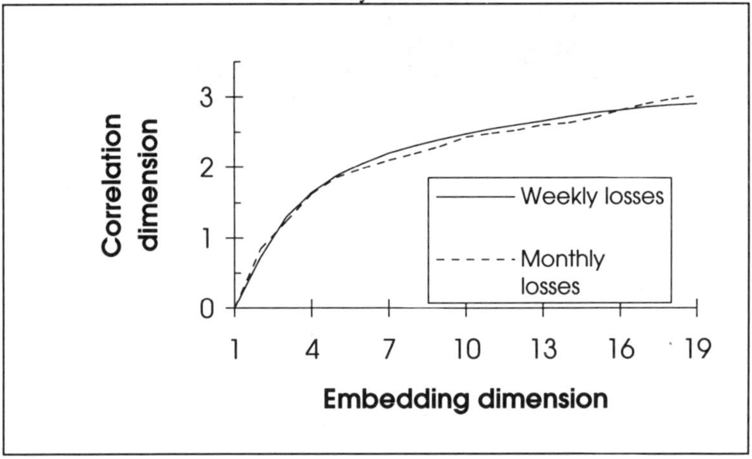

2.3 *Operational level*

At the operational level of war two examples were examined. Figure 6 shows the convergence of the estimated correlation dimension for soldiers killed in action (3.2), all personnel losses (2.75), and the losses normalized by the number of troops (1.2). An unexpected discovery is that with the increase of factors taken into account (in the order presented), the fractal dimension decreases.

Figure 6: Convergence of the estimated fractal dimensions for three sets of 196 data points for W.W.II casualties

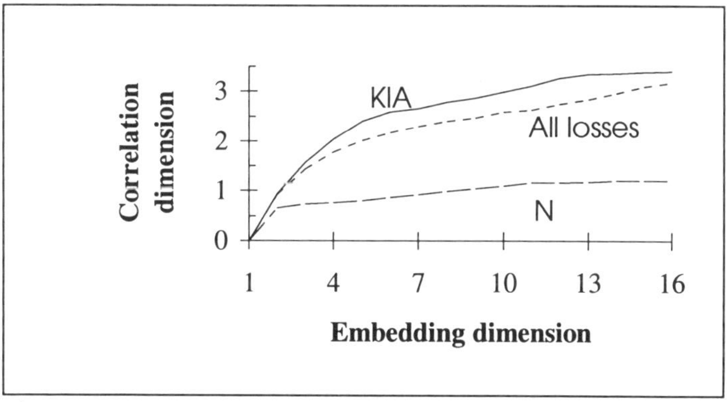

Figure 7 presents the two-dimensional Poincare map for the normalized casualties.

The time series is too short to allow us to discover patterns and attractors, but obviously the map is confined in the phase space, which is one of the qualitative signs for the presence of chaos[15].

Figure 7. Poincare map for the normalized W.W.II casualties. The two-dimensional cross section is of a three-dimensional phase space constructed by a time delay of one and eight days.

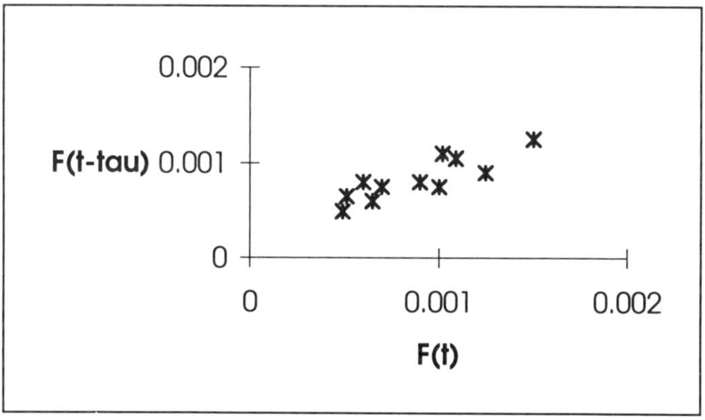

Dimension algorithms have low precision and it is difficult to judge whether the dimension is an integer [16]. Furthermore, in the 'chaos theory' sense the available data sets are very short, which makes the precision of our estimations questionable. Nevertheless, the examined data sets pass the qualitative tests, as well as one quantitative test for identification of chaotic behavior.

3. Implications of the Presence of Chaos in Warfare

Some chaos researchers have attempted to find chaotic characteristics in warfare to demonstrate that warfare is chaotic. While this approach has significant value, it has the disadvantage that observation of some of the characteristics of chaos in warfare does not prove that warfare is chaotic. For example, Beyerchen[6] has effectively argued that war is nonlinear. However, not all nonlinear systems are chaotic and therefore observation of nonlinearity in warfare does not actually prove that warfare is chaotic. In contrast, having previously focused on a mathematical approach to show that chaos is present in war, we will now reverse the above logic. That is, we can say that if war is chaotic, then it must have the characteristics of a chaotic system. We will now attempt to list the characteristics of chaotic systems and define what they mean in the context of warfare.

3.1 Chaotic systems are deterministic

All chaotic systems must be at least partly deterministic[1]. Therefore, if warfare is chaotic, then it too must be partly deterministic and must also obey a set of rules. The particular set of rules that are obeyed would depend upon the aspect of war being examined. For example, at the tactical level of war, bombing effectiveness may be governed by rules such as "a B-52 dropping twenty 500 pound bombs from 10,000 feet has a 50% chance of hitting a 10-foot by 10-foot target." At the operational level of war, the outcome is determined by the decisions and the behaviors of the participants. Therefore, if operational warfare is chaotic, the decisions and behaviors of the individuals and groups participating in the war must follow a set of rules.

In fact, if human behavior did not have any rules, if it were completely random, we would not be able to predict behavior at all. In reality, the better we know someone, the better we are able to predict their behavior. This is because we understand the rules that govern their behavior. In particular, the process of strategic decision making, which is a key element in operational warfare, has been shown to demonstrate chaotic behavior[5]. The presence of chaos in warfare means that it should sometimes be possible to predict enemy response if we take the trouble to learn the rules that govern his behavior. Put another way, this is Sun Tzu's conclusion of "know your enemy"[17]. The following sections will build on this foundation by suggesting ways that chaos theory can help us develop and use models of warfare.

3.2 Computer simulation can greatly enhance our understanding of chaotic systems

Computer numerical modeling or simulation has greatly increased our understanding of physical chaotic systems and might reasonable be expected to do the same for warfare. Chaos theory, however, cannot be used by itself to derive a theory of warfare. As with any other theory that describes a phenomenon, a theory of warfare must be based upon observation, hypothesis and testing. Specifically, development of a model of warfare would require the development of the structure of the model, determination of the number and type of variables, and determination of the form of the equations. In addition, system parameters and control factors would have to be identified as well as sources for noise. This is a very difficult task which is complicated by the possibility that different models might apply for different antagonists.

Chaos theory can help us by suggesting ways to develop our model and ways to use the model once developed. For example, observation of a chaotic system can be used to determine the dimension of the system. The number of variables needed to describe the system must at least equal the dimension of the system. Therefore, chaos theory can be used to define the minimum number of variables required in our computer model. Chaos theory also suggests that computer models of warfare must contain some nonlinear relationships between system variables so that it can reflect the chaotic nature of warfare. This may actually prove to be advantageous as the fractal nature of chaotic systems may allow relatively small and simple wargames to accurately simulate warfare. Realistic wargames that could be run on a desktop computer would have significant educational and operational advantages. Finally, the Kolmogorov entropy can be calculated for a chaotic system. The Kolmogorov entropy is a measure of how fast information is lost in a chaotic system and thus indicates how far into the future predictions can reasonably be made.

The ways in which computers have been used to understand chaotic behavior in physical systems also suggest ways to use the computer to model warfare. For example, chaos has given weather forecasters a means to determine whether their forecasts are likely to be accurate. Chaotic systems are highly dependent upon initial conditions but they are not always equally so. If a chaotic system is in a portion of its phase space where the initial conditions are critical, then uncertainty in determining the initial conditions makes a large number of outcomes possible. If a chaotic system is in a region of its phase space where the initial conditions are not critical, then only one outcome (prediction) is likely. In practice, weather forecasters use this behavior by inputting small changes in initial conditions into their model. If the small changes produce small variations in the prediction then they have shown that the system is in a portion of phase space where the initial conditions are not critical and their prediction is likely to be true. If the minor changes in initial conditions produce large deviations in future behavior, they know that their prediction is likely to be in error.

The same approach could be taken to understand when predictions in warfare are likely to be accurate. This in itself would be a valuable contribution of computer simulation to understanding warfare. There are, however, two additional reasons why this approach may be even more applicable to warfare than it is to weather. First, unlike weather forecasters, we have some ability to change the initial conditions. Specifically, if we find ourselves in a region of great uncertainty, we could determine which conditions would have to be changed to move the system to a position where the outcome was predictable and desirable. The quantity and type of forces are an example of initial conditions that we might be able to change. Second, we could use our model to determine which initial conditions and which variables had the most profound effect on our predictions. This would aid in identifying centers of gravity (COGs) and identify information that we needed to know precisely. That is, it would tell us where to concentrate our attack and what intelligence information was most critical.

3.3 Chaotic systems are dynamic

All chaotic systems are dynamic. So, if war is chaotic, then it too must be dynamic (this also agrees with experience). If a system is dynamic then the important parameters that describe the state of the system are constantly changing. In the physical example of a pendulum, potential energy, a maximum when the pendulum is at its height, is converted into kinetic energy which reaches a maximum at the bottom of the swing. As the pendulum rises, the kinetic energy is once again converted into potential energy. Understanding these sources of power is crucial in attempting to understand a dynamic system. Indeed, Pentland[18] has argued that these sources of power and the processes for their transformation will often be COGs for operational systems.

For societies, Pentland describes the sources of power as being the armed forces, the government bureaucracy, industry and natural resources, society and culture, and the value system. He relates these sources of power to the forces (or the instruments of power) that they produce--military, political and diplomatic, economic, social-cultural, and ideological. For societies, one example of the flow of power from one measure of the system to another is the conversion of economic power into military power. Pentland also points out that these sources of power, in combination with their linkage to instruments of power, and feedback control loops, can be the COGs for the system.

By definition, the values that describe a dynamic system are constantly changing. Even in a dynamic system, however, there are things that don't change. In a pendulum, the mass and the length of the pendulum are constant. Also, even though kinetic and potential energies are constantly changing, the sum of kinetic and potential energies remains constant. Analyses of dynamic systems rely on identifying aspects of the system that are constant and identifying processes for change--such as potential energy into kinetic energy for a pendulum. This suggests that similar approaches should be applicable to warfare and that emphasis should be placed not on the current values of the system (e.g. the number of tanks at a particular location) but on the processes (such as how the logistics operation moves tanks) and on the things that are relatively constant (such as industrial capacity).

3.4 Chaotic systems are nonlinear

All chaotic systems are nonlinear. Among other things, nonlinearity means that a small effort can have a disproportionate effect. This concept, in fact, is central to the whole military notion of a COG. Nonlinearity also means that the future behavior of a chaotic system cannot be predicted based upon initial conditions. This observation suggests that the campaign planner should concentrate on processes in an enemy system rather than data on its current condition. In particular, these observations suggest that nonlinear processes in an enemy's system should be fruitful areas to find COG's. The following paragraphs will describe some of the many sources for nonlinearity in warfare.

Feedback loops are one process that can introduce nonlinear effects in many systems. One example of such feedback is the rate of attrition. High attrition rates could force a commander to change his tactics. For example, the loss rates of 16% experienced by the US in the daylight bombing raids over Schweinfurt in 1943 were enough to stop the bombing raids for four months until a long range fighter was developed. Warden[9] used this and other historical examples to argue that the maximum acceptable rate was about 10%. He continued, however, by pointing out that the effect of one mission with a 10% attrition rate and nine missions with negligible casualties was much greater than a steady 1% attrition rate over ten missions. In a linear system there would be no difference between the two -- the additive effects would be the same. The fact that there is a difference shows that the feedback is nonlinear.

A second source for nonlinearity in warfare is the psychology associated with interpreting enemy actions. Because of the danger of a wartime environment, small actions on the part of your enemy often assume larger significance in a commander's mind than they deserve. According to Liddell Hart[19] this nonlinear effect occurred in World War I before the first Battle of the Marne. The Germans, aware of a possible seam in their dispositions, had been ordered to retreat if the British Army advanced over the Marne. As it happened, a British division sent out a reconnaissance patrol. The Germans misinterpreted this as a general advance and so they retreated when the way lay open for victory.

A third source for nonlinearity in warfare is the role of mass. Warden[9] showed that, for air power, losses vary disproportionately with the ratio of the forces involved. For example, in 1944, 287 American aircraft attacked a target defended by 207 German fighters. The Americans lost 34 aircraft. A month later, when 1,641 American aircraft were opposed by 250 German fighters, America lost 21 aircraft which is a lower percentage and a lower absolute number.

A fourth source of nonlinearity in warfare is Clausewitzian friction[6]. Basically, there will be events in war, perhaps as a result of chance, that have an effect out of all proportion to their apparent importance. By its very nature, this is an exceedingly difficult form of nonlinearity to anticipate but it can be taken advantage of once it happens.

Finally, the process of decision making itself can be a source for nonlinearity. Sometimes the decision is clear cut. Often, however, the decision can depend upon relatively minor circumstances at the time. One source[20] suggests that the steam engine lost out to the gasoline internal combustion engine largely as a result of an outbreak of hoof and mouth disease. Because of this outbreak, many horse troughs, which steam engines had used to top up on water, were removed. Once the decision is made, it is often irreversible because of the drive for standardization. Any major decision, including those made in wartime, can be nonlinearly based on such relatively minor factors.

3.5 Fractal geometries apply

If warfare is chaotic, then aspects of it must be fractal. This has implications for the analysis of an enemy system. First, the attractor for a chaotic system is fractal and so is infinitely complex. Efforts to analyze every aspect of an enemy's system are, therefore, bound to be in vain as there will always be some finer level to analyze. Second, behaviors at the tactical, operational, and strategic levels are linked. If a technique is successful at one level then we can expect it to be successful at all levels. This suggests that we should, when possible, try out strategies on a small scale where the consequences of losing are inconsequential. Third, if the small scale is similar in behavior to the large scale, then we can use observation of the small scale to predict the behavior of the large scale. For example, Admiral Yamamoto was fond of playing Shogi. In his biography of Admiral Yamamoto, Agawa[21] noted that his style of playing this game was to risk everything on a bold, early stroke. If that failed, then Admiral Yamamoto would often lose the game. Agawa suggests that this philosophy was behind the way in which Admiral Yamamoto planned his large campaigns -- such as Pearl Harbor and Midway.

The fractal nature of war may also have implications for the way we should organize for war. Sun Tzu[17] implied a fractal nature of war when he said, "Generally, management of many is the same as management of few." This indicates that he thought that the principles of organizing to fight were essentially the same regardless of the scale of the fight. Some principles, such as span of control, appear to be similar regardless of organizational level.

In addition, it is possible that "fractal" organizations may be more effective. By this, we mean that an organization where each unit has sufficient autonomy to conduct operations independently as the whole does. The German "Auftragstaktik" (mission oriented tactics) is an example of this notion in practice. In 1940, the Germans developed a system of independently operating units characterized by mobility and movement in conjunction with fire power. This system allowed both the search for and exploitation of tactical as well as operational opportunities. The command objective became coordination through communication rather than actual deployment and direct control of movement. This approach contrasted greatly with the French approach where subordinate units were controlled completely by centralized headquarters. The difference in approaches was one of the major factors in Germany's resounding victories. Decentralized organizational approaches, such as Auftragstaktik, can increase system abilities in the face of an unknown and unpredictable environment, can enable systems to form and change strategies on all levels and to generate diverse behavior. Preserving self-consistency in achieving a communicated vision becomes critical in such approaches as they are dependent upon subordinate units understanding what it is they should be doing and why they are doing it.

3.6 Multiple attractors are possible

Multiple attractors are possible in a chaotic system. In warfare, armed forces can drastically change their organization and means of fighting a war. The People's War of Mao Tse Tung is an example of this. Mao divided the phases of war up into different stages. In some stages, his army fought a guerrilla war as small units. Only later, when conditions were right (i.e. the opposing armies had been sufficiently weakened), did he combine his units into a conventional force. If warfare is chaotic, then chaos theory warns us that enemy systems can exist in different states. The implications are that we must be aware of these possible states and, if necessary, be capable of changing our own system state to counter the enemy strategy. Chaos theory also warns us that the transition from one state to another can be very fast.

4. Conclusions

Several qualitative and quantitative tests for identification of chaotic behavior were applied to data generated in war. The results show that chaos is present on the grand strategic, strategic, and operational levels of war. It should be pointed out that the amount of data was severely limited and that therefore we do not claim to have calculated exact values for such descriptors as the fractal dimension. Nevertheless, the examined data sets pass the qualitative tests as well as one quantitative test for identification of chaotic behavior.

In addition, the characteristics of chaotic systems were examined for their implications in the context of war. These implications include ways to use computer simulations to find centers of gravity, a need to place emphasis on nonlinear processes if seeking to leverage effects, some advantages in creating fractal organizations and the possibility of multiple attractors. Finally, several sources for nonlinearity in warfare were described.

Acknowledgements

This work was largely the result of research conducted at the US Air Force's Air Command and Staff College. We are grateful for the contributions of Major Peter Axup, Major Randal Franklin and Major Michael Dolgov (Russian Air Force). In addition, we are thankful to our research advisors Lt Col Gus Liby and Major Gary Burg. We also thank Commander Sarigul-Klijn, formerly at the Naval Postgraduate School, who confirmed the validity of our software and provided us with Poincare maps, and historian George Kuhn of the US Army's Logistics Management Institute, who provided us with the casualty data for the 12th Army Group in World War II.

References

1. Gleick, J., Chaos: Making a New Science. New York: Penguin Books (1987).
2. Shinbrot, T., Grebogi, C., Ott, E., & Yorke, J.A., "Small Perturbations to Control Chaos. *Nature*, 363, 411-415 (1993).
3. Jubak, J., "Can Chaos Beat the Market?" *Worth*, 3 , 66-70 (1993).
4. Weiss, G., "Chaos Hits Wall Street--The Theory That Is" *Business Week*, 2, 138-140 (1992).
5. Richards, D., "Is Strategic Decision Making Chaotic?" *Behavioral Science*, 35, 219-232 (1990).
6. Beyerchen, A. Clausewitz, Nonlinearity, and the Unpredictability of War. *International Security*, 17(3), 59-90 (1992).
7. Dewar, J.A., Gillogly, J.J., & Juncosa, M.L., "*Non-Monotonicity, Chaos, and Combat Models*" (Report R-3995-RC). Santa Monica, CA: RAND Corporation (1991).

8. Grossmann, S. & Mayer-Kress, G., "Chaos in the International Arms Race". *Nature*, 337, 701-704 (1989).
9. Warden, J. A. The Air Campaign: Planning for Combat. Washington: Pergamon-Brassey's (1989).
10. US Air Force. Basic Aerospace Doctrine of the United States Air Force (Vol.2). Washington, DC (1992).
11. US Government. Budget of the United States Government: Fiscal Year 1993 (Supplement). Washington, DC (1992).
12. McCrea, M.M., *U.S. Navy, Marine Corps, and Air Force Fixed-wing Aircraft Losses and Damage in Southeast Asia (1962-1973)* (Report CRC 305). Arlington, VA: Office of Naval Research. (1976).
13. Takens, F.,"Detecting Strange Attractors in Turbulence". *Lecture Notes in Mathematics*, 898, 366-381 (1981).
14. Grassberger, P. & Procaccia, I., "Measuring the Strangeness of Strange Attractors." Physica D, 189-208 (1983).
15. Schuster, H. G., Deterministic Chaos: An Introduction. New York: VCH Publishers (1989).
16. Parker, T. S., & Chua, L.O. Practical Numerical Algorithms for Chaotic Systems. New York: Springer-Verlag (1989).
17. Sun Tzu, The Art of War, Oxford University Press (1963)
18. Pentland, P. A., Center of Gravity Analysis and Chaos Theory, or How Societies Form, Function, and Fail. Maxwell Air Force Base, AL: Air University Press (1993).
19. Waldrop, M.M., Complexity: The Emerging Science at the Edge of Order & Disorder. New York: Simon & Schuster (1992
20. Agawa, H., The Reluctant Admiral: Yamamoto and the Imperial Navy. New York: Kodansha International (1979).
21. Hart, B.H.L., The Real War: 1914-1918. Boston: Little, Brown and Company (1964).

FRACTAL STRUCTURE IN BUSINESS CYCLE DURATIONS

(JEL classification: E32)

ALI ABDERREZAK

Department of Economics,The Pennsylvania State University,College Place, DuBois, PA 15801 E-mail: axa17@psuvm.psu.edu

Paper presented at the 1994 annual conference of the Society for Chaos Theory in Psychology and the Life Sciences, Johns Hopkins University, Baltimore, MD, June 24-27, 1994.

To the extent that many natural phenomena with long-term dependence inherently impact economic activity, it is reasonable to suspect that some of this dependence would reflect on key macroeconomic time series despite an apparent lack of orderly patterns in their behavior. To test this hypothesis, Rescaled Range (R/S) analysis is applied to U.S. business cycle durations. R/S is also modified to an "adaptive" rescaled range technique. The findings indicate that business cycles tend to have a long-term memory. Antipersistence is revealed as what links cyclical durations; longer cycles precede shorter ones, and vice-versa. This finding translates into a rejection of the alternate hypothesis of randomness in cyclical turning points.

Keywords: Antipersistence, business cycle, fractal, Hurst coefficient, persistence.

1. Introduction

The purpose of this paper is to investigate the issue of long-term irregular fluctuations in business cycles from a different perspective, that of what has come to be known as 'Rescaled Range' Analysis. To the extent that many natural phenomena (e.g. weather and sunspots) with seemingly erratic behavior which inherently impact economic activities exhibit a biased random process and so long-term dependence, it is reasonable to expect these phenomena to impact key macroeconomic time series to a certain extent.

Though the presence of long-term dependence in natural phenomena is relatively well-documented, it has only begun to be considered for macroeconomic and financial time series. Among the pioneers in this field, Mandelbrot (1971), relying on long-forgotten mathematical techniques, finds speculative markets to be fractal.[1] Asset prices in these markets, he argues, are characterized by persistent long-term statistical de-

[1] Benoit Mandelbrot was interested in natural objects that could not be defined by Euclidian geometry because of their lack of asymmetry and roughness. He theorized fractal geometry.

pendence.[2] The implication of this dependence is of theoretical and practical importance. Because it implies patterns and trends in capital markets, it calls into question the Efficient Market Hypothesis.[3] Studies by Greene and Fielitz (1977), and Peters (1989, 1991a, 1991b), among others, do support Mandelbrot's findings. Other studies, by Lo (1989) for instance, indicate otherwise. Using macroeconomic time series, Diebold (1989) finds evidence of long memory, which induces persistence in U.S. aggregate output.

The analysis of long memory in business cycles can be traced to Fisher (1925) who hypothesized that cycles were only a myth, thereby rejecting the notion of cyclical duration dependence. More recent empirical work—though limited to short-run analysis—has refined Fisher's initial proposition to suggest that cyclical fluctuations are Monte Carlo cycles, thus making the termination probability of a cyclical phase independent of its age.[4] If this hypothesis is correct, the analysis of business cycles would become of a dubious value, and the only economic policy to pursue is that of pure "laissez-faire." Such findings are advanced by McCulloch (1975), Savin (1977), and Diebold and Rudebusch (1987).

This paper focuses on the test of long memory in business cycles. This hypothesis is investigated using a statistical technique known as the rescaled range (R/S) or range over standard deviation method. R/S was first proposed by Hurst (1951), and later refined by Mandelbrot (1972). The original statistic was developed to analyze questions associated with water storage and management of the Nile river in Egypt. The theory defines a policy of discharge such that the reservoir never overflows or empties given a naturally irregular water influx. Subsequently, Hurst extended the analysis to various other natural phenomena. Analogous to the condition of river discharges, he found that these phenomena followed is known as a "biased Hurst walk" which indicates the existence of long-term dependence.

In the present context, the results from the R/S analysis indicate that durations of business cycles have a biased distribution. Business cycles are found to have long memory. Longer cycles tend to precede shorter ones and vise-versa. The results are also supported when the modified model, the "Adaptive" Rescaled Range, is tested.

---

[2]Mandelbrot and Wallis (1968) use the term "Joseph Effect" to describe long-term dependence. In the biblical story, Joseph interpreted Pharaoh's dream to indicate that seven years of plenty would be followed by seven years of famine.

[3]A market in which prices always "fully reflect" all available information is called "efficient," Fama (1970).

[4]The term "Monte Carlo cycles" is used to suggest cycles "superstitious gamblers believe they can discern in their luck at casinos like the one at Monte Carlo," McCullock (1975)

The organization of this paper is as follows. Section 2 consists of an introduction of orderliness (or lack of it) in business cycles. Section 3 is a brief technical presentation of the theory of the rescaled range technique and its potentially useful application to cyclical duration data. In Section 4 an adaptation of the NBER-business cycle chronology is performed to create a time series that would allow testing for cyclical duration long-term dependence. Section 5 summarizes the principal statistical findings. In section 6, the R/S is modified to formulate the "adaptive" Rescaled Range (R/S_i) model. Finally, section 7 summarizes the main points and the findings of the paper.

2. Orderliness and Business Cycles

Just as dams are an attempt to control potential destruction from a natural phenomenon, sound economic policies are perceived as means to manage the impact of business cycle fluctuations. Although the term "business cycle" implies steadiness and predictability in advanced capitalist markets, neither has proven to be the case. In fact, the evolution of these markets has been characterized by seemingly uncontrollable and recurrent ups and downs.

Monetary and fiscal policies for activists are instituted to stabilize cyclical durations and fluctuations, thus exploiting an inherent orderliness in market economies. For these economic policy activists, the hidden orderliness within the volatile fluctuations in business cycles is what justifies their interventionism. For many other economists, the apparent absence of systematic patterns and trends makes cyclical fluctuations Monte Carlo cycles, and thus renders fiscal and monetary activism inefficacious and even counterproductive.

Perhaps the adoption of a different approach which combines both views from above (activists and non-activists), would be more successful. This alternative would treat business cycles as complex and evolving structures driven by a deterministic process. But to do so, we first must test for long memory in business cycles.

The purpose of this article is then a preamble to the alternative approach. It consists of examining whether cyclical duration really does exhibit a random walk in the long run. Such random behavior of business cycle activities would of course reflect the absence of duration dependence, and thus previous cyclical duration would be irrelevant to present and future durations. But if long memory is established for business cycles, then previous durations become instrumental in predicting future ones.

3. Rescaled Range Analysis

As mentioned earlier, the R/S analysis was first intended to investigate water storage. Only Mt, water discharge, may be manipulated to regulate reservoir capacity. The water influx, et, is uncontrollable; its fluctuations stem from the natural difficulty in predicting rain patterns. The ideal reservoir would then call for the efficient management of Mt. When doing so, the reservoir would never empty or overflow. To achieve this optimal policy, we need to consider the cumulative deviation of et from M_t over N, the number of years.

We continue this brief technical introduction to R/S in the context of business cycles instead of water storage. To examine the behavior of cyclical fluctuations, we consider the cumulative deviation between the actual and anticipated cyclical durations over consecutive business cycles.[5]

Let d_t = the actual business cycle duration in months,

M_N = the anticipated cyclical duration over N cycles.

M_N is the arithmetic mean duration over a longer time horizon of N consecutive business cycles. $X_{t,N}$, the accumulated departure from the average M_N may be defined by the expression:

$$X_{t,N} = (d_t - M_N), \quad t=1,...,N \tag{1}$$

The range from the accumulated departures from the average, M_N, is defined by the following difference:

$$R_N = \underset{1 \le t \le N}{\text{Max}}[X_{t,N}] - \underset{1 \le t \le N}{\text{Min}}[X_{t,N}] \tag{2}$$

R_N depends on the time period under consideration; it would likely increase as the historical time series (Business-cycle chronology) increases.

Hurst standardized the different ranges R_N by dividing them by S, the standard deviations of the e_t (cyclical durations for our purpose) under consideration. The use of this dimensionless ratio permits the comparison of observed ranges of various phenomena.

The essence of *Hurst Empirical Law* simply consists of arguing that (R_N/S) may

---

[5]The anticipated cyclical duration may be as simple as the arithmetic mean from past business cycles.

be estimated by $(N/2)^H$. H, the Hurst exponent, may now be estimated by (RN/S). Thus,

$$(RN/S) = (N/2)^H, \quad 0 \leq H \leq 1 \tag{3}$$

Statistical results collected by Hurst indicate that many natural phenomena have H coefficients that are larger than 1/2.

A random walk distribution, indicating the absence of long-term statistical dependence, in the behavior of cyclical duration is achieved when the parameter estimate of the Hurst's exponent is 1/2. If H is less than 1/2, then a case of antipersistent behavior is realized. Antipersistence indicates that shorter business cycles precede longer ones, and vice-versa. Finally, if H is larger than 1/2 then a case of persistent behavior is established. Persistence in business cycles indicates that longer business cycles follow longer ones, and vice-versa.

Persistence and antipersistence also associate with the long-term memory effect of business cycles. A statistically significant Hurst coefficient estimate translates into a long memory effect which in turn indicates that the age (and thus the turning-point) of a business cycle depends not only on currents economic events but also on earlier economic conditions through the age of the cycle itself.

4. The Data

The data used in this study are constructed from the National Bureau of Economic Research (NBER) business-cycle chronology which dates back to 1854 for monthly observations. A business cycle is defined as a sequence of two consecutive cyclical phases connecting either two peaks or two troughs. Business cycles depict recurrent sequences of synchronized upturns and downturns in a great number of economic activities. An expansion (contraction) consists of a period of a sustained increase (decline) in aggregate economic activity. The observed cyclical fluctuations do vary in amplitude and duration. These fluctuations are specified in the time frame defined by the NBER, which would influence the outcome of studies that use the NBER chronology.[6]

---

[6]The NBER chronology has been questioned by some economists who raise legitimate concerns. Romer (1986), for instance, attributes the relative post World War II stability in economic activity to the prewar measurement. However, the NBER chronology remains the most comprehensive one that is used in macroeconomic studies.

U.S. business cycles may be subdivided into two categories: major cycles which mainly coincide with wartime, and minor cycles associated with 'normal' aggregate economic activity in a market economy. Minor cycles are important to consider because they do not involve major reallocation of long-term investment opportunities in the economy. Because of its importance, this distinction between minor and major cycles will be used to examine whether the long-term dependence in cyclical duration depends on the nature of the cycles used.

U.S. business cycles may also be subdivided into two groups based on another criterion: prewar cycles (1854-1948) and postwar cycles. Postwar business cycles have been less severe than in the prewar period. The duration data indicate that while the average length of recession has fallen by one-half, the average length of an expansion doubled. The pre- and postwar division in business cycles will also be used when testing for long memory.

Using the NBER monthly business-cycle chronology of Table1 below, the cyclical duration is set up to coincide with the age of business cycles. At each cyclical turning point (peak or trough) the duration of a business cycle is one month. In the succeeding month's period this duration becomes two (months) and so forth. Thus, a cycle whose duration or age is 48 months will then be represented by a time series that ranges from one to 48 (months), with increments of one (month). For our purpose here, a time series (business-cycle duration) is random if it is influenced by a *large* number of events that are equally likely to occur. A nonrandom time series will reflect the non-randomness of its influences. This time series will then be fractal.

5. Estimating The Hurst Coefficient

To estimate the Hurst coefficient, equation (2) may be rewritten as follows:

$$\log(R/S) = H.\log(N/2) \tag{4}$$

Equation (4) can then be estimated using a regression analysis.

The correlation between consecutive business cycles is depicted by the coefficient C_N. Over period N C_N may be computed as:

$$CN = 2^{(2H-1)} - 1 \tag{5}$$

From Equation (4) it is obvious that a Hurst exponent of 1/2 makes the correlation coefficient zero. The correlation of past and future increments vanishes over a period or cycle N as implied for an independent random process. The case of persistent behavior ($0.5 \leq H \leq 1.0$) will translate into a positive correlation coefficient, while that of antipersistence behavior ($0 \leq H \leq 0.5$) will reflect a negative correlation coefficient.

Equation (4) is estimated using the Ordinary Least Squares (OLS) technique. Because of low D-W statistics (indicating the presence of serial correlation), the Cochrane-Orcutt transformation is performed. The regression results are summarized in Table 2. The findings do support the hypothesis of antipersistence in business cycle duration. Except for the Peak-to-Peak definition, the Hurst's coefficient estimates are less then one-half and are statistically significant in all scenarios.[7] The negative correlation coefficient estimates from Table 2 also indicate support for the hypothesis that the memory effect does not dissipate in the context of cyclical durations.

Because business cycles do vary in duration and magnitude over time, minor cyclical duration dependence is analyzed in a abridged sample. Table 4 below summarizes the findings for minor business cycles only. The results reveal that major cycles do not seem to alter the findings; the long-term duration dependence hypothesis is still tenable. However, the degree of dependence does seem to be impacted by the time period under consideration.

6. An "Adaptive" Rescaled Range Statistic

To consider the hypothesis from a different angle, leading to perhaps a more effective scheme, an adaptive weighing function is proposed. The ratio R/S is modified to become an "adaptive" rescaled range statistic, R/Si [(i=1,2,...t), t being the age of the business cycle]. Instead of using one value per cycle, S varies every month over t, thus allowing the standard deviation to reflect a learning process in which the rescaled ratio captures the monthly updates in the variability of cyclical durations. The model to estimate then becomes:

$$\log(R/Si) = H.\log(N/2 \qquad (6)$$

The regression results for the "adaptive" Rescaled Range analysis are summa-

[7]The definition used for business cycles (peak-to-peak versus trough-to-trough) does seem to slightly impact the size and the significance of the estimated Hurst coefficient. One possible explanation may reside in what the preceding cyclical phase may have on the following one. In the trough-to-trough definition, expansions may impact contractions in a way that is quantitatively and/or qualitatively different from the way contractions impact expansions in the peak-to-peak definition.

rized in Tables 3 and 5. The findings support the hypothesis of antipersistence in cyclical duration using both major and minor cycles. The Hurst coefficient estimates as indicated in Tables 3 and 5 are less than one-half and are statistically significant.

7. Summary and Concluding Remarks

The purpose of this paper was to gain an insight into long-term cyclical duration dependence using NBER business-cycle chronology. Using both standard and "adaptive" Rescaled Range analyses, long memory in the form of antipersistence in U.S. business cycles was discernable throughout. The established long-term association indicates that shorter business cycles precede longer ones, and vice-versa. Moreover, the size of the estimated Hurst coefficient suggests that the degree of dependence between cycles has increased for postwar business cycles.

Antipersistent behavior in business cycle duration may now be viewed as an element underlying a deterministic process that, if uncovered, could improve our ability to make drastically better economic forecasts. If this deterministic process happens to derive from natural phenomena which inherently impact economic activities, it is important to reconcile cyclical antipersistence with persistence in natural phenomena.[8] It is within this apparent contradiction that we may have an answer to the long memory in business cycles. Finally, in terms of policy implications, since business cycles and thus cyclical turning points are not generated by a random process, decision-makers may successfully utilize the above deterministic process to implement appropriate and timely monetary and fiscal policies.[9]

Acknowledgments

The author would like to thank F.K. Vinlove and Bill McCarthy for their generous assistance.

---

[8]The Hurst exponent H for natural phenomena is found to be more or less symmetrically distributed about a mean of 0.73, with a standard deviation of about 0.09.

[9]Keeping in mind that, in chaos, even with a deterministic system in place, we may not realize accurate and dependable predictions. As a result, economic policies may prove ineffective or even counterproductive yet cyclical durations are not random.

References

1. Diebold, F. and G. Rudebusch (1987), "Does the Business Cycle Have Duration Memory?" *Board of Governors of the Federal Reserve System, Special Studies*, Paper no. 3.
2. Diebold, F.X. (1989), "Long Memory and Persistence in Aggregate Output," *Journal of Monetary Economics*, 24, pp. 189-209.
3. Fama, E. (1970), "Efficient Capital Markets: A Review of Theory and Empirical Work," *Journal of Finance*, 25, pp. 383-417.
4. Fisher, I. (1925), "Our Unstable Dollar and the so-called Business Cycle," *Journal of the American Statistical Association*, 20, pp. 179-202.
5. Hurst, H. (1951), "Long Term Storage Capacity of Reservoirs," *Transactions of the American Society of Civil Engineers*, 116, pp. 770-799.
6. Lo, A.W. (1989), *Long-Term Memory in Stock Market Prices*. NBER Working Paper No. 2984.
7. Mandelbrot, B. and J. Wallis (1968), "Noah, Joseph and Operational Hydrology," *Water Resources Research* 4, pp. 909-918.
8. Mandelbrot, B. (1971), "When Can Prices Be Arbitrage Efficiently? A Limit to the Random Walk and Martingale Models," *Review of Economics and Statistics*, 53, pp. 225-236.
9. ____________, (1972), "Statistical Methodology for Non-periodic Cycles: From Covariance to R/S Analysis," *Annals of Economic and Social Measurement*, 1, pp. 259-290.
10. McCullock, J.H. (1975), "The Monte Carlo Cycle in Business Activity," *Economic Inquiry*, 13 pp. 303-321.
11. Peters, E. (1989), "Fractal Structure in Capital Markets," *Financial Analysts Journal,"* July/August, pp. 32-37.
12. __________,(1991a), "R/S Analysis Using Logarithmic Returns: A Technical Note," *Financial Analysts Journal*, July/August, pp. 81-82.
13. __________,(1991b), *Chaos and Order in the Capital Market.* New York: John Wiley & Sons.
14. Romer, Christina (1986), "Is the Stabilization of the Postwar Economy a Figment of the Data?" *American Economic Review*, 1, pp. 1-37.
15. Savin, N.E. (1977), "A Test of the Monte Carlo Hypothesis: Comment," *Economic Inquiry*, 15, pp. 613-617.

Table 1: The NBER Business-Cycle Chronology

Cycle number	Expansion Begins	Length (months)	Full Cycle (TT)[1]	Contraction Begins	Length (months)	Full Cycle (PP)[2]
Prewar						
1	Dec. 1854	30	.	June 1857	18	.
2*	Dec. 1858	22	48	Oct. 1860	8	40
3*	June 1861	46	30	Apr. 1865	32	54
4	Dec. 1867	18	78	June 1869	18	50
5*	Dec. 1870	34	36	Oct. 1873	65	52
6*	Mar. 1879	36	99	Mar. 1882	38	101
7	May 1885	22	74	Mar 1887	13	60
8	Apr. 1888	27	35	July 1890	10	40
9	May 1891	20	37	Jan 1893	17	30
10	June 1894	18	37	Dec. 1895	18	35
11	June 1897	24	36	June 1899	18	42
12	Dec. 1900	21	42	Sep. 1902	23	39
13	Aug. 1904	33	44	May 1907	13	56
14	June 1908	19	46	Jan. 1910	24	32
15	Jan. 1912	12	43	Jan. 1913	23	36
16	Dec. 1914	44	35	Aug. 1918	7	67
17*	Mar. 1919	10	51	Jan. 1920	18	17
18	July 1921	22	28	May. 1923	14	40
19	July 1924	27	36	Oct. 1926	13	41
20*	Nov. 1927	21	40	Aug. 1929	43	34
21*	Mar. 1933	50	64	May 1937	13	93
22*	June 1938	80	63	Feb. 1945	8	93
23	Oct. 1945	37	88	Nov. 1948	11	45
Postwar						
24	Oct. 1949	45	48	Oct. 1953	10	56
25	May 1954	39	55	May 1957	8	49
26	Apr. 1958	24	47	Apr. 1960	10	116
27*	Feb. 1961	106	34	Feb. 1969	11	47
28	Nov. 1970	36	117	Nov. 1973	16	74
29	Mar. 1975	58	52	Mar. 1980	6	18
30	July 1980	12	64	July 1981	16	108
31	Nov. 1982	104	28	Nov. 1990	8	?
32	Mar. 1991	?	112			

Notes: (*) refers to a "major" cycle,
(1) signifies trough to trough
(2) signifies peak to peak.18

Table 2: (Regression Estimates of eq. (4): log(R/S) = a0 + H.log(N/2)
(All Business Cycles)

Business Cycle	a0	H	R^2	SSR	C_N
Peak-to-Peak					
Full Sample	6.18	0.091	.998	1.24	-.433
	(43.3)*	(13.3)			
Prewar	5.86	0.106	.998	1.20	-422
	(38.6)	(11.8)			
Postwar	7.02	-0.034	.946	11.8	-.523
	(48.8)	(-0.38)			
Trough-to-Trough					
Full Sample	6.14	0.095	.999	.997	-.430
	(41.2)	(13.8)			
Prewar	5.82	0.102	.998	.978	-.424
	(34.9)	(11.5)			
Postwar	7.16	0.043	.997	.013	-.469
	(228.)	(11.2)			

Notes:(*) the t-statistics are between parentheses,
C_N is the correlation coefficient,
November 1948 (peak) is the pre- and postwar dividing point.

Table 3: Regression Estimates of eq. (6): $\log(R/S_i) = a_0 + H.\log(N/2)$
(All Business Cycles)

(Adaptive Standard Deviation Model)

Business Cycle	a0	H	R^2	SSR	C_N
Peak-to-Peak					
Full Sample	4.83	0.031	.999	.859	-.478
	(2.36)*	(29.6)			
Prewar	5.25	0.037	.998	.787	-.474
	(3.40)	(27.1)			
Postwar	6.53	0.019	.946	11.3	-.487
	(9.25)	(1.46)			
Trough-to-Trough					
Full Sample	5.56	0.029	.999	.678	-.479
	(3.51)	(32.0)			
Prewar	5.34	0.035	.999	.592	-.475
	(3.88)	(29.5)			
Postwar	7.08	0.009	.999	.004	-.494
	(51.9)	(40.2)			

Notes:(*) the t-statistics are between parentheses,
C_N is the correlation coefficient.
November 1948 (peak) is the pre- and postwar dividing point.

EXHIBIT 4: Regression Estimates of eq. (4): $\log(R/S) = a_0 + H.\log(N/2)$
(Minor Business Cycles)

Business Cycle	a0	H	R^2	SSR	C_N
Peak-to-Peak					
Full Sample	6.47	0.083	.997	.924	-.439
	(45.6)*	(10.7)			
Prewar	6.12	0.100	.993	.892	-.426
	(52.7)	(9.01)			
Postwar	7.22	0.006	.995	.002	-.496
	(178.)	(1.37)			
Trough-to-Trough					
Full Sample	6.46	0.040	.997	1.10	-.471
	(64.8)	(4.49)			
Prewar	6.07	0.047	.994	1.10	-.466
	(60.1)	(3.59)			
Postwar	7.19	0.044	.996	.016	-.469
	(200.)	(1.16)			

Notes:(*) the t-statistics are between parentheses,
C_N is the correlation coefficient,
November 1948 (peak) is the pre- and postwar dividing point.

EXHIBIT 5: Regression Estimates of eq. (6): $\log(R/S_i) = a_0 + H.\log(N/2)$
(Minor Business Cycles)
(Adaptive Standard Deviation Model)

Business Cycle	a_0	H	R^2	SSR	C_N
Peak-to-Peak					
Full Sample	6.14	0.029	.998	.704	-.479
	(5.86)*	(21.1)			
Prewar	5.90	0.037	.995	.654	-.474
	(7.31)	(17.4)			
Postwar	7.18	0.011	.998	.008	-.497
	(46.3)	(27.9)			
Trough-to-Trough					
Full Sample	5.63	0.023	.998	.847	-.484
	(4.16)	(15.8)			
Prewar	5.64	0.030	.995	.808	-.479
	(5.04)	(13.0)			
Postwar	7.04	0.009	.998	.007	-.494
	(34.0)	(22.9)			

Notes: (*) the t-statistics are between parentheses.
C_N is the correlation coefficient,
November 1948 (peak) is the pre- and postwar dividing point.

Part II

THEORETICAL STUDIES

CAUSALITY AND EMERGENCE IN CHAOS AND COMPLEXITY THEORIES

Jeffrey Goldstein, Ph.D.
Adelphi University
Garden City, NY 11530

The concept of causality is revised in the light of the phenomenon of emergence as explored in chaos and complexity theories. Emergence is examined by taking a close look at the arising of new, more complex attractors in the logistic equation as the parameter k is increased. The "qualitative dynamics" of these attractors are understood in terms of a pattern-based causality. Philosophical issues associated with this revised view of causality are then discussed.

> *From where I sit it looks as if most of the behavioral scientists...tried to hook up their stuff to the wrong half of the fundamentals of science. That is, they followed the Ionians in asking questions about substance "What is it made of? How much of it is there?" They should have followed the Pythagoreans in asking about forms and patterns.* (Letter from Gregory Bateson to George Sarant 6/25/1969).

1 Introduction

1.1 The End of Causality?

Besides their various applications in the sciences, chaos and complexity theories are presenting intriguing implications for certain age-old philosophical issues such as determinism and causality. For example, the philosopher of science Stephen Kellert in his influential work *In The Wake of Chaos*[1] has claimed that chaos theory has supplanted the need for causal explanation:

> ...chaos theory does indeed give us understanding by showing us the

> mechanisms responsible for unpredictable behavior, but that these are not causal processes. Instead, these mechanisms are best understood as geometrical, and chaos theory tells us how they function (pp. 104ff).
>
> ...an account of scientific understanding as the disclosure of hidden causal processes is not only inadequate for the biological and social sciences...but it is inadequate for the physics of nonlinear systems as well (p. 105).
>
> Chaos theory does not reveal the causal process at work because it does not need to...Furthermore, it is impossible in principle to trace out the workings of the actual causal mechanism in a chaotic system. To do this would require calculations using the evolution equations of the system (p. 105).
>
> The theoretical hypotheses of chaos theory assert relationships of topological similarity, not congruence of physical causes, between its exemplary models and actual systems (p. 107).
>
> So chaos theory provides us with understanding that is holistic, historical, and qualitative, eschewing deductive systems and causal mechanism and laws...(p. 114).

According to Kellert, chaos theory can abandon causality because of the property of sensitivity to initial conditions whereby the unpredictability of chaotic systems is a function of a supposed sundering of the initial from the final state of the system[a]. In this fairly typical line of thinking about the implications of chaos theory, the "holistic" and top-down approach of "qualitative dynamics" supercedes causal mechanisms considered as merely bottom-up and reductionistic.

The claim that modern science should eschew causal explanation is not new. As far back as 1912, Bertrand Russell[2] declared that philosophy should follow what

---

[a] Kellert doesn't mention Robert Shaw's very significant insight that chaotic systems can show a gain of experimental information when the present state of the system is contrasted wtih an earlier state so that there is an increase in accuracy in backward determining the earlier state. This process can act as a constraint on unpredictability due to sensitivity to initial conditions.

science had putatively already done and leave causality behind as an anthropomorphic relic of the past. Indeed, in physics there has been a general tendency towards mathematical formalisms which, in some cases, seem to replace causal, mechanistic explanations[3]. However, even though the eminent philosopher of science Ernest Nagel[4], in his perusal of scientific literature, conceded that explicit use of the term "cause" was rarely found, he, nevertheless, found that the idea behind the term "cause" continued to have wide currency in the accounts scientists gave of their experiments, as well in their interpretations of the explanatory power of various mathematical formalisms. Nagel also made a very prescient observation about the relation of causal explanations to a science's abiltity to predict: explicit discussions of causality usually occur when a scientific development creates doubts about the adequacy of typically used procedures of inquiry such as the finding of a lack of predictive power. Nagel's remarks pertain very much to the current situation surrounding the new type of unpredictability discovered in chaos theory.

1.2 Emergence

It needs to be pointed out, however, that in an earlier paper[5] I found that despite Kellert's protestations to the contrary, natural scientific investigations utilizing chaos theory did use causal explanations even in the midst of a great deal of "qualitative dynamics." In the current paper, I want to show that the role of causal explanation in chaos and complexity theory is indeed problematic but not primarily having to do with the unpredictability associated with sensitivity to initial conditions. Instead, I am claiming that causality is problematic because of the phenomenon of *emergence* or the arising of global, complex order in chaotic or complex systems. But, rather than emergence calling for the end of causality, I am suggesting, instead, a revision of causal explanation in the light of emergence.

Emergence, as the arising of complex, global patterns presents a challenge to our established notions of causality because it represents the coming forth of something radically new. Emergent phenomena seem to just pop-out of the blue similar to the three dimensional images that suddenly appear in Magic Eye[6] artwork as your focus shifts and all of a sudden a three dimensional image emerges that you didn't see before when your eyes were focussing on the details of the picture. This

3-D image doesn't appear connected to the details of the surface of the picture, but seems to hover there in a space of its own somewhat like the phase space of nonlinear dynamics. (In fact, if you begin to focus on the details in the picture, you may lose the 3-D image.)

Similarly, the phenomenon of emergence in a complex system seems to emerge out of the blue, and, as such, emergence seems to violate our commonsensical notion of causality whereby something becomes what it is by a process or a causal chain involving something that was there before. That is, in our conventional notion of causality, the cause is understood as that which was there before, somehow like the effect, at least to the extent that the effect is seen to naturally follow from the cause. But, emergence as an effect arises in such a startling manner that it seems to disrupt the chain of cause and effect. This reminds me of a cartoon by Sidney Harris in which two scientists are standing next to a blackboard, one explaining to the other the equations written on the board. This scientist points to a certain section of an equation and declares "Then a miracle occurs!" Of course, one consequence of believing that a miracle occurred would be to stop asking causal questions about the miraculous phenomenon. Analogously, it is the startling phenomenon of emergence, in my opinion, which strains and threatens to crack causal explanation. In focussing exclusively on whatever it is that has emerged, in Kellert's case the chaotic attractor depicted as a "geometric" figure on a phase portrait, and ignoring how it arose, that is, the causal sequence, is like believing emergence is a miracle and therefore requires no further inquiry.

1.3 Mythologies About Chaos and Complexity Theories

This way of proceeding by ignoring the causal chain leading to emergence is related to one of the problems encountered in trying to explore the phenomena of emergence--the way that certain presumptions, what I will call "mythologies," have been developing about what is actually going on in chaotic and complex systems. One such mythology concerns the relation of simplicity to complexity. This is the notion that the complexity of chaotic behavior emerges en masse, as a whole, all at once out of simplicity like Athena from the head of Zeus. Thus, one hears it touted that chaos demonstrates that amazing complexity can come forth from simple rules or equations, as if there was no preparation for how simplicity is transformed into complexity. Certainly, for the original explorers of the fantastic realms of chaotic or complex systems, the emergence of complex phenomena out of seemingly simple systems seemed so extraordinary, even counterintuitive that the emphasis was on the chasm between the emergent phenomena and the simple rules defining the initial situation. But, just because an effect is so startingly novel doesn't entail there is nothing responsible for it.

It isn't a far step from this emphasis on the chasm between simplicity and complexity to the impression that the causal chain between the simple rules and the later qualitative dynamics has been unhinged. Of course, Kellert and other qualitative dynamicists would not claim that chaos is uncaused, it is, after all, *deterministic* chaos. However, by so emphasizing the so-called "geometric" intricacies of phase space to the detriment of exploring the causal chain bringing about emergence, this approach neglects the extremely crucial role played by what I have discerned as four interrelated factors in the causal chain responsible for emergence. First, I will give an example of these four causal factors in operation and then later show how they lead to a revising of our conventional notion of causality.

But, before I proceed with an examination of these four causal factors, I want to refer back to the opening quote from Gregory Bateson. I believe our conventional notions of causality have been built-up over the years from our ideas of the concept of substance, which Bateson attributes to the ancient Ionian questions concerning "What is it made of?" or "How much of it is there?" That is, much of what underlies our idea of causality is a *substantialist* causality, meaning that it is a view of how this or that substance comes about through the action of this or that substance on it. In my opinion, the new view of causality demanded by chaos and complexity theories must be an understanding of causality not in terms of substance but in terms of *pattern*, what Bateson calls the Pythagorean approach. Such a reframed view of a pattern-based causal explanation would be concerned with questions like: How do the new patterns shown in emergent phenomena relate to previous patterns in the system? How are these patterns transformed? What in the new patterns is novel and what is a carryover from earlier states of the system? I will return to this issue of pattern- based causality later on in the paper. For now, let's turn to an example of emergence.

2 An Example of Emergence

2.1 The Logistic Equation

As an example of emergence, I propose we take a careful look at the period doubling route to chaos exemplified in the evolution of the logistic equation as the parameter value is increased. Since, the explorations of this route to chaos in the early work of Robert May[7] and Mitchell Feigenbaum[8], the logistic map has become somewhat paradigmatic for how we think of dynamical systems, particularly ones exhibiting a route to chaos. Emergence is encountered as this purely mathematical system undergoes bifurcation and the transition from simpler to more complex attractors. The logistic map is also instructive because how May and Feigenbaum have described the route to chaos (Feigenbaum's account building on May's) has been a main source of some of the popular mythologies about chaos that I mentioned earlier, e.g., the impression that complexity arises en masse somewhat mysteriously out of a simple substrate with simple rules. Consider, for example, a title from one of May's articles: "Simple mathematical models with very complicated dynamics." I'm not saying May's article actually commits what this impression suggests. After all, May's and Feigenbaum's schemes are purely mathematical and asking about causes is not customarily what mathematicians do. Yet, the way complex dynamics emerge out of simple rules cannot but provoke curiousity as to where all this complexity comes from.

The logistic equation is a first-order difference equation often used to model a system in which populations change from one generation to the next in such a way that the population increases when it is small and decreases when it is large, with a parameter k expressing birth or death rates. According to May's terminology, this parameter k "tune's" the degree of the curvilinearity of the resulting parabola. The logistic equation assumes this form:

$$X_{t+1} = kX_t(1-X_t) \qquad (1)$$

It is mapped on the unit interval, with X on the x axis and X_{t+1} on the y axis (see Figure 1).

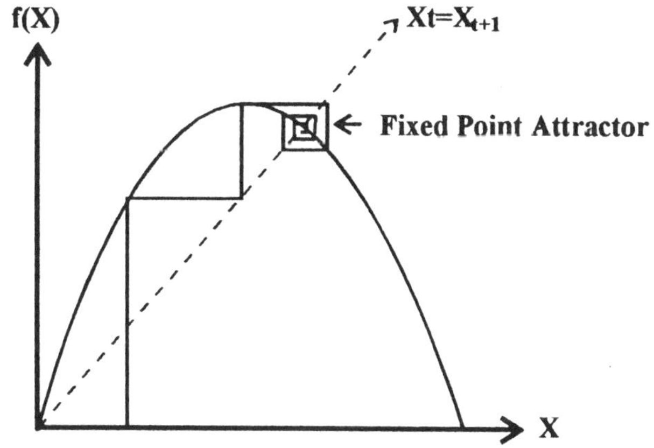

Figure 1: Functional Iteration--Logistic Equation

This is, of course, a nonlinear equation because X is squared, and as a difference equation, there is the operation of what Feigenbaum calls "functional iteration," that is, the value of X for each later generation is found by replacing the value of X for the preceeding generation into the equation and successively iterating the function (a kind of feedback).

Figure 1 shows that, as functional iteration occurs, at low values of k the dynamics of the equation are rather simple with a fixed point attractor. Emergent phenomena occur as the value of k is increased and the system undergoes bifurcation into more complex attractors. I am interested in the causal process responsible for this emergence of complexity. First, we must recognize what exactly is involved in functional iteration. Because the logistic equation looks so simple, the impression is that functional iteration is a relatively simple operation, but, in my opinion, this impression is deceptive. For example, the algebraic transformation in merely 4 generations shows a rather complicated evolution:

$$X_{t+1} = kX_t(1-X_t) \tag{2}$$

$$X_{t+2} = k(kX_t(1-X_t))(1-(kX_t(1-X_t))) \tag{3}$$

$$X_{t+3} = k(k(k(kX_t(1-X_t)))(1-(kX_t(1-X_t)) \\ (1-(kX_t(1-X_t)))) \tag{4}$$

$$X_{t+3} = k(k(k(kX_t(1-X_t))(1-(kX_t(1-X_t)))) \\ (1-(kX_t(1-X_t))(1-(kX_t(1-X_t)))) \\ (1-k(k(kX_t(1-X_t))(1-(kX_t(1-X_t)))) \\ (1-(k(kX_t(1-X_t))(1-(kX_t(1-X_t)))) \tag{5}$$

Let's see how this process of functional iteration looks on the mapping of the function (see Figure 1). Functional iteration, here, is demonstrated by starting at any initial value of X_0 and moving up to the curve $f(X)$, then moving horizontally to the 45^0 line representing the case when $X_t=X_{t+1}$ and then moving vertically back to the curve $f(X)$ to find the new value of X. This vertical-horizontal-vertical stepping is just a quick way to expedite the process of functional iteration by replacing X at t with X at $t+1$. Let's follow the sequence of X's that are generated by functional iteration (Figures 2 & 3).

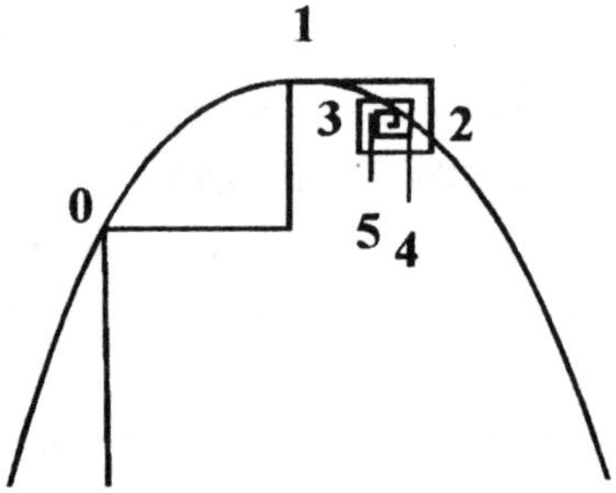

Figure 2: Sequence from Functional Iteration

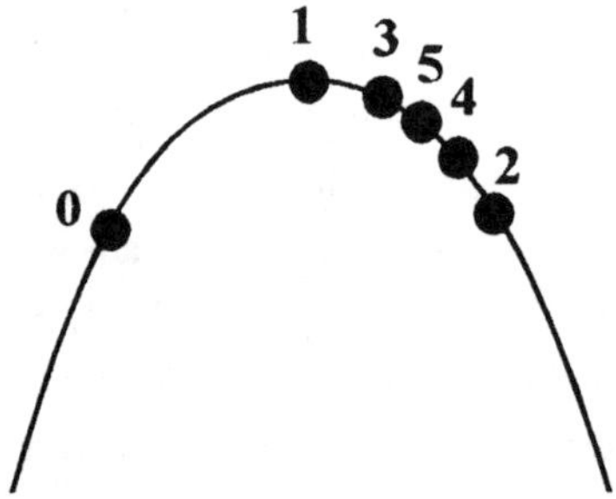

Figure 3: Functional Iteration as a Complexifying Operation

2.2 *Pattern Complexifying Operations*

It is clear that this so-called simple process of functional iteration leads to a more convoluted route taken by X as it traverses each succeeding generation than the route taken by X simply following the parabolic curve of $f(X)$. As Figures 2 and 3 show, the sequence on the parabola $f(X)$, at first, as we would expect, is from the points labelled 0 to 1 to 2, but then it folds back so that 3 is between 1 and 2, and then another fold as 4 is between 3 and 2, and then still another fold as 5 is between 3 and 4. To be sure this pattern ultimately spirals in to the fixed point attractor, and we don't yet see the phenomena of emergence. But my point is that functional iteration is a *pattern complexifying operation* which generates a more complex pattern than that found on the original, nonfunctionally iterated function represented by the parabola. My use of "more complex" here is deliberate, following from the etymologically root of "complex": "com"--"together with" as in a plurality and "plex"--"fold" or "crease." "Complexity" is the occurrence of a plurality of folds. "Pattern" refers to the sequential relationships of the points representing the succeeding generations of X to one another. A noncomplexified, i.e., nonfolded, sequence on the parabolic function would show that as X increases in value, one can follow the corresponding point on $f(X)$ as a noncomplex sequence on the parabola.

This same *pattern complexifying operation* can be seen in the algebraic expansion of functional iteration. At this juncture, two points need emphasis. First, eventhough there is a complexification of pattern, at the same time the original pattern is sustained, albeit in a transformed form. That is, this is not pattern destruction, it is pattern reorganization, with elements of the previous pattern showing up in the later generations. For example, in equations (2) through (5), the original pattern of $kX_t(1-X_t)$, is also found in the next generation of X, and then the next generation of X and so on, as long as functional iteration continues. The same holds true for the equation for generation X_{t+2} continued into generation X_{t+3} and then into X_{t+4}, and so on. What I am trying to highlight here is that patterns are being sustained in their very transformation--that is, there is a built-in redundancy in pattern complexifying operation.

This redundancy is a crucial factor in the eventual emergence. This can be understood in two ways. One, Stuart Kauffman's work[9] on the origin of emergent order in his so-called random Boolean Networks, can be interpreted to show how emergent order relies, at least in part, on the propagation of redundant patterns in his networks. In Kauffman's scenario, Boolean rules which are canalyzing (i.e., generating the same values in connected nodes) plus feedback loops connecting the nodes propagate the canalyzed values, resulting in the establishment of "walls of constancy." These "walls of constancy" are the foundations of the emergent order. In other words, redundancy is generated through the sustaining and propagation of patterns of on-off activity in the components of the network. Two, Henri Atlan[10] has pointed out how, from an information theoretic perspective, self-organization requires redundancy in a system. That is, the noise which self-organization incorporates into its reorganized patterns "feeds" off of the sustained and regular patterns in the system.

The second point that needs to be made about the pattern complexifying operation is that even though it is given a crucial boost as the parameter value is increased, complexification is already occurring with the mere act of functional iteration. A look at the metaphors used to describe nonlinear transformations in the literature indicates the complexifying nature of these operations and how markedly different they are from simple "increase," "expansion," or "exponential growth"[11]." Instead, we find: "folding and stretching," "mixing," and, "kneading dough." Thus, Prigogine[12] pointed out that the nonlinear operation of the "Baker" transformation resembles the process of kneading dough -- flattening, cutting, and then reassembling the transformed rectangles. Therefore, the complexity of these emergent phenomena have complex precursors so that the chasm between emergent complexity and initial simplicity is starting to close.

2.3 Criticalization of Parameter Values

The pattern complexifying operation is one part of the causal chain leading to the onset of emergence in the evolution of the logistic equation. A second causal process is what May terms the "tuning" of the nonlinearity of the system, referring to how an increase in the parameter *k* steepens the parabolic curve. It is obvious that as the curvilinearity of the parabola $f(X)$ steepens, there is a corresponding change in the slope of the tangent where $f(X)$ meets the 45^0 line $X_t = X_{t+1}$ (Figure 4).

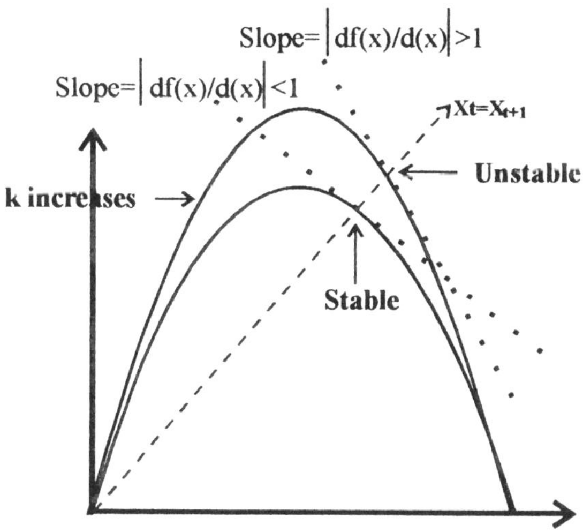

Figure 4: Instability and Emergence

Figure 4 shows the steepening of the parabola in the logistic map as *k* is increased. This change in slope is used by May to explain the onset of instability around the fixed-attractor point. His reasoning is as follows: The slope, of course, is the derivative *df(X)/dX*. As long as the slope lies between 45° and -45°, i.e., absolute value of the derivative is less than 1, the fixed attractor point is stable. But as the parameter *k* is "tuned" to steepen the curve of the function *f(X)* then, the slope at the attractor point steepens beyond -45° or the absolute value of the derivative is greater than 1. Consequently, the attractor point becomes unstable and bifurcation to a new attractor occurs, in this case the emergence of a limit cycle attractor. This is the first period doubling event during the route to chaos as the parameter is increased. In a little while I will come back to further explore what is going on with the onset of instability.

As can be seen from this scenario, emergence requires a *criticalization of parameter values*. This insures the instability necessary for bifurcation and the transition to a more complex attractor. Near bifurcation, a small change in the parameter leads to not just a large quantitative change in the sytem, but a qualititative jump to a new realm of dynamical behavior in the system. The criticalization of parameter values involved in emergence shows that near bifurcation, every little increase in the parameter value does not have the same effect, i.e., it is not like a linear additive process. This is analogous to the Sorites Paradox discussed by the philosopher of mathematics Brian Rotman[13]: subtracting a single grain from a heap ("sorites") of grains leaves in place a heap--such is the meaning of heap. Each further subtraction will likewise leave a heap. But this means there is a critical point where there is a qualitative change from heap to nonheap, contrary to the original supposition that a heap remains after each subtraction of a grain. This is similar to the metaphor of the the straw that broke the camel's back. The last grain has an effect or potency different than what each of the previous grains have when there still is a sizable heap. This is not just nonlinear disproportionality between cause and effect since there is not just a quantitative but a qualitative change. In a sense, it is at the critical point that quantity changes into quality.

According to Rotman, "Such a recognition opens up a gap inside the logic of statically conceived, context-free "equality," since decrements and increments are seen to be not equal in their cumulative effects even if they are "equal" as signs designating the "same" activity." Rotman uses the Sorites Paradox as a way into his revision of the meaning of the mathematical concept of infinity according to what he calls a "Non-euclidian Arithmetic," but my reference to his ideas is to mark how the criticalization of parameters is a crucial factor whereby a quantitative shift in parameter values causes, in part, a qualitative change represented by the emergence of complexity.

2.4 Appropriation of Randomness

The criticalization of parameter values is related to the third causal element, how emergence signifies an *appropriation of randomness*. This can be seen by going back to the way the change in slope showed that the attractor point had become unstable (see Figure 4). Remember that the slope at the tangent is the derivative of $f(X)$. When $|df(X)/dX| < 1$ the attractor is stable because changes in $f(X)$ are less than changes in X. Stability means that random fluctuations or perturbations are dampened. That is, random fluctuation increasing the value of X lead to an increase in X faster than the accumulating increases in $f(X)$. Therefore, these random fluctuations are dampened, and the system does not diverge from its current attractor. Bifurcation does not occur and more complexity does not emerge. However, when $|df(X)/dX| > 1$, changes in $f(X)$ are greater than changes in X. Here, random fluctuations increasing X cause an increase in $f(X)$ faster than in X. Therefore, random fluctuations are amplified and lead to a divergence from the attactor. The system at that critical value has become unstable leading to a shift to a new attractor (period doubling) and the emergence of greater complexity.

In this way, emergence as the consequence of instability is an appropriation of randomness. It is because of this appropriation of chance, that emergence can demonstrate novelty since departures from behavior consonant with the previous attractor are amplified and incorporated into the new organization of the system. As Eigen and Winkler[14] put it, "We need a model that incorporates instability, for it is instability that forces a clear selection from available alternatives and prevents the return to an already tested state" (p. 51). This kind of utilization of randomness in a causal chain mandates that our conventional notion or causality as not permitting the presence of randomness be reframed. That is, our conventional notion would have it that system is either determinate or indeterminate, caused or stochastic. But, emergence suggests that causality and randomness are interwoven in an intriguing new fashion.

2.5 Actualization of Potentials

Finally, the fourth factor in complex causality leading to emergence in the evolution of the logistic equation concerns actualizing potentials that are, in an important sense, already inherent as a possibility in the logistic equation. I think one of the most amazing and significant discoveries of dynamical systems theory is that the bifurcation points and the various attractor regimes exist as potentials in nonlinear systems, potentials which are actualized under the appropriate operations and parameter values. These potentials of complexity already exist in the seemingly simple nature of the equations. The pattern complexifying operations and the tuning of the nonlinearity bring out and release this inner potential. This actualization of inherent complexity is a causal factor similar to switching on a light bulb. The switch is a causal factor not because it generates the electricity, but because it releases the potential electric current to do its work in heating the filament. The switch is the prompt that allows the causal process to proceed. The criticalization of the parameter value is like turning on a light switch--there is already a potential instability and bifurcation potential inherent in the logistic equation which is "turned on" by the critical parameter value. However, this potential comlexity is not apparent in the initial lower ranges of the parameter. As Nicolis[15] put it when describing how far-from-equilibrium conditions can bring about self-organization in physical systems, the inherent nonlinearity, which is hidden and appears linear at equilibrium conditions, is brought out of hiding and revealed at far-from-equilibrium conditions. Far-from-equilibrium conditions correspond to the criticalization of the parameter values in the logistic equation when undergoing the operation of functional iteration.

To summarize our examination of the evolution of the logistic equation: when we ask what is responsible for the emergence of complexity, that is, what causes emergence we need to include all four factors discussed above-- a pattern complexifying operation; a criticalization of the parameter values; an appropriation of randomness; and the actualization of potentials. There is a pattern complexifying operation on a system which has built into it instability and consequent appropriation of the random in traversing into more complex attractor regimes when the parameter reaches a critical value.

3 The Law of Requisite Complexity in Causality

3.1 Emergence and Causality

When considering causality in relation to emergence, at first sight there seems to be only two options. First, one can put one's entire focus on the emergent phenomenon itself, disregarding the local causality that leads to the emergence. To do this requires an uncoupling of the emergent phenomena from the local properties out of which it emerged. This is analogous to the uncoupling of levels that occurs with coarse graining in quantum mechanics to account for classical dynamics. The coarse graining ignores the fine grained quantum effects by "summing them over" in favor of coming up with real probabilities applicable on the classical level[16]. Kellert and others choose this first option when they uncouple the qualitative dynamics of phase space. A second option is to do just the opposite, namely, to focus only on the local level and ignore the emergent phenomena as an epiphenomenon. This option is the heart of a strong kind of reductionism, often found in physics, which disparages more complex levels as not fundamental enough. However, there is a third option which I am suggesting: both the local and the global level are important, the causal chain leading and the emergent complexity. But this third option requires a view of local causality that is commensurate with the complex effect.

According to the philosopher of science R. Schlegel[17], "Causality is not an a priori principle, or set of principles, but is rather a general characteristic of scientific knowledge. As such, it partly sets the form of science; but also, the history of causality amply demonstrates that the content of science responds back as a determinant of causal principles (his emphasis; p. 14)." Such an interaction between our conceiving of causality and our scientific findings is happening in our own era in the study of chaotic and complex systems. But there is an important factor complicating the issue: as the confluence of mathematics and natural science, chaos and complexity theories harbor a potentially troublesome clash since, as mentioned earlier, mathematical formalisms tend to replace causal explanation with functional relations. This potential clash need not reach actuality if it is recognized that causal assumptions underlie even the most abstract mathematical formalisms. This is seen in chaos theory in the caveat offered by several mathematical physicists[18] when they pointed out that attention must be carefully paid to the selection of the right delay parameters in state space reconstruction following the Takens Theorem. If one chooses a delay parameter that is too small, each coordinate is about the same and the trajectories are squeezed along the identity lines; however, if the delay parameter is too large, "...in the presence of chaos and noise the dynamics at that time become effectively causally disconnected from the dynamics at a later time, so that even simple geometrical objects look extremely complicated" (p. 339). This need to keep causality in mind in mathematical formalisms is echoed in the warning of the theoretical biologist Robert Rosen[19] concerning overly mathematical interpretations of complexity: "The very fact that the same mathematical language does not (in fact, cannot) distinguish between essentially distinct categories of causation means that the mathematical language we have been using is itself somehow fundamentlly deficient and that it must be extended by means of supplementary structures to eliminate those deficiences" (p. 200).

The approach advocated by Kellert puts all eggs in the mathematical basket of the "geometry" of phase portraits. Instead, I think what is needed is a view of causal explanation that certainly uses mathematical methods and insights, but, heeding Rosen's caution, supplements those formalisms with what I propose to call complex, pattern causality. This view of causal explanation would follow the physicist A. D'abro's[20] determination of causality, "...we must include under the applelation 'cause' all the items of information that are required to render the future sequence of events determinate" (p. 52). Borrowing from H. Ross Ashby's famous "Law of Requisite Variety," I suggest we need a Law of Requisite Complexity in Causality.

3.2 Pattern Complexification and The Commensurability of Cause and Effect

According to Ernst Nagel[21], the notion of cause had a legal origin in ancient Greece denoting actions accountable for wrongs for which there were legal remedies. This legal usage was extended by analogy so that a cause of an occurrence was taken as that which was believed to be either partly or wholly "responsible" for that occurrence" p. 17). As that which is responsible for the effect, causes must be commensurate with and connected to their effects so that there can be a direct line of inquiry from effect to cause. Although he criticized the idea of causality in general, Russell[22] referred to the accepted philosophical maxim that "cause and effect must more or less resemble each other" (p.182). Descartes[23], considering such a principle as self-evident, relied on it in his notion that a cause must be as great as its effect since a lesser cannot produce a greater. In a similar vein, Leibniz[24] stressed that an effect involves its cause to the extent that a perfect understanding of an effect would automatically lead to a perfect understanding of its cause. The scientist and philosopher of science Isabelle Stengers[25] interprets Leibniz's idea of sufficient reason in a similar way as laying the foundation for the balance of cause and effect found in various conservation laws. William James[26] decried such a commensurability of cause and effect in the ideas of the great scientist and mathematician William Rowan Hamilton who wrote, "...that all that we at present know as an effect must previously have existed in its causes; though what these causes are we may perhaps be altogether unable to surmise"(p. 194, footnote 1). James found this abominable because he thought it denied the possibility of novelty where the effect contains more than its cause.

The study of nonlinear systems echoes James' critique of commensurability of cause and effect since effects can be of a magnititude out of all proportion to their causes. This was noted in a prescient fashion by the philosopher of science Mario Bunge[27]:

> ...the association of causes (in a linear fashion) will not constitute a novelty in itself. In general, such a conjunction will produce a quantitative change only, that is, an increase or decrease of certain effects, not however, the emergence of new traits; this is the hypothesis of the summative character of causes...The additive character of causation is involved in the very notion of the *causal analysis* of a given situation. In fact, such an analysis consists in the decomposition of a complex determinant into a sum of causes, or in the separation of a given result into the separate contributions of the individual components or aspects of a system or process" (p. 166).

For Bunge, linear causality remains trapped in the assymmetry of a unidirectional influence from cause to effect, thereby leaving out reciprocal causation or feedback.

However, just adding nonlinearity into the picture doesn't get us very far in accounting for emergence since an effect disproportionate to its cause is not the same as an effect showing a qualitative difference with its cause. In emergence, properties on a given level do not appear in the level beneath, and therefore there is required, in the words of Diner[28] "explicit passage from one level to the other...to disclose the appearance of these properties" (p. 276).

This need to distinguish levels and thereby point to the discrepancy between local quantities and global qualities has been addressed by many thinkers in the new nonlinear domain. Thus, Pagels[29] (1988) characterizes the discrepancy between levels found in emergent phenomena in terms of a "causal decoupling" indicating again the chasm between the properties on the global versus the local level. Rod Swenson[30] (1989) highlights this difference of level by referring to emergent phenomena as a process of "entification" pointing to the kind of ontological irreducibility inherent in emergence. And, challenging the philosopher Stephen Pepper's disavowal of the sui generis reality of emergent properties, Meehl and Sellars[31] point to what they call "supervenient regularities" at higher levels not reducible to regularities or laws formulated at lower levels. The issue highlighted in these various characterizations of emergence is the radical novelty shown in emergence. At bifurcation something does appear that wasn't there before: a new solution path of the system. In self-organization, for example, there is the emergence of correlation across the system.

A new view of causality will need both a commensurability of complexity between cause and effect, while at the same time enabling for the emergence of a kind of novelty that is not purely a matter of increased numerical magnitude. The four causal factors described above to account for emergence in the period-doubling route to chaos satisfy both of these conditions: the pattern complexifying operation and the actualization of potentials satisfies the commensurability condition, while the criticalization of parameter values and the appropriation of randomness satisfies the novelty condition.

But, notice that we are talking about causality now in terms of patterns, not substance. The change in pattern shown in emergence is of a different sort than the change in motion of one billiard ball hitting another, which is a typical model of causality in the substance view. The mathematician Ben Goertzel[32] has proposed a "pattern dynamics" to understand self-organization as a replacement of the Prigogine School's thermodynamic approach. My reconceptualization of causality follows Goertzel's "pattern dynamics." It is interested in emergence as the emergence of new patterns, patterns of complexity and coherence that differ from the previous, local patterns in the system. The transition into a new attractor realm signifies a transformation of patterns in the behavior of the system. According to Goertzel, "...it is appropriate to look at the flow of patterns rather than the flow of physical quantities" (p. 96). Such a pattern-based causality calls for a revision, but not an elimination, of "causal mechanisms," a subject to which we now turn.

3.3 A Renewed Mechanics of Causality

The English analytical philosopher Peter Strawson[33] has pointed to the mechanical transactions underpinning the idea of a cause. Criticizing Hume's grounding of causality in a subjective impression of association or necessary connection, Strawson cites examples of how we constantly see causality at work in our everyday life: "...we put our shoulders to the wheel, our hands to the plough, push a pen or a button, pull a lever or a trigger. Entering into them ourselves, we find in them a source of the idea of power and force, compulsion and constraint....even when it is most clearly metaphorical, the language of mechanism pervades the language of cause in general..." (pp. 118, 119). Strawson suggests that the sense of mechanical interaction as being "paradigmatically explanatory" accounts for the reluctance to accept the idea of action at a distance and instead draw out some medium through which an influence or cause is propagated (p. 119). We demand that something had to happen for this effect to have occurred, some efficient, mechanical answer to the question of "why." According to Mackie[34], we are ready to attribute causes to those times that can be described by clauses introduced by "because." Strawson concludes that the search for causal explanation in science is a seeking for those modes of action and reaction not observable at the ordinary level but hypothesized and which are considered intelligible because they are modelled on our experience of mechanical interaction in our everyday life. The natural scientist "... wants to know the detail of the link, the inner mechanism of the connection. He wants to know *how* it works" (p. 123).

Similarly the philosopher of science Wesley Salmon[35] claims that the best-developed explanations in science are attempts to "*elucidate the mechanisms at work*" (p. 262; italics in the original). Salmon adds, however, that these mechanical explanation need not be

be a crude atomism, but can allow for fields, and other more sophisticated, but nevertheless mechanical notions. This call for an "mechanical" explanation of how things work is as true for complex systems as simple systems. Thus, Dellatre[36] says of complex systems, "...any real evolution problem, often irreversible, raises afresh questions of time sequences and therefore causality" (p. 150).

It is crucial to note, however, that causal explanation as "elucidating the mechanisms at work" has in the past been equivalent to getting rid of random error, accidents, and other chance occurrences as much as possible in order to get to the root causes. In this context, Robert Rosen[37] has made the intriguing observation that the empirical concept of error arises from the discrepancy between a system conceived abstractly as an ideal model and the same system as it actually behaves. Since, the former is really a system which for the sake of abstraction is conceptualized as a closed system and the latter is always an open system, error consequently arises when a system is open to its environment(s). But, "open" systems are precisely what was ignored by the abstract ideal systems predominant in classical science. When systems are "open" to their environment, the dynamics are nonlinear as pointed out by the physicist Bruce West[38]:

> The notion of isolating a process (system) is an intrinsically linear one, that is to say, the coupling of a process to the environment can only be made arbitrarily weak when the coupling is independent of the system. If the coupling is dependent on the process then for some values of the quantities characterizing the system the coupling to the background cannot be ignored. Such a system-dependent coupling is therefore nonlinear (p.67).

In a similar vein, the philosopher M. Brand[39] has pointed out that when a system is closed, as is assumed for the causal laws of Newtonian dynamics, there is an exhaustive formulation of the ceteris paribus clause ("other things being equal"). Only in a closed system, can the initial conditions and only the initial conditions be exhaustively given by assigning a value to every state variable (p. 14). However, in "open" systems, there is no informative and systematic way to state the ceteris paribus clauses.

Furthermore, mechanical causality as explaining "how things work" has mostly been restricted to what Aristotle termed "efficient causality" or that by which some effect is wrought, e.g. how one billiard ball moves because of the efficient causation of another billiard hitting it with enough force. This limited view of causality as merely efficient causality is related to an often heard criticism that mechanical explanations reduce causal linkages to machine-like contrivances not commensurate with the nature of the effect. For example, certain of Freud's ideas on the displacement or sublimation of "psychic energy" have been disparaged as too much like a mechanical hydraulic device, an explanatory scheme that does not seem appropriate, in other words, is incommensurate, for describing human feelings, thinking, or behavior. Associated with this conventional notion of efficient causality is constraining causal influence to an external action of one thing on another--again, billiard balls hitting one another. In such a notion there is no place for a cause as the prompt or actualization of a potential. For example, it doesn't make much sense to say that the causative billiard ball prompts the effected billiard ball to actualize its potential to move. However, it does make sense to talk about the actualization of potentials when one is talking about, say, the developmental stages of a person's life, e.g., puberty is a potential actualized by certain internal and external prompts, including biochemical changes in sex hormones, peer pressure, and cultural expectations.

We need to heed another warning from Robert Rosen that the concept of causality is a complicated notion, a fact recognized in Aristotle's need to postulate four categories of causality, corresponding to the need to answer the "why" question in four different ways. That is what a Law of Requisite Complexity of Causality must do, provide a way to answer the complexity of the "why" question. In this context, there is not only one way to take "mechanical" in relation to causality. According to the *Oxford English Dictionary*, the etymological roots of "mechanical" or "machine" is related to similar cognate words in other language groups such as the words "may" (meaning "to be able") or "might" (power). "Mechanical" causality, then, could be conceived as an explanation in terms of whatever it is that has the ability or the power to bring about the effect. It is in the light of these kinds of new ways of thinking about "mechanical" that S. Diner can write that chaos and complexity theory have led to a "renewal of mechanism" in which efficient and final causation are reunited. As Sally Goerner[40] has pointed out, a limitation of a purely "mechanistic" is an overemphasis on explaining wholes by only considering the parts, and leaving out the powerful idea of how a system can be attracted to possible future states.

A Renewed Mechanics of Causality would include: the appropriation of randomness and the actualization of a potential complexity. The appropriation of randomness in the emergence of complexity is connected to the unpredictability found in chaotic and complex systems. To be sure, aspects of emergent structures are unpredictable, e.g., the direction of rotation of the convection currents in the Benard Cells. They are unpredictable because at bifurcation the system is unstable and therefore will amplify random currents, the amplification of these random currents determining the directionality of the convection cells. And of course there is the unpredictability of future states of a chaotic system due to sensitivity to initial conditions.

But what ramifications on causality does unpredictability imply? According to thethe philosopher A. Roque[41], etymologically, "explanation" implies a spreading-out or "un-folding" on a planar surface of something convoluted or "com-plicated." This suggests to Roque, that the "explanadum" or what needs to be explained must be able to be laid out all at once, indicating that the explanadum is something static and ahistorical. In Roque's view, predictability is a core of this view of explanation, an ability to lay out the explanadum all at once. But the unpredictability in nonlinear systems, in Roque's opinion, means that such systems are not accessible to this kind of causal explanation, "Non-linear, essentially historical phenomena cannot be explained deductive-nomologically because the phenomenon is not already there waiting to be explicated, explained; the creativity (emergence) that result from the bifurcation is irreducible, in principle unpredictable" (p. 248).

I want to make two points about these claims concerning the connection of unpredictability to causality. First, unpredictability is proving to be *not* the last word in chaotic or complex dynamics. If it were, then why have two of the leading chaoticians formed a company called "Prediction, Inc." Moreover, Ben Goertzel[42] has pointed out that even though future states of a complex system may be unpredictable there is nevertheless a great deal of what he calls "structural predictability." For example, as chaotic as the weather may be, it is still very predictable that it will be hot in August in New York, even though the specific temperature on a specific day is not. I don't see why a causal scheme requires state rather than structural predictability.

The second point has to do with Roque's alternative scenario for explanation: a hermeneutic explanation, borrowing from the great German philosopher Hans Georg Gadamer's idea of a narrative: "In true narrative the telling of the tale explains by interweaving sequential but overlapping threads such that a temporal pattern, the *meaning* that has flowed throughout the unique sequence of events and has bound them into a whole, emerges" (p. 251). According to Roque, explanatory narrative can capture the interlevel dialectic involved in self-organization. This narrative explanation, consisting of temporal patterns, fits nicely with the four causal factors of the Law of Requisite Complexity in Causality that I have been expounding. Complexifying operations are a kind of temporal pattern transformation, not unlike the complexities of a narrative in a story. In conclusion, I don't see any reason that unpredictability cannot be incorporated into a Renewed Mechanics of Causality.

3.4 A Revolution in Causal Explanation

To summarize how chaos and complexity theories are revolutionizing the concept of causal explanation in science, it would be helpful to turn to Paul Thagard's[43] examination of the ideational changes occurring in a scientific revolution. First, there is coalescence or the bringing together of two or more previously separated items under a new, superordinate classification. For example, Newton's theory of gravitation linked together the previously separate realms of the earthly and the celestial, and Einstein's theory of equivalence in General Relativity coalesced acceleration and gravitation into a new Riemannian space-time nexus. I can think of two salient coalescences that happen to causality in the light of chaos and complexity theories: one, the bringing together of mechanical linkage and the appropriation of randomness in the same causal chain; and, two, the coalescence of efficient causality with the other categories of causation, particularly final causation. As Dellatre[44] indicates, efficient causation has to do with local properties and final causation with global emergence seen under the rubric of a pull by an attractor. Sally Goerner's remarks highlight this coalescence of causality, "Chaos suggests that effects may be thought of as coming from either the top or the bottom and actually should be thought of as coming from both" (p. 49).

In an important sense, which category of causation, efficient or final, we choose to focus on has to do with what our causal interest is in a particular situation. As the philosopher Hilary Putnam[45] has indicated, how we attribute causality is context-bound and interest-dependent. He gives the example of a man suffering a heart attack. If the man has high blood pressure and has not followed a doctor's advice on diet and exercise and then suffers a heart attack, we are likely to say that the bad nutritional habits and lack of exercise caused the heart attack. Whereas, if we are interested in what would have happened to him if he had not had high blood pressure, we are likely to say that the high blood pressure caused the heart attack. According to Putnam, it is important to recognize that these two ways of looking at causation are not mutually exclusive, today you may be interested in blood pressure, tomorrow in lack of exercise. Similarly, efficient causation and final causation are not mutually exclusive, it depends on what your causal context and interests are.

The converse of coalescence is Thagard's conceptual category of differentiation, a process whereby making a distinction brings about two or more new separate things where before there was only one. For instance, Newton differentiated mass from weight to make it possible to understand weight as the combined result of mass and the force of gravity, and, Einstein differentiated mass into rest mass and relativistic mass. Differentiation occurs in chaos and complexity theories by the unhinging of predictability from determinism. As Isabelle Stengers put it, "The infinite point of view in which causal, deterministic, and foreseeable are synonymous now appears as the night in which all the cats are gray whereas it is precisely and irrestibly the different shades which interest us which gives rise to questions and essential distinctions" (p. 271). Furthermore, one can say that chaos and complexity theories differentiate causality according to quantitative and qualitative dynamics. One of my points above was that an crucial causal factor in emergence is how the criticalization of paramter values at bifurcation changes quantity into quality. Finally, chaos theory has distinguished deterministic chaos or noise from randomness, a distinction that has vast repercussions for causal explanation. This is seen, for example, on how the finding of low dimensional chaos means the system is not totally stochastic, which is, therefore, a goad to looking for the nonlinear dynamics in the system causing the chaos.

Branch jumping is another conceptual change suggested by Thagard, having to do with the moving of a concept from one classification scheme or branch to another. For example, Darwin reclassified humans as a type of animal rather than the previous separate set existing for humans alone. Newton moved gravity away from the Aristotelian idea of an intrinsic property of a body to gravity being a force between one body and another. In my revision of causality, the very putting randomness under the category of causality at all is an example of branch jumping. Also, including the factor of actualizing potential under the category of causality is a type of branch jumping. This branch jumping shows a great departure from the Newtonian transformation of gravity from the category of intrinsic property to external force. Of course, actualizing a potential in this context is different than Aristotle's intrinsic property, for the potentials being talked about in a nonlinear system are not some kind of Aristotelian impetus, but the potential complexity inbuilt into the system, only requiring the appropriate conditions to be realized.

Finally, Thagard's concept of tree switching refers to changing the very organizing principle involved in a conceptual framework. For example, Darwin not only reclassified humans as animals, he changed the very notion of classification so it referred to not just similarities in form, but also being of common descent, a historical concept. In quantum mechanics tree switching occurs with the new classification system of particles based on their quark composition. Robert Rosen suggests a type of tree switching in relation to complexity. Instead, of physics being the tree trunk with biology being one of the branches, he views complexity as the trunk and biology and physics as two separate branches. Such a tree switching has important ramifications for a model of causality by indicating that complex systems and their dynamics call for a requisite complexity in causal explanations.

4 Conclusion

Certainly causal explanation is not the only form of scientific explanation. Abraham Kaplan[46] lists other types of explanations: taxonomic, mathematical, statistical, temporal laws, and goal-seeking. Much of what is included under Kaplan's list of noncausal explanations can find a home here in our reconceptualization of causality in the light of chaos and complexity theories: the need to devise new taxonomies of pattern transformations; a place for mathematical formalisms; a role for stochastic processes; temporal laws whereby phases or complexity are actualized; and, goal-seeking explanation in the form of the finalistic causation implied by the concept of attractors.

The claim that eventually science would become purified of the atavistic notion of causality has been an ideal right from the birth of modern science. Descartes, for instance, exclaimed that he hoped "All of my physics is but geometry." In our twentieth century, the great mathematician Hermann Weyl thought such a "geometrization" of physics had taken place to a large degree in Einstein's formulation of General Relativity in which the force of attraction at a distance is dispensed with in favor of a system of inertial movements of trajectories following a geodesic in a Riemannian geometry of space-time[47]. When Kellert, though, talks about a "geometry" of phase space superceeding and eliminating causal explanation, that is another matter altogether. Phase space is not an actual "geometry" of the world, it is an abstract mathematical space for visualizing time series data from a dynamical system. As such, it is not providing a picture or model of a real world "geometry" of space-time and what moves in that space-time. The finding of fractal dimensionality or some other mathematical indicator of low dimensional chaos in the "geometry" of a phase portrait doesn't obviate causality, but indicates that some kind of nonlinear dynamics are going on in the system. In other words, the "geometry" of phase space is an indicator of when and where to start looking for causal processes that are leading to the type of behavior consonant with that kind of low dimensional chaos. This reveals an expansion, not a diminunition of the role of causality, a reframed view of causality.

In philosophy, the concept of emergence was first drafted as an antidote to scientific reductionism in the work of George H. Lewes, C. Lloyd Morgan, Henri Bergson, and so on[48]. Part of the problem, though, with these early ideas on emergence was their lack of a clear scientific demonstration. The advent, however, of emergent phenomena in chaos and complexity theories has provided just the right kind of scientific and mathematical grounding that emergence needs. But to focus on emergence simply as a sui generis phenomena, while neglecting the causal mechanisms required for emergence to occur is to do injustice to the scientific enterprise. Emergence has determinate and indeterminate roots. In this sense, it provides a different view of the relation of causal and random processes than quantum mechanics. In the latter, the basic substrate of reality is assumed to be stochastic, throwing out causality altogether at that level. But, the emergence of chaos and complexity take place in the macroscopic universe, not the subatomic realm, and it is in that macroscopic world where causality is still operative and in which we need to frame our research and our questions guiding our research. Therefore, the task confronting an exploration of emergent complexity consists of a search for the way determinate and indeterminate processes are interwoven.

We are inclined to ask "why" questions and search for causes when we notice something odd, something that stands out, that is different. The emergence of complexity naturally leads to asking "why" questions and seeking causes. I suggest we are at the stage of the sciences of chaos and complexity in the social and life sciences where indeed the why questions are especially relevant, and therefore, causal inquiries are what is called-for. But it has to be a reframed type of causality, so what we have learned so far can be fedback into our notions of causality and transform them into more adequate tools of understanding. As Robert Rosen[49] asserts, "It is the entire task of theoretical science to establish congruence between causal order in the external world and implicative order in the formal world, which embodies the very idea of natural order" (p. 179).

References

1. S. Kellert, *In the Wake of Chaos* (Univ. of Chicago Press, Chicago, 1993).
2. B. Russell, *Mysticism and Logic* (Doubleday, Garden City, NY, 1912).
3. G. von Wright, *Causality and Determinism* (Columbia University Press, NY, 1974).
4. E. Nagel in *Cause and Effect: The Hayden Colloquium on Scientific Method and Concepts,* ed. D. Lerner (The Free Press, NY, 1965).
5. J. Goldstein, *So You've Found Chaos in Your data. So What? Scientific Explanation in Chaos Theory*. Society for Chaos Theory in Psychology and the Life Sciences, Second Annual Winter Conference-1994, New Haven, Connecticut, Feb 6, 1994.
6. N.E. Thing, N.E., *Magic Eye: A New Way of Looking at the World* (Andrews and McMeel, Kansas City, MO: 1993).
7. R. May, *Nature*. **261**, 459-467 (1976).
8. M. Feigenbaum in *Order in Chaos*, ed. D. Campbell and H. Rose (North-Holland Publishing Co, Amsterdam, 1983).

9. J. Goldstein, *The Chaos Network*, **4,** 5, 7-11 (1993).
10. J. Goldstein, in *Proceedings of the Third Annual Chaos Network Conference: From Chaos to Complexity--New Tools for a Complex World*, ed. M. Michaels (People Technologies, Urbana, Ill., 1994).
11. J. Goldstein, in *The Chaos Network Annual Conference--the Second Iteration*, ed. M. Michaels (People Technologies, Urbana, Ill., 1992).
12. I. Prigogine, *From Being to Becoming: Time and Complexity in the Physical Sciences* (W.H. Freeman and Company, NY, 1980).
13. B/ Rotman, B., *Ad Infinitum: The Ghost In Turing's Machine* (Standord University Press, Stanford, CA: 1993).
14. M. Eigen, & R. Winkler, R. *Laws of the Game: How the Principles of Nature Govern Chance* (Harper & Row, NY, 1981).
15. G. Nicolis, G. in *The New Physics*, ed. P. Davies (Cambridge Univ. Press, Cambridge, England, 1989)
16. M. Gell-mann, *The Quark and the Jaguar: Adventures in the Simple and the Complex* (W.H. Freeman & Co, NY, 1994).
17. R. Schlegel, R., in *Causality and Physical Theories*, ed. W. Rolnick (American Institute of Physics, NY, 1974).
18. M. Casdagli, D. Des Jardins, S. Eubank, J. D. Farmer, J. Gibson, & J. Theiler, in *Applied Chaos*, ed. J.H. Kim & J. Stringer (John Wiley & Sons, NY, 1992).
19. R. Rosen, in *Theoretical Biology and Complexity: Three Essays on the Natural Philosophy of Complex Systems,* ed. R. Rosen (Academic Press, Orland, Florida, 1985).
20. A. D'abro,. *The Rise of the New Physics* (Volume One) (Dover Publications, NY, 1951).
21. E. Nagel in *Cause and Effect: The Hayden Colloquium on Scientific Method and Concepts,* ed. D. Lerner (The Free Press, NY, 1965)
22. B. Russell, *Mysticism and Logic* (Doubleday, Garden City, NY, 1912).
23. R. Taylor, in *The Encyclopedia of Philosophy*, Volume 2, ed. P. Edwards (MacMillan, NY, 1967).
24. L. J. Russell, *The Encyclopedia of Philosophy*, Volume 4, ed. P. Edwards (MacMillan, NY, 1967)
25. I. Stengers, in *Dynamical Systems: A Renewal of Mechanism*, ed. S. Diner, D.

Fargue, & G. Lochak (World Scientific Publishing, Singapore, 19860.
26. W. James, *Some Problems in Philosophy* (Greenwood Press, NY, 1968)
27. M. Bunge, *Causality: The Place of the Causal Principle in Modern Science* (Harvard University Press, Cambridge, MA, 1959).
28. S. Diner, in *Dynamical Systems: A Renewal of Mechanism*, ed. S. Diner, D. Fargue, & G. Lochak (World Scientific Publishing, Singapore, 1986).
29. H. Pagels, *The Dreams of Reason: The Computer and the Rise of the Sciences of Complexity* (NY: Simon and Schuster, NY, 1988).
30. R. Swenson, *Systems Research.* **6**, 3, 187-197 (1989).
31. P. E. Meehl, & W. Sellars, in *The Foundations of Science and the Concepts of Psychology and Psychoanalysis*, ed. H. Feigl & M. Scriven (University of Minnesota Press, Minneapolis, 1956).
32. B. Goertzel, *The Evolving Mind* (Gordon & Breach, Langhorne, Pa., 1993).
33. P. Strawson, *Analysis and Metaphysics* (Oxford University Press, Oxford, England, 1992).
34. J. L. Mackie, *The Cement of the Universe: A Study of Causation* (Oxford University Press, Oxford, England, 1974).
35. W. Salmon, *Scientific Explanation and the Causal Structure of the World* (Princeton Univ. Press, Princeton, NJ, 1984).
36. P. Dellatre in *Dynamical Systems: A Renewal of Mechanism*, ed. S. Diner, D. Fargue, & G. Lochak(World Scientific Publishing, Singapore, 1986).
37. R. Rosen, R. *Anticipatory Systems* (London: Pergamon, London, 1986).
38. B. West, *An Essay on the Importance of Being Nonlinear* (Springer-Verlag, Berlin, 1985).
39. M. Brand, in *The Nature of Causation*, ed. M. Brand (University of Illinois Press, Urbana, Illinois, 1976).
40. S. Goerner, S. *Chaos and the Evolving Ecological Universe* (Gordon & Breach, Langhorne, Pa., 1994).
41. A. J. Roque, *International Philosophical Quarterly*, **XXVIII**, 3, 247- 255 (1988).
42. B. Goertzel, *Chaotic Logic* (Plenum Press, NY, In Press).
43. P. Thagard, *Conceptual Revolutions* (Princeton University Press, Princeton, NJ., 1992).
44. P. Dellatre, in *Dynamical Systems: A Renewal of Mechanism*, ed. S. Diner, D.

Fargue, & G. Lochak(World Scientific Publishing, Singapore, 1986).

45. H. Putnam, H. *Renewing Philosophy* (Harvard University Press, Cambridge, MA, 1992). 46. A. Kaplan, in *Cause and Effect: The Hayden Colloquium on Scientific Method* and Concept, ed. D. Lerner (The Free Press, NY, 1965).

47. J. Garcia, & J. Piaget, in *Understanding Causality*, (W.W. Norton & Co, NY, 1974).

48. S. Jaki, *The Purpose of It All* (Regnery Gateway, Washington, D.C., 1990).

49. R. Rosen, in *Theoretical Biology and Complexity: Three Essays on the Natural Philosophy of Complex Systems,* ed. R. Rosen (Academic Press, Orland, Florida, 1985).

CHAOS AND THE DOUBLE FUNCTION OF COMMUNICATION

P. S. AULA
Department of Communication, P.O. Box 54 (Unioninkatu 37), FIN-00014 University of Helsinki

Since at least the needle model age, communication researchers have systematically sought means to explain, control and predict communication behavior between people. For many reasons, the accuracy of constructed models and the studies based upon them has not risen very high. It can be claimed that the reasons for the inaccuracy of communication models, and thus the poor predictability of everyday action, originate from the processes' innate chaos, apparent beneath their behavior. This leads to the argument that communication systems, which appear stable and have precisely identical starting points and identical operating environments, can nevertheless behave in an exceptional and completely different manner, despite the fact that their behavior is ruled or directed by the same rules or laws.[1]

'Changes aren't permanent, but change is'
- from the song Tom Sawyer by Rush -

1 Introduction

Finnish communication researcher Osmo A. Wiio's first law of communication,[2] according to which communication fails except by chance, is the stochastic model of communication dynamics. Many phenomena of communication include factors which seem random, and a researcher with a classical communication model background might regard these factors as stemming from errors in measurement or merely intractable disturbances in the process, and thus may leave them outside the results which he considers as characteristic of the utilized linear sender-centered models. A process in which event A affects event B which affects event C which in turn affects D, and so on, has the characteristics of a linear deterministic system. If D leads to event A, the system forms a circle which behaves in a completely different fashion. Instead of a linear interpretation, we can examine communication as an occurrence which is a bifurcation formed by different unfulfillable possibilities, in other words as a chain formed by potential or realizable alternatives.

Communication is ruled by nonlinear characteristics when the present cannot be returned to the past. Development is thus dissipative or dispersive. Conclusions cannot be drawn from events happening at this moment nor can the past be predicted. The communication process develops with the help of non-continuity. It is always broken by a factor considered a disturbance, whereby earlier explanations and the understanding of the state of things no longer hold true. Chaos theory is that part of the new research domain which studies nonlinear systems, believed to be stable but whose behavior can nevertheless change by surprise from dynamic balance to imbalance and develop towards growing disorder. As with the concept of communication, the use of chaos is also ruled by imprecise meanings. In chaos research, what is meant by a system's chaos is its apparently complicated and random behavior, which is nevertheless ruled by fundamental order and its properties.

Also, in connection with communication research it is important that the concept of chaos is separated from its everyday meaning. The scientific content of chaos shares very little in common with its generally understood meaning, which implies disorder or confusion. In science, chaos is better understood as a new theory and research method, which is used to find a higher degree of order behind seemingly disordered behavior. Chaos is also not merely an analogy for a system's imbalance or the irregularity of its behavior. Chaos has to be separated from chance. A chaotic phenomenon is unpredictable, and precise information about its development or future final state cannot be obtained. The process of chaos however is not dependent on chance. Instead it develops within a unique chaotic path (i.e. a strange attractor) which has determined boundaries. Chaos theory provides a dynamic conception of a reality which understands the behavior of phenomena as complicated and unpredictable, but still law-obeying.

Now it is well known that a large group of simple deterministic systems can produce behavior which appears coincidental, but which is by nature in fact chaotic.[3,4,5,6,1] The clearest examples of chaotic systems are mathematical models which have an extremely simple, perfectly specified deterministic description, but which produce completely unpredictable behavior. In mathematics, physics, meteorology, ecology and other scientific fields outside of the social sciences, the existence of chaotic systems has been quite well established.

Proof of chaotic behavior in social science research began to appear at the end of the 1980s, and today chaos research is growing also in this domain.[1] References to chaos theory to the social sciences have grown in the recent past as evidenced by a steadily increasing international literature.[7,8,9,1,10,11,12] Unfortunately, proper, widely accepted social science applications remain however scarce. For example in communication studies, research that makes use of chaos theory concepts [13,14,15,16,17,18] is very young.

According to Buder and Ritchie,[5] the epistemology of chaos research helps us to understand and explain communication processes. Using nonlinear dynamic systems we can construct essential base-characteristics of social systems. Richter[19] has also shown that by utilizing chaos theory, we can validly construct different behavior patterns of human dynamic relations. By using chaos in communication research, we can pay more attention to the complexity of human behavior and its long-term significance, which is necessary if we want to fully understand human social functioning.

2 The Purpose of this Article and Concepts Used

The basis of this paper is the observation that communication models, based on the traditional linear sender- oriented communication main paradigm, are not sufficiently able to describe, explain and predict communication behavior between people. My hypothesis is that the reasons for the insufficiency of communication models arise from the inherent complexity of communication phenomena and processes, which often appear to behave in a chaotic manner. The purpose of this paper is to consider the possibilities of chaos-theoretical modelling of communication behavior. Furthermore, I attempt to clarify whether chaos models can describe and explain those properties of the communication system which traditional models, complying with the standard paradigm, have had to leave outside the discussion. These properties are: (1) dynamics, (2) nonlinearity, (3) sensitivity to changes and (4) unpredictability. As an example of a comprehensive communication system I'll present an organization.

As we know, chaos models are not based on any one single explanatory chaos theory, but instead on a group of theories of diverse phenomena using equally diverse models and which arise relatively independently in different scientific fields. In chaos models one finds several abstract structures which have in my mind a special significance in the modelling of communication processes and systems. In my work I have stressed especially the following: (1) a sensitive dependence on initial conditions (i.e. the butterfly effect), (2) bifurcation and (3) bifurcation tree.

The myth of the tornado caused by the flap of a butterfly's wing is known far and wide, and its telling has spread beyond the boundaries of the academic world. The butterfly effect refers to that property of a nonlinear system, according to which a small change in the condition of the system at a certain moment can lead to large changes in the future. It is a basic property of the system, which we can define as its extreme sensitive dependence on initial conditions or initial value sensitivity.[20,5,6]

Bifurcation refers to a system's condition or behavior's suddenly dividing or branching into two different or merging part-behaviors. Consecutive bifurcations form a development chain (i.e. bifurcation tree) which can lead the system to chaos. The first bifurcation is the first sign of chaos, from which a threat of potential chaotic development leading to a system's breakdown can be determined. If you cannot control the bifurcation process, you cannot control chaos.

3 The Legacy of Shannon and Weaver

The Mathematical Theory of Communication of Shannon and Weaver [21] has been one of the most important and authoritative in the development of later communication process models and theories. Regardless of the model's omissions and deficiencies, it has remained as perhaps the longest-lasting description of the communication process.[22] The model is, by nature, a linear sender-centered communication model, and its effect on the development of communication models as well as the way by which communication is understood, has been monumental.

In spite of its notable contributions to the study of communications (including the measurement of the quantity of information by bits and the understanding of entropy as applied to communication), Shannon and Weaver's model[23,24] is in itself a straight derivative of communication's main linear paradigm:

$$A \rightarrow B = X \tag{1}$$

(A communicates something to B, with X resulting.)

"Several western communication models and theories are based on this model, when first we strip from it surface understanding and development whose purpose is to differentiate the models from each other. No matter what type of communication we talk about, this model directs -whether in its appearance or by its support - a large part of research and literature. It also determines how we relate to everyday communication - the model is linear, mechanical, atomic, reminiscent of algebraic formulas and appears official.[23]"

Furthermore, Shannon and Weaver's theory is associated with the balancing and strengthening of a system's structure. Only one direction is possible, that is towards balance. Similarly, new information is also in the first place centered on the strengthening of already existing information structures. The amount of information is taken as given; due to inevitable disturbances it can only diminish.[25]

As heirs to the theory of Shannon and Weaver, we can mention, for example Westley and MacLean's model, Rogers and Shoemaker's diffusion model as well as mass communication's newsflow, gatekeeper and uses-and-gratification models. The models that describe these communication processes are linear and predictable. They are more rational than a human being himself.[23] Residual values, stochastic errors or human factors are abstract explanations of deviations from the linear model and most often they seem to be the receiver's mistake. The visual presentation of the communication processes' models define the matter. Traditional communication models, which are more often than not on the box and arrow level, cannot be used to quantify the processes they portray. Causality appears in the boxes' order and in the direction of the arrows; nonlinear factors are framed within the disturbance or noise square.

When we examine these traditional communication theories, whose models include the linearity of existing order, we can see that these theories' verbal descriptions often attempt to handle prevailing nonlinear relations within the communication system. Regardless of the nonlinearity of this concept, the theories' verbal description based hypotheses, visual models and empirical studies return generally to linear knowledge claims. One reason for this is the absence of methods that comply to the aims. Several statistical techniques in use, for example Anova and regression analyses, have been developed for the analysis of linear relationships. We can say that communication studies suffer from a certain degree of methodological determinism, when available methodological tools determine how a researcher develops a theory and research hypothesis.[26]

As Åberg[31] already stated in 1979, the research methods of communication processes are quite primitive. However, they still partly direct the development of linear communication thought. Another - and more noteworthy - reason for the dominance of the linear communication paradigm is intellectual. We are not necessarily able to intelligently understand nonlinear relations and their long-term influences. That is one of the reasons why communication's linear main paradigms still hold a position of authority in western thinking.

Due to these reasons, it is important to try to develop alternative communication models so that we can focus our understanding of nonlinear human communication systems and their short- and long-term influences. The phenomenon of communication is unalterable. It does not change; only our understanding of it does. Communication is not itself problematic. The human understanding of it is.[27]

An attempt is often made to understand the behavior of empirical communication research objectives causally, using deterministic models for the expected reasons. These models normally include the possibility of error-terms, which nevertheless cannot be explained. A recent study of the nonlinear behavior of dynamic systems has shown that the problem of these models is perhaps not, for example, the precision of explainable behavior observations, but rather the properties of the models used for explaining.[5] Shannon and Weaver's model has long been regarded as insufficient for the description of complex human communication processes. The model attracted many followers, and during the subsequent decades produced communication models actually developed componentially but built upon linear thinking. Even though models are just models and not reality, linear models that describe communication processes are by nature insufficient abstractions for describing the complexity of human communication dynamics.

4 Organizational Communication

According to Robbins[28] "An organization is a consciously coordinated social unit, composed of two or more people, that functions on a relatively continuous basis to achieve a common goal or set of goals". Organizational communication is two-way interaction that makes possible in different situations the realization of the goals of the organization and its members. Communication joins the parts of the organization to each other and also to the organization's environment. The organization's communication system is a system which is formed in use from given communication channels, communication structures and communication rules. Organizations are maintained with the help of management and communication. From these specifications I consider an organization as an example of a comprehensive communication system.

As Buder and Ritchie stated, the butterfly effect "is one property of chaos which we believe to be of particular importance for social science".[5] Processes can be found among the different fields of organizational communication, which are exposed to the butterfly effect. The constructing of communication relations, the spreading of unofficial information, changes in public opinion, chains of association, the adoption of new innovations, and so on, are in their many forms and in their unexplainability comparable, for example, to the unpredictability of the weather system. Many communication events can be presented as periodic systems, in which the output of the last event is the input of the following. Small and seemingly insignificant changes on the micro-level appearing in communication can grow in strength, which effects the macro- level organization. It is important that any system which is susceptible to the aforementioned effect, can in a short space of time, change its nature unpredictably. In particular, there is reason to emphasize that chaoticity is an inherent property of a system. Although in reality the causes of chaotic behavior are usually changing conditions in the environment, chaos does not necessarily come from external factors whose effects are often impossible to differentiate from internal chaos factors.

5 Is Organization Chaotic?

Organizations working in society often change due to environment-changing factors and it is apparent that an organization is often subject to pressures of change from the environment. For example, according to McCann and Selsky [29] an organization's change is usually effected primarily by environmental factors together with other secondary factors. The system's growing structural instabilities, the loss of control or the absence of resources due to environmental turbulence can cause a critical state or conditions for the organization, which can be called the bifurcation point. Normally, a rational organization tries to stabilize or balance its behavior by different methods. It attempts, for example, to cultivate its structure and to increase its capacity to process information, or to increase its openness to the possibilities offered by the changed environmental conditions. If the change cannot be controlled, the organization's instability may begin to grow cumulatively toward chaos; actions made within the organization can surpass the next effective operational level, causing again a new wave of change.

So, it is self-evident that the predictability of the behavior of an organization is vital for effective planning and management. There is thus reason to ponder on what conditions an organization can be put on a level with a potentially chaotic- behaving system. We can identify certain features which all chaotic systems have in common:

1. Chaos appears in nonlinear, dynamic systems,
2. Chaos appears in systems which are recursive, by which is meant cycles in which a system's output at a given moment is used in the input of the following moment.

It is easy to see the primary characteristics of these features, for example in present-day firms and other organizations and for their functioning environments. In the functioning of an organization, recursiveness plays a large role. Prepared plans are based by rule on the previous period's achievements and lessons are sought from the past. In practice, nearly all methods are composed of activities which the members of the organization try to improve. Sudden irregular and unforeseeable changes in the functioning of an organization can be, for example, the failure of the operation plans, the interruption of production, a surprise strike and internal conflicts. In other words an organization can be determined as a comprehensive communication system whose behavior is at all times susceptible to chaotic phenomena. Bearing in mind these features, one can examine an organization's ability to control change caused by environmental or internal factors.

As a simple model for discerning changes one can use, for example, the familiar formula:

$$A(t+1) = E(t) - E(t)^2 \tag{2}$$

Symbol A presents the organization's ability to control complexity, and E the change in environmental factors occurring in time (t). The organization thus has certain possibilities A to control environmental turbulence during a specific period of time. As environmental pressures increase, that is as E increases, the organization's behavior also acquires new forms. Controlled change can, however, get out of hand; the functioning change may be chaotic thus paralyzing and possibly destroying the system completely (fig. 1-4).

In practice, one can specify the system's oscillations between different values by examining it as an expression of the changed behavior in relation to the environment. The behavior of the organization represented in figure 1 is stable. Its operations are in balance with the new conditions and demands of the environment, and random conditional changes occurring in the environment do not cause changes in the organization's behavior. In the situations of figures 2 and 3, the effect of the environment causes periodic change in the system's behavior, but which is nevertheless controllable. The system in figure 4 never achieves a stable level of operation in relation to its environment, but the variation still remains within the specific limits of the chaotic attractor.

A better understanding of the pressures surrounding an organization can be achieved if we use the bifurcation diagram to describe the system's behavior produced by the formula. I shall call the diagram in figure 5 the developmental behavior model of the chaotic (communication) system.

Again the organization's ability to manage change is presented by the y-axis and the changes in environmental conditions by the x- axis. For every E value a point is drawn on the graph, which presents the state corresponding to the system's balance. The pressures due to the environment being few, the system's structure is not changed. With growing pressure the system moves towards the first doubling, which is the first bifurcation point. Behavior changes from stability and balance to disturbance and variability. If the pressures of the environment continue to increase, the strength of the disturbance factors increase and management becomes more difficult, leading to the situation in which behavior becomes even more complicated and hard to control. The organization is headed towards chaos.

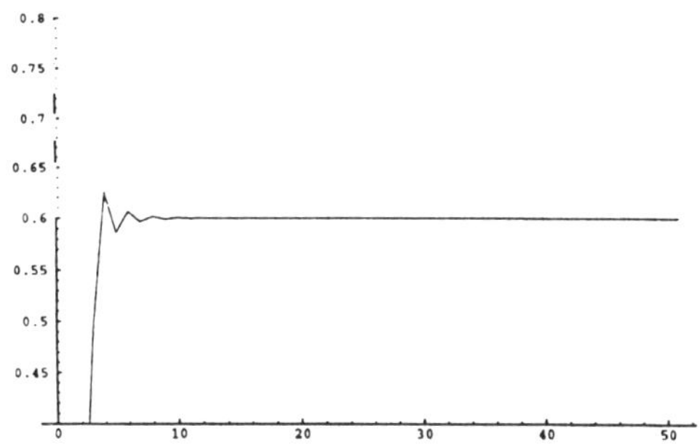

Figure 1: At the value E = 2.5 the system maintains its balance.

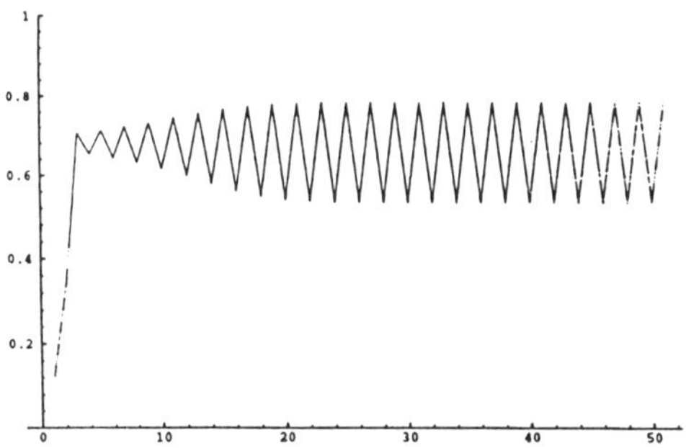

Figure 2: At the value E = 3.15 behavior divides and the system oscillates between two different values.

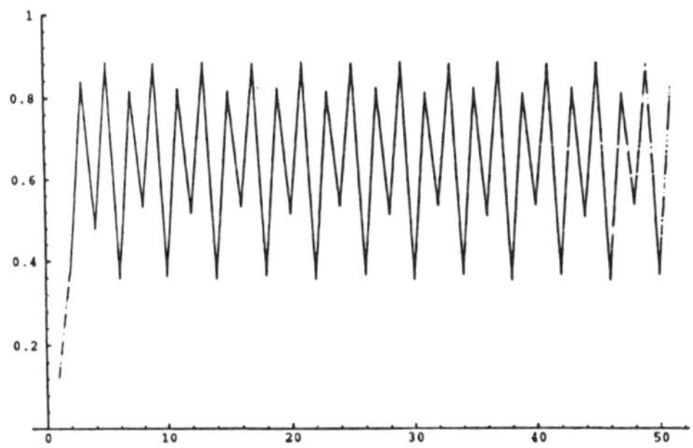

Figure 3: At the value E = 3.55 behavior divides again and oscillates between four different values.

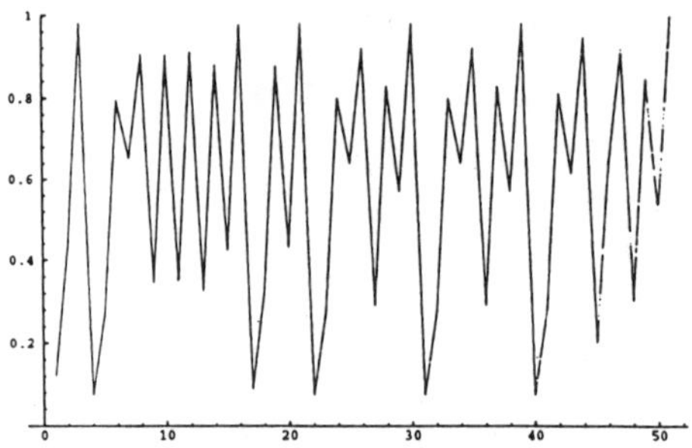

Figure 4: At the value E = 4 the system's behavior has divided many times and become chaotic.

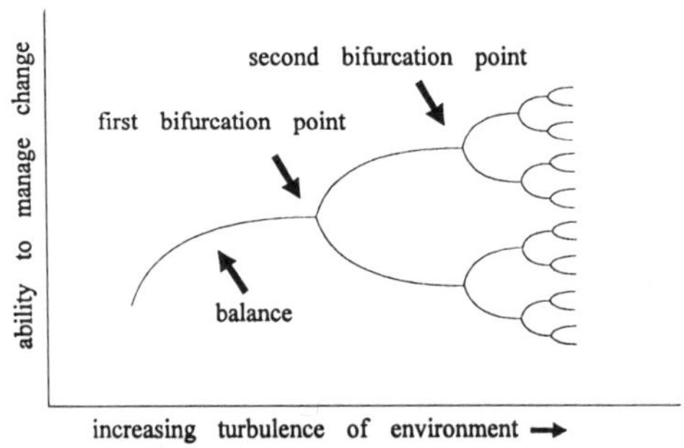

Figure 5: Behavior model of a chaotic system. The representation of the bifurcation tree is simplified due to observability

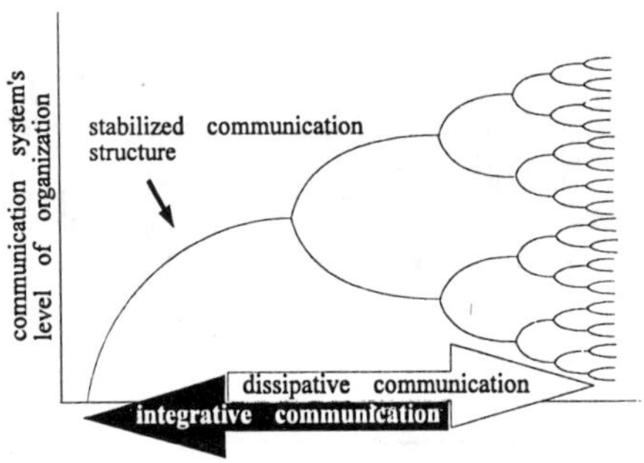

Figure 6: The significance of intergrative and dissipative communication for the structuring of the organizations's communication system.

6 Integrative and Dissipative Communication The Double Function

The organization's communication system's relationships to the organization's behavior can be pictured so that the organization (i.e. superstructure) tries to minimize instabilities at the bifurcation point with the help of communication or control (i.e. the system's operational subsystem) when it chooses the right path of development. In connection with every realized choice, the organization moves towards new points of communication space (phase space) along the path it has created.

Communication as a substructure of an organization can be an effective tool in the safeguarding of correct choices and the realization of favorable development. Since control is not an independent system and can itself be unbalanced or chaotic, it is clear that not just any communication is system-productive or integrative. Communication can be regarded as a qualitatively bisectional system in which one part directs the superstructure towards order and the other towards growing disorder. By its qualitative properties I have named these two areas or functions of communication as:

1. integrative communication and,
2. dissipative communication.

When we describe the organization's development by the bifurcation diagram, the integrative communication directs the organization towards order and the dissipative towards growing disorder (fig. 6).

Effective integrative communication would thus be the organization's important instrument when it meets with pressure for change from the environment. On the other hand, dissipative communication works in reverse; it makes harder the organization's choice made in bifurcation. It can even become damaging and direct the organization towards the wrong development path. It can thus be claimed that every communication act realized within an organization is by its nature bipartite, and that communication can simultaneously carry out simultaneously two different functions. The thesis is that every communication event, regardless of its original meaning or appearance, is simultaneously both integrative (constructive) and dissipative (destructive). At the optimum level, an organization's communication would be directed so that the communication's integrative properties would be maximized and its dissipative properties minimized. On the other hand, existing structures and functions can be broken with controlled dissipative communication and changed forward into the desired direction.

It is with regard to the improvement or rationalizing of the organization's operations that this concept of double function has its advantage. Its realization, that for example all communication occurring in every stage of a delegation process can work in two different ways regardless of what meaning it had designed, offers superior possibilities to better control communication. A talented superior constructs his message in such a way that its area of effect is as broad as possible. The appearance of the communication is not decisive, but rather the relation between the integrative and dissipative functions, which also depends on the receiver of the message, on his worldview. The interpretation of the message is facilitated by the realization of its double function. This idea applies in both individual as well as group delegations. It could be said that in delegating tasks to a group, the significance of the realization of communication's double function grows. There are more people in a group and thus more chances of a wrong interpretation. In extreme circumstances, the message can be integrating to one of the group members and to another dissipative, again regardless of the message's nominal meaning or content.

In practical group situations, the benefit of communication's double function could give room to the free forming of the group's structure. Successful free communication, which is a necessary condition for the synergy rising from the group, makes possible a lively group and opens the road to change. Ossified structures are broken and self- organization surfaces as the main property of the group's functioning. In such a case, the significance of interactional communication grows; continual interactive exchange between members is necessary, if the double function in a communication situation is to work in the desired manner. The common knowledge of the actors of the communication process, shared meanings, empathy, cultural interaction, all are important to group communication and thus also to the group's functioning.

Communication and a control system's structuring to the level that makes possible the effective functioning of the main system (organization as a whole), is not identical with the traditional balance concept of the communication system. As an independent system, the communication system continuously changes, for example in its relation to its environment at least on the microlevel; neither are the dynamic communication relations ever structurally-stable. The traditional balance between an organization and the environment is an artificial concept used as an aid to modelling; its appearance in organization reality is impossible. The traditional balance concept of organization models builds a protecting shield around the organizations, in the sense that even after the dramatic change it ensures a return to a safe and often more effective state of stability. Chaotic balance makes an organization susceptible to disturbances and sudden changes when the organization itself seems to stand on a firm surface. It is important to note that the seemingly random disturbance or bias in an organization's operation perhaps does not have to be a deviation from the state of balance which can be corrected, but instead a characteristic of the organization's fundamental nature.

7 Chaotic Knowledge Claims

From chaos-theoretical phenomena, and examples that arise from them, let us present a few general assumptions about the organization's characteristics and thus about the comprehensive communication system's nature and behavior:

1. A communication system develops from dynamic and nonlinear elements which are in a recursive interactional relationship with each other,
2. From the communication system's development we can distinguish stable behavioral stages (the chaotic balance) and crisis points (the bifurcation points). The changes of the stable period are to a certain degree predictable, whereas this is impossible at the crisis points,
3. In the development of communication systems, sudden and unexpected changes appear which change it qualitatively and lead the development to a completely new level. The uncontrollable development of a communication system can lead to chaos. The significance of qualitative changes are as important if not more important than quantitative changes,
4. Integrative communication realized in a communication system directs the system towards stable forms and dissipative towards disintegration. The double function of communication works in a way as a balancer and director of behavior in a system's development between order and disorder.

In an organization's communication research, chaos theory highlights the problems involving the measurement of the communication's effectiveness and influence. Observations, measurements and calculations are always imprecise, which also leads to inaccuracies in conclusions and predictions. According to the butterfly effect, mistakes and nonspecifics can grow cumulatively and arbitrarily rapidly. However, the behavior of chaotic systems is at all times completely predetermined. Unpredictability is merely the result of the fact that no finite information is enough to describe the system so precisely that the prediction will not in a short time become worthless. We can say that the more precisely the system's condition is known, the more precisely and long-term a prediction can be made, albeit only to an extent. What can be done? Although the behavior of communication systems cannot be predicted absolutely, it can still be influenced in that direction. With regard to chaos in the development of a system, we can limit it by controlling the parameters. Similarly, by changing the system's structure and properties we can stretch the plan of chaos and create so-called slow chaos. In place of single predictions and precise measurement, for the behavior of the organization we can construct alternative scenarios which are checked and corrected as conditions dictate.

Traditional empirical organizational communication research, which tries to find linear causal relationships and numerical indicators, might find chaos theory very frustrating. According to the theory, the precise prediction of the behavior of chaotic systems is lost over time. The only possibility is continual checking, calculating and correcting of prepared predictions.

According to London and Thorngate [30], the most interesting dynamic expressions of social systems are in traditional social science models, often lost through incorrect terminology. In chaos theory, these characteristics are part of reality and can be understood by using simple mathematical models. Due to nonlinear dynamic properties, it is possible that what traditionally in communication models was determined as random disturbances, might in reality be the appearance of chaos. Chaos can be described by deterministic low-dimensional models, whose explainable power is in its universal principles.[5]

Gregersen and Sailer [1] have stated that the chaotic behavior of systems, verified in physical reality, is probably also quite common in the social world surrounding us. In these uncertain times, it is nevertheless absurd to speak of chaos as the Savior dressed in fractals' clothing, who solves all of the problems of communication research. All communication systems surely do not behave chaotically. Even part of the objectives of empirical communication research is doubtlessly governed by norms, when long-term operation itself may be linear and regular. A linear communication understanding can also be the only possible way of thinking which can be used to estimate and govern certain forms of communication. The verification of chaos by simulation and hypothetical models in actual communication circumstances, without valid methods, is still largely unknown terrain, whose exploration demands time and specific equipment. It is not enough that the nonlinear dynamic communication models themselves, used in description, produce chaotic behavior. It must also be possible to draw a connection between these models and the surrounding communication reality. Until now, we have had only a foggy idea of their connection. For the extension of the theory, we need concrete applications. Then again, what do we do if a traditional paradigm and the prolific theories about it do not stand up. Chaos is hard to handle, and its application to social science is problematic. Still, already in the mirror of chaos' short history, it appears that it is, in its innovative significance, perhaps the strongest primus motor in decades that has pushed itself into academic quarters true to traditions. As Pagels [20] has said: "Life is nonlinear, and so is just about everything else of interest."

Notes

The formula is the same as x(t+1) = rx(t)(1-x(t)), which is known as the so-called Verhulst's equation. As early as 1976, May[3] noted that it is also possible to use similar functions in the social sciences. Equation 2 is in the form $x(t+1) = rx(t) - rx(t)^2$, in which x is replaced by the symbol A and rx by the symbol E.

References

1. H. Gregersen and L. Sailer, *Human Relations*, **46**(7), 777-802, (1993).
2. O.A. Wiio, *Wiion lait ja vhn muidenkin*, (Weilin+Göös, Espoo, 1978).
3. R.M. May, *Nature*, **261**, 459-467, (1976).
4. D. Ruelle, *Mathematical Intelligencer*, 126-137, (1980).
5. E.H. Buder and L.D. Ritchie, *Are the Data Random or Merely Chaotic?: Chaos Theory and the Limits to Prediction*, (A paper presented at the annual conference of the International Communication Association in San Francisco, 1989).
6. J. Gleick, *Kaaos*, (Art House, Jyvskyl, 1990).
7. L. Bailyn, *Sloan Management Review*, 77-83, (1993).
8. G. Feichtinger and M. Kopel, *European Journal of Operational Research*, **68**, 145- 159, (1993).
9. Y. Friedman, *Leonardo*, **26**(4), 359-363, (1993).
10. T.J. Kiresuk, *Journal of the American Society for Information Science*, **44**(4), 235-241, (1993).
11. D. Richards, *International Studies Quarterly*, **37**, 55-72, (1993).
12. M. Somma, *Social Science Information*, **32**, 371-385, (1993).
13. N.O. Contractor, *Unpublished manuscript*, (1992).
14. N.O. Contractor and D.R. Seibold, *Unpublished manuscript*, (1992).
15. P. Aula, *Viestinnn kaaos - kaaosteoria ja viestinnn eplineaarinen mallintaminen. (Communication and Chaos)*, Master's Thesis, (Department of Communication, University of Helsinki, Helsinki, 1993).
16. P. Aula, *Kaaos - uusi paradigma organisaatioviestinnn tutkimuksessa? (Chaos - New Paradigm for the Study of Organizational Communication?)*, (Paper presented at The Conference for Communication Studies, Helsinki, 1994).
17. P. Aula, *Tiedotustutkimus*, **1**, 33-46, (1994).
18. P. Aula, P. *Kaaos - viestinnn uusi paradigma vai tehoton metafora? Kaaosteoria osana viestinttutkimuksen viitekehyst (Chaos - Communication's New Paradigm or Ineffective Metaphor? The Chaos Theory as a*

Part of Communication Research's Framework), (Licentiate Thesis, Department of Communication, University of Helsinki, Helsinki, 1995).
19. F. Richter, in *Metatheory in Social Science: Pluralisms and Subjectivities*, eds. D. W. Fiske and R. A. Shweder, (University of Chicago Press, Chicago, 1986).
20. H.R. Pagels, *The Dreams of Reason; The Computer and the Rise of the Sciences of Complexity*, (Simon and Schuster, New York, 1988).
21. C. Shannon and W. Weaver, *The Mathematical Theory of Communication*, (University of Illinois Press, Urbana, 1949).
22. J.P. Bowman and A.S. Targowski, *The Journal of Business Communication*, **24**, 21-34, (1987).
23. L. Thayer, in *Viestinnn virtauksia*, eds. E. Erholm and L. Åberg, (Otava, Keuruu, 1978).
24. S.W. Littlejohn, *Theories of Human Communication*, (Charles E. Merrill Publishing Company, Ohio, 1978).
25. E. Jantsch, *The Self-Organizing Universe; Scientific and Human Implications of the Emerging Paradigm of Evolution*, (Pergamon, New York, 1980).
26. P.R. Monge, *Unpublished manuscript*, (1990).
27. B.A. Fisher, *Perspectives on Human Communication*, (Macmillan Publishing, New York, 1978).
28. S.P. Robbins, *Organizational Behavior; Consepts, Controversies and Applications (5th ed.)*, (Simon and Schuster, New Jersey, 1991).
29. J. McCann and J. Selsky, J. *Academy of Management Review*, **9**, 460-470, (1984).
30. I. London and W. Thorngate, *Psychological Reports*, **48**, 203-228, (1981).
31. L. Åberg, in *Viestinnn virtauksia*, eds. E. Erholm and L. Åberg, (Otava, Keuruu, 1978).

THE APPLICATION OF TOPOLOGICAL MODELS TO SOCIAL SYSTEM DYNAMICS

H.R. ERWIN
Department of Computer Science
George Mason University
Fairfax, Virginia 22030
USA

Accurate measurement of parameters describing social systems is difficult, and the model input data resulting from such measurements are likely to be in error. An understanding of when noisy data can invalidate model results is critical to the successful use of non-linear models to analyze social systems. This paper discusses modelling approaches for non-linear systems, including the use of topological modeling and uses case studies to provide some insight into factors affecting model validity and stability.

1 Introduction

"The purpose of computing is insight, not numbers." [1]

Our understanding of human systems is in the middle of a revolution. Many are now known to be non-linear, becoming irregular or aperiodic when stressed. This paper examines the limitations of modeling for non-linear systems and uses case studies to suggest valid approaches to addressing their evolutionary dynamics.

An important problem the social scientist should be aware of is that not all deterministic non-linear systems can be modelled — some have behavior that cannot be predicted even in theory. The phenomenon underlying this — sensitive dependence on initial conditions — is a key idea in modern systems theory. A system with sensitive dependence on initial conditions is a difficult modeling problem - there is an absolute limit to the accuracy with which any model (including the system itself) can replicate observed system behavior given unavoidable inaccuracies in measurement. This strange property is basic to how chaos occurs in deterministic systems and usually reflects a 'folding' of the system dynamics (that is, trajectories with different histories passing through the same point) and a simultaneous loss of information about the initial system state as the evolutionary paths of nearby points diverge exponentially. In nonconservative or dissipative systems this process results in fractal patterns, since an initial neighborhood diverges in some directions while simultaneously decreasing in total volume. Deterministic chaos can occur in

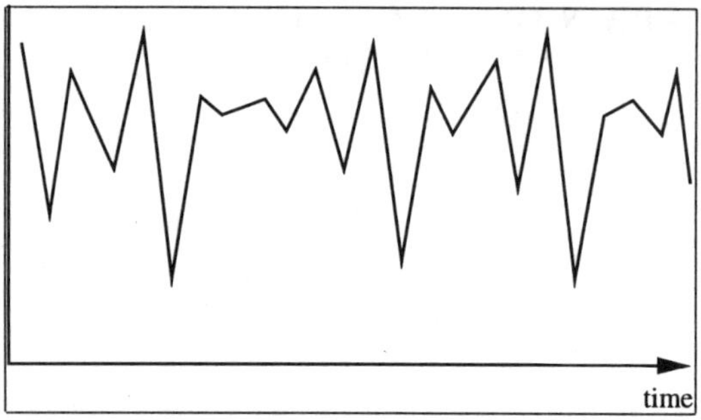

Figure 1: Discrete chaos for a strongly non-linear system

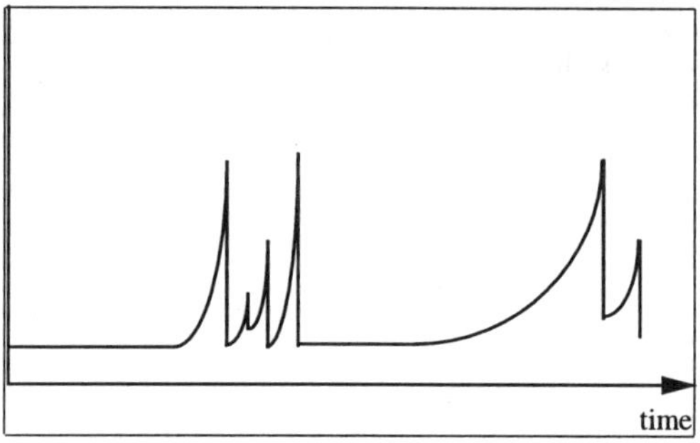

Figure 2: Intermittancy

discrete and continuous dynamic systems in the following ways:

- Strongly non-linear dynamics — seen, for instance, in difference equation models of population dynamics in ecology [2]. Strong non-linearity results in dynamic processes that merge distinct initial states into the same terminal state, while spreading nearby initial states over the entire state space. Figure 1 illustrates the typical pattern seen.

- Intermittency — dynamic systems that flow smoothly at most times, but sometimes show rapid fluctuation and divergence [3]. The periods of rapid fluctuation inflate small neighborhoods so that the eventual return to smooth flow finds that system state information has been lost. Figure 2 illustrates a sample of this in a continuous flow.

- Multiple interfering periodicities — where periodic or cyclic processes interact non-linearly. The periodicities add and the non-linearity then

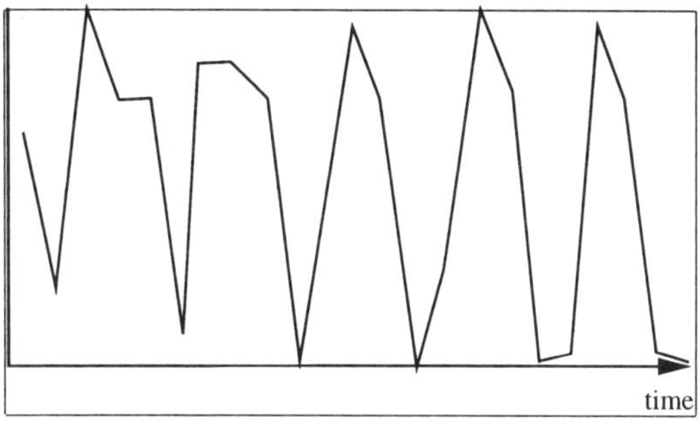

Figure 3: Chaos due to multiple interfering periodicities

folds the interaction so that state history is lost [4]. Figure 3 illustrates this for a discrete flow.

- Non-dissipative chaos, e.g., Arnold's 'Cat maps' [5] and circle maps:

$$f(x) = (\alpha x + \beta) \; modulo \; 1$$

for $\alpha > 1$. Information is lost at each iteration as points an integer distance apart are folded together by the modulo operation. Pseudo-random number generators belong to this category.

Chaotic processes now appear to be ubiquitous, and mathematical dynamicists suspect they include nearly all macroscopic low-dimensional "random" processes [6]. Even in random systems, interaction between statistical noise and non-linear dynamics can be remarkably complex.

2 Non-linearity and Model Validation

The key issue for any model is how valid a represention it is of the behavior and characteristics of the system being simulated [7,8]. The traditional approach to validating a model is by comparing the results to static analysis, the results of other simulation models, and data taken from the actual system. That heuristic approach has the advantage of avoiding the difficulties that can be encountered applying formal validation techniques to non-linear models, but it also requires many validation points to adequately address the issues of correctness, local accuracy, dynamic stability, and transient response. The costs of multiple models and data collection usually limit the amount of feasible validation, even when possible. Comparative model validation must be paired with formal methods.

In recent years, the computational resources available to the social system modeler have grown to the point where formal methods are becoming computationally feasible. Although comparative techniques continue to be important, formal methods may now be used as well. This allows the modeler to address issues of system stability and transient response, where the comparative method was always weak.

2.1 Formal Validation of Models

In computational science, a model is formally validated based on criteria of correctness, accuracy, and stability. These are defined as follows:

- Correctness — does the model represent the behavior of the system correctly as the finite approximation mesh size is taken to zero? In models where the approximations are qualitative simplifications of individual submodels, rather than numerical approximations, the question changes to: "Does the model represent the behavior of the system correctly in the limit as simplified submodels are individually replaced by fully correct ones?"

- Accuracy — how accurate are the estimates derived from the model as a function of the local mesh size? For models with simplified submodels, the question becomes: "How accurate are the estimates of the model as a function of the numerical errors resulting from the qualitative simplifications in the individual submodels?"

- Stability — do small roundoff and statistical errors eventually disappear, or do they grow to dominate the final results? A related question is whether the modeled system is stable or divergent in the region of interest.

In computational modeling, correctness is addressed by demonstrating that the approximations used in the model converge to the correct equations as the mesh size goes to zero. This corresponds to the following notional equation:

$$\lim_{(\Delta x \to 0, \Delta t \to 0)} \{model(\Delta x, \Delta t)\} = reality$$

and is usually addressed using the differential calculus. Accuracy is usually addressed by estimating the size of the prediction error as a product of powers of the spatial and temporal mesh dimensions. This is usually worked out using an approximation formula such as the Taylor series. Stability is normally addressed through an eigenvalue analysis. The model is fed a representative error distribution and its tendency to diverge or converge on the $error = 0$ state is determined.

2.2 *Informal Validation of Models*

Informal calibration and validation of models is the comparison of model predictions and parameters with data from other sources, particularly data and models that are likely to be independent of the preconceptions that went into the design of the model being validated. There are typically four sources of comparison data:

- Benchmarks
- Static analysis results
- Other dynamics models, and
- Detailed submodel results.

Benchmark Data

Benchmark data are measurement data from the real system. These are often unavailable, usually expensive to collect, and almost always noisy, but they reflect the 'ground truth' about the system being modeled. There are at least three dangers associated with their use:

- A complex model may have more free parameters than there are calibration data points available. A fit then only reflects a specific selection of parameter values by the modeler, rather than a 'ground truth' about system dynamics.
- The benchmark data often reflect a single point of system behavior, when the real interest lies in understanding system sensitivity to parameter variation.
- The elements of the benchmark data may be erroneously associated, leading to a model calibration to an imaginary state of the real system.

Static Analysis

A model may also be compared to the results of a static analysis of the system. This is particularly important when concerned with detecting behavior that is an artifact of the model design, rather than present in the real system. An example of this might be a discrete event simulation of a factory with a hidden limit on the number of simulated processes that can be active. This would result in unexpected queue build-ups or job losses whenever the number of jobs to be simulated exceeded the hidden limit. A static queueing theory analysis would be able to identify this problem. Static analysis is weak, however, in predicting transient system dynamics and the presence of stable states of the system that result from the interaction of dynamic processes.

Dynamic Models on Varied Time and Distance Scales

Multiple models with varied temporal and spatial meshes and differing architectures can be used to gain insight into limitations of the model under study and its sensitivity to parameter variations. The weaknesses of this approach are, of course, the added cost and the risk that the models would have the same blind spots.

Submodels

Individual components of the model under study can be validated separately. This is commonly done in commercial simulation work and provides the user a high degree of confidence in major elements of the model. This may not be feasible or cost-effective in modeling human systems, and there is a significant risk that system interactions between submodels are still not correctly modeled.

2.3 Problems Modeling Social Systems

Experience has shown that there are a number of other problems in modeling social systems, including:

- poor choice of modeling technique,
- noise,
- lack of data, and
- non-linearity and chaos.

Poor Choice of Modeling Techniques

There is a broad literature on the strengths and weaknesses of various modeling techniques. Stability is often unexpectedly important[9,10]. The popular choices of explicit compartmental models and discrete event models have particularly poor stability. On the other hand, implicit methods are usually stable, and so should be preferred unless the computational cost of the matrix operations required is too great.

Noise

How closely can or should you calibrate your model when you know the data are noisy? The presence of noisy data should encourage the modeler to investigate the sensitivity of his or her results to variation in free parameters.

Limited Data

Given enough free parameters, you can fit *any* model to your data. The model should be sufficiently complex to investigate the issues of interest, but not more than can be calibrated using available data. A low dimensional model is better than one of too high a dimension. If calibration data are unavailable, the behavior of the model should be studied throughout the region of possible values.

Non-linearity and Chaos

Non-linear systems are often chaotic, with sensitive dependence on initial conditions. This implies the model, to be correct, must be unstable. How do you model an inherently unstable system? Do you trust your results? What types of results should you ask for? These issues must be addressed.

2.4 An Approach — Topological Thinking

No model should be studied at just one point. Instead, the sensitivity of the model and the system to changes in parameter values must be investigated. In some cases, it may turn out that no single-point model is valid, and some more generalized approach to modeling must be used. One such approach, which I call 'topological modeling'[7,11,12], is to study the topological dynamics of the system in the abstract. Abraham and Shaw's visual introduction[13] is perhaps the best discussion of topological modeling, although F. D. Abraham's[14] book discussing applications to psychology is also excellent.

3 Catalog of Topological Models

Two types of topological model are of particular interest:

- dynamic models — models of how the state of a system evolves over time, and
- bifurcation models — models of how the state of a system can change with parametric variation.

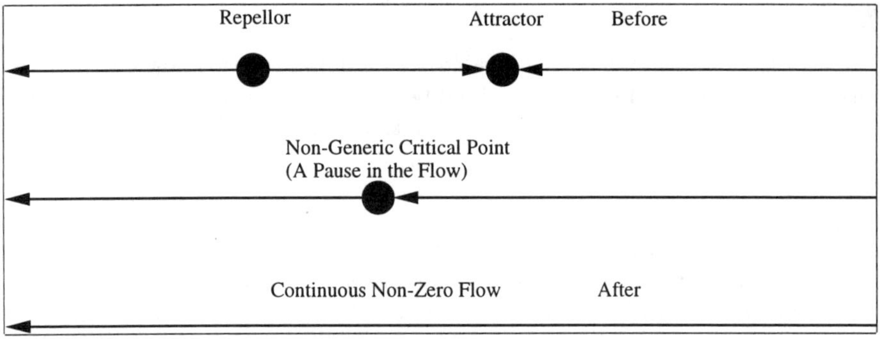

Figure 4: Static fold bifurcation

Terminology is important in this analysis. First a model should be 'generic' — having a non-zero probability of happening so that it does not need a miracle to represent the system. Secondly, dynamics are defined in terms of 'limit points' and 'limit sets.' A limit point of a dynamical flow is a point fixed under the action of the flow. Similarly, a limit set is a collection of points mapped into itself. Attractive limit sets are called 'attractors,' while limit sets that repel nearby points are called 'repellers.' (Note that repellers cannot be easily detected in simulations.) The 'inset' of a limit set is the set of points that converge to the limit set under the action of the flow. If a flow is time-reversed, with attractors becoming repellers, the result is also a flow, and the 'outset' for the original flow is defined as the inset for the time-reversed flow.

With this terminology in hand, a number of examples can be investigated.

3.1 One Dimension

In one dimension, all generic limit points of dynamic flows are either attractive or repelling. Bifurcations in one dimension usually consist of a single limit point splitting into an array with two limit points of the original type on the outside and one of the inverse type in between. (The reverse process can also occur.) A second bifurcation of some interest in dimension one is the static fold. This dynamic system starts with a flow that moves to the right or left as it approaches infinity, but with two limit points defining a finite segment with the opposite direction of flow. Figure 4 illustrates this.

In the figure, the left hand point is a repeller and the right hand point an attractor. The middle segment is a barrier to leftward flow. The bifurcation occurs when the length of the middle segment becomes zero. First, there is a momentary pause in the flow as the limit points merge. Then the pause

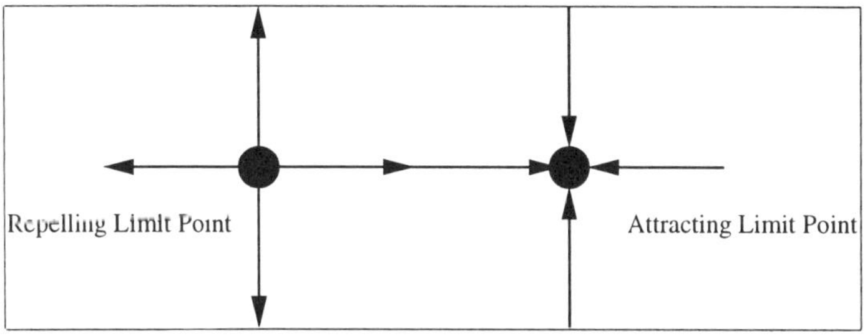

Figure 5: Attractive and repelling limit points in two dimensions

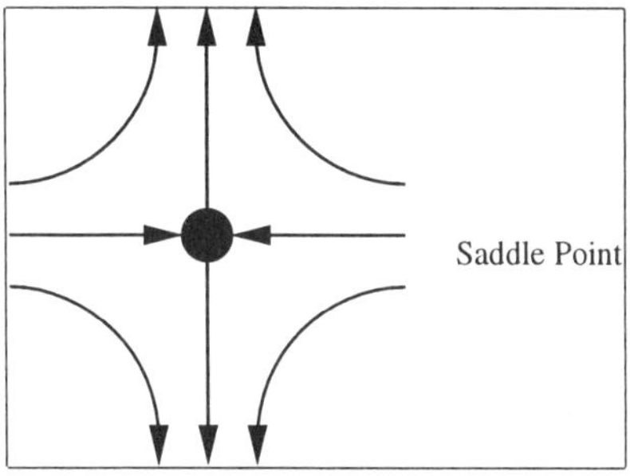

Figure 6: Saddle point in two dimensions

disappears, and a non-zero flow to the left exists throughout the real line.

This dynamic system describes the disappearance of a local stable point and is one possible model for technological innovation or the disappearance of a barrier to trade or migration. In real systems, the barrier region has to be fairly wide; otherwise, random fluctuations will 'tunnel' through it.

3.2 Two Dimensions

In two dimensions, there are several additional types of generic limit sets. (See Figures 5–8.) The attractive and repelling limit points in one dimension generalize naturally, and attractive and repelling limit cycles appear. There is also a new type of limit point, the saddle point, which attracts in some directions and repels in others.

In dimension two, we see the appearence of the first excitation or Hopf bifurcation, characterized by an attractive fixed point spawning a small attractive limit cycle that grows gradually as the control parameter is increased (Figure 9). The Hopf bifurcation is one model for the gradual emergence of periodic dynamics from a fixed point. In economic and governmental systems,

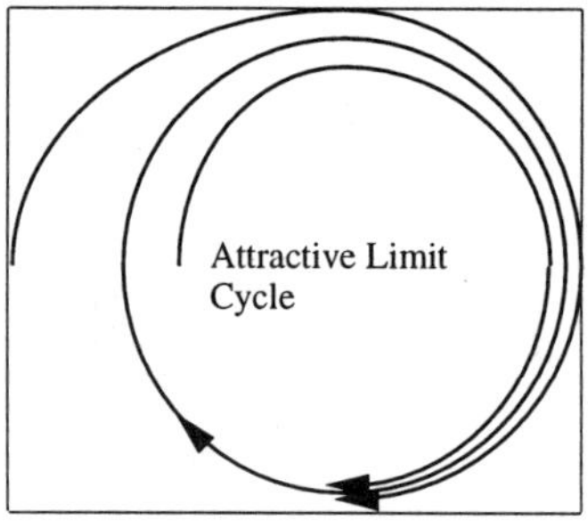

Figure 7: Attractive limit cycle in two dimensions

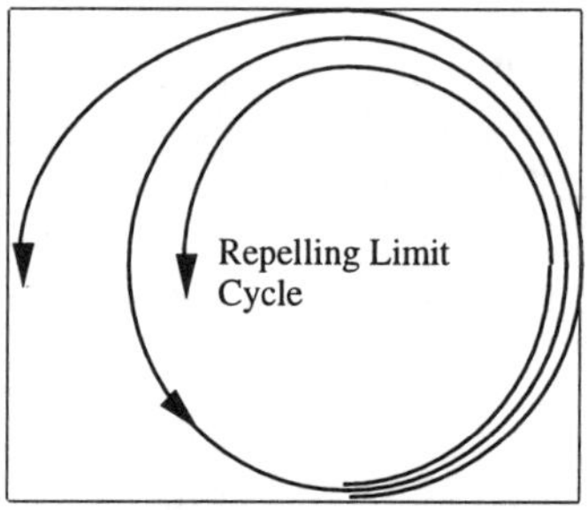

Figure 8: Repelling limit cycle in two dimensions

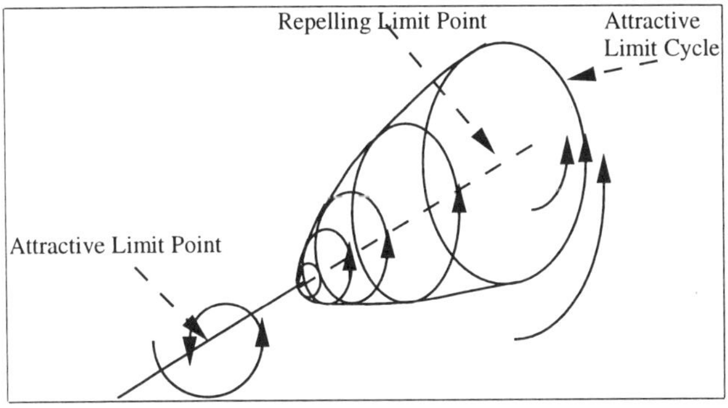

Figure 9: Hopf bifurcation in two dimensions

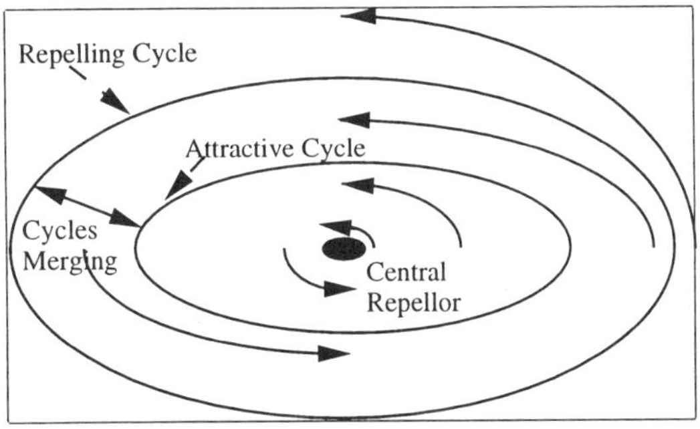

Figure 10: Initial state of the periodic fold

this can occur as the time lag between information acquisition and policy action grows. Dynamics change gradually as a basin of attraction slowly grows from a point attractor. These models are robust and control theory can be used to analyze the transition process.

The periodic fold in two dimensions is a bifurcation of particular application to the evolution of human systems. (See Figure 10.) Here, there are two concentric limit cycles around a central fixed point, one cycle attracting and the other repelling. An intermediate flow passes between the cycles. As one limit cycle contracts or expands to approach the other, the intervening flow disappears resulting in a single fixed point. (See Figure 11.) This process describes how a migratory adaptation to a cycle of environments or seasons can be replaced with a sedentary adaptation.

The basin bifurcation in two dimensions is an interesting pattern with relevance to historical patterns. First some terminology: if a trajectory leaving

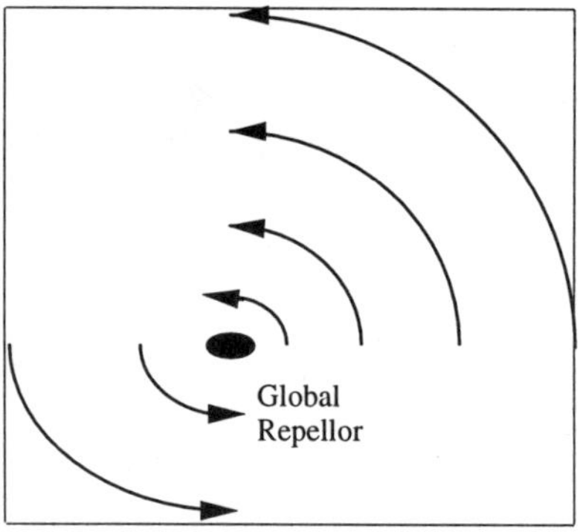

Figure 11: Final state of the periodic fold

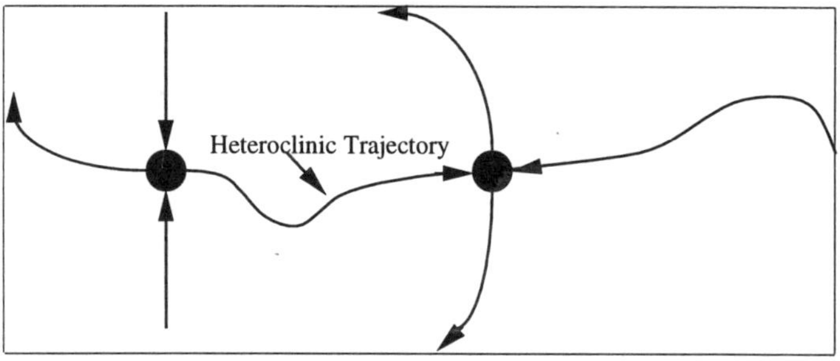

Figure 12: Heteroclinic trajectory

one saddle point arrives at another saddle point, it is termed a 'heteroclinic' trajectory (Figure 12). If it forms a loop, returning to the saddle point, it is called a 'homoclinic' trajectory (Figure 13).

The process where one saddle point passes another momentarily is illustrated in Figure 14. This results in a right-left swap of basins of attraction between the saddle points — the basin bifurcation. An historical example of this process was the Arab capture of Syria, breaking Constantinople's land communications to Egypt, and eventually detaching the whole of Northern Africa from Byzantine rule. The key pattern to be recognized is the sudden transfer of a large region between capitals where there had been no long-term encroachment in the area.

The blue loop is also an interesting bifurcation that shows how a cycle may appear in a region that previously had only fixed points. Consider the attractive fixed point at the right in Figure 15. Also present is a repelling fixed point at the center and a saddle point below. Both trajectories leaving

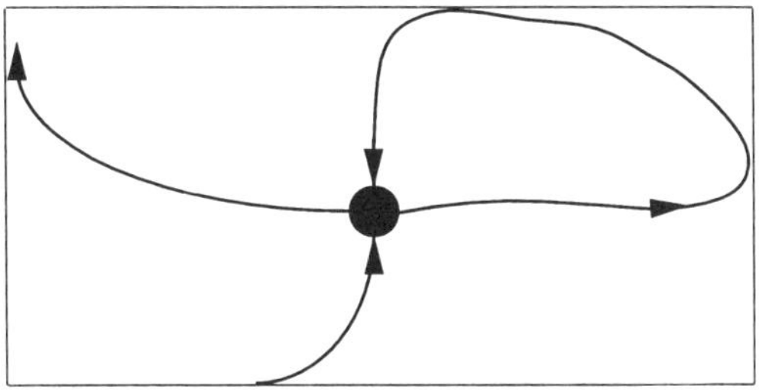

Figure 13: Homoclinic trajectory

the saddle point eventually arrive at the attractive fixed point. Allow the attractive fixed point to approach the saddle point. Eventually the points will merge, leaving an attractive periodic trajectory surrounding the repelling fixed point (Figure 16).

3.3 Three Dimensions

In three dimensions, the potential dynamics become more complicated, but the processes by which limit points generalized for two dimensions still apply. The various types of two-dimensional limit points, in particular, now spawn collections of three-dimensional attracting, repelling, and saddle cycles, with the insets and outsets spiraling an integral number of times as the flow passes once around the cycle. If the integer is odd, the insets and outsets for those cycles form Möbius strips. (See Figure 17.)

An important way higher dimensional dynamics can be spawned is the 'suspension' operator. A suspension of a dynamic flow is a dynamic flow of one dimension higher which substitutes a circular flow for each point in the original space. If we do this gradually for a limit cycle, we get the three-dimensional bifurcation called the 'solenoid.' This is also called the second excitation, where a process similar to a Hopf bifurcation spawns doubly periodic dynamics (See Figure 18).

Another bifurcation that first appears in dimension three is the octave jump in two dimensions or period doubling bifurcation. This occurs when a periodic attractor at the middle of a Möbius band splits into two. The attractor is replaced by a repeller at the middle of the band and a single attractor twice as long on each side of the repeller. Experiment by cutting a Möbius band in half lengthwise to get insight into this (See Figure 19). This bifurcation is a stage in the emergence of chaos.

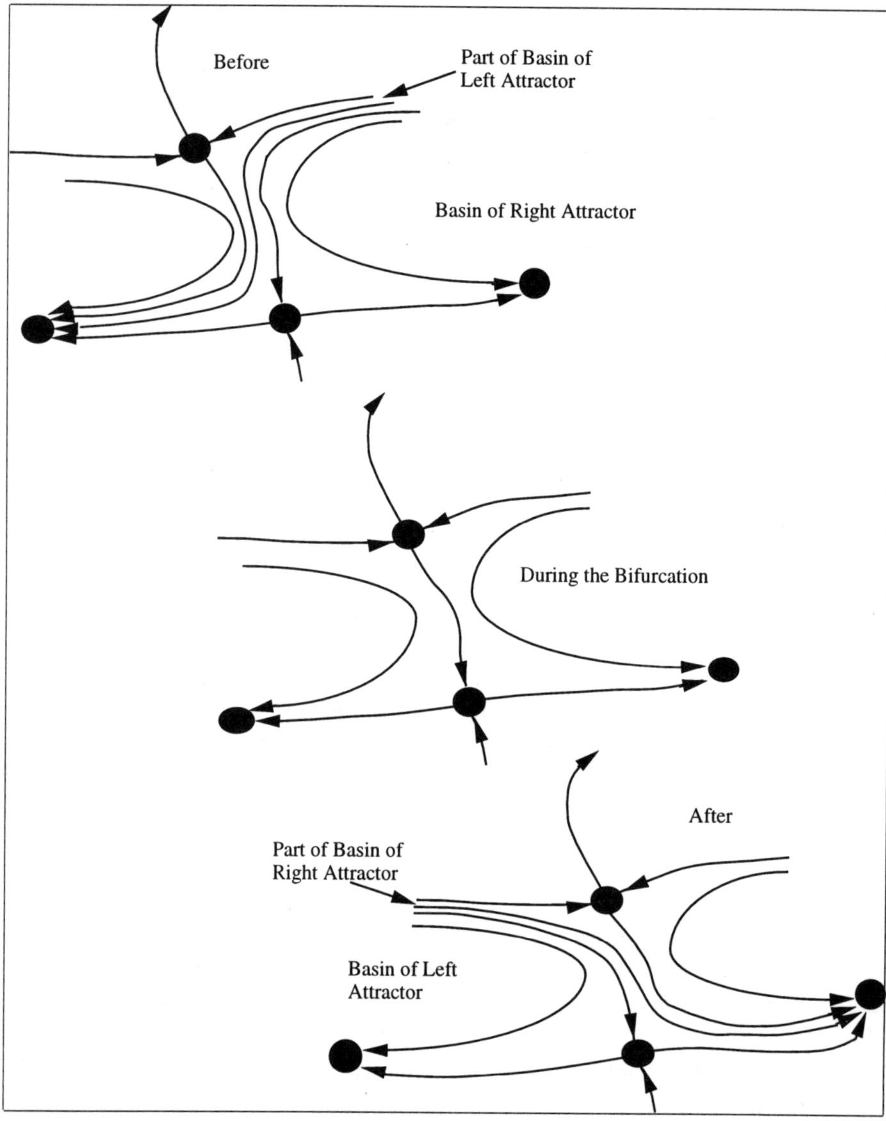

Figure 14: Basin bifurcation

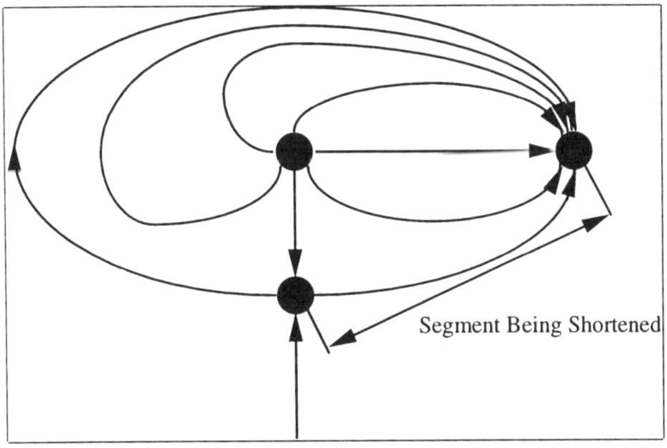

Figure 15: Initial stage of the blue loop bifurcation

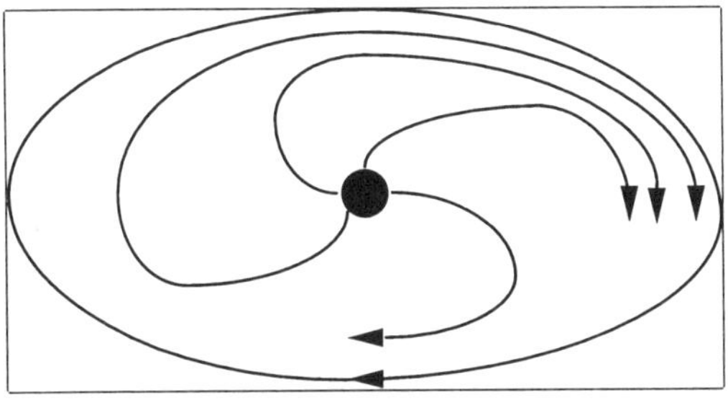

Figure 16: Final stage of the blue loop bifurcation

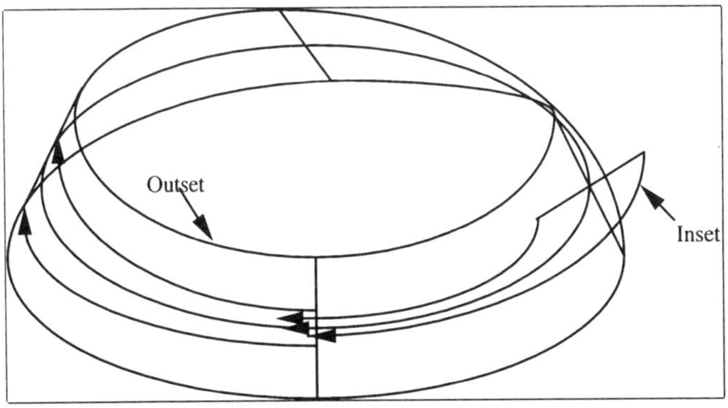

Figure 17: Möbius strip

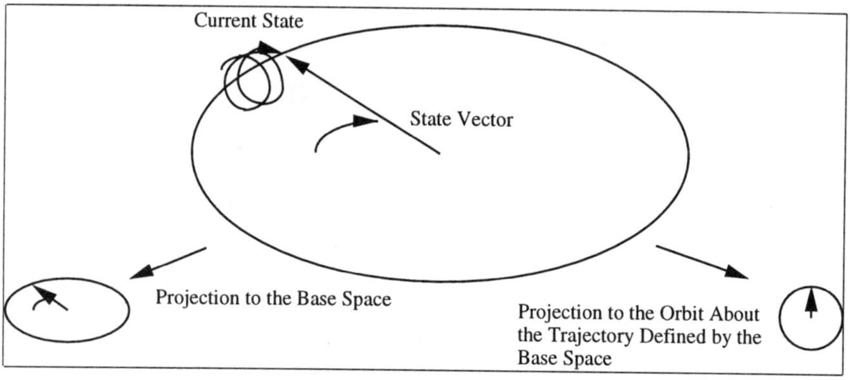

Figure 18: Suspension of a cyclic flow

4 Case Studies

A pair of case studies provide some insight into the effectiveness of topological modeling. The first examines the origins of agriculture, while the second considers the collapse of complex cultures.

4.1 Modeling the Origins of Agriculture

The following model is an interesting example of the periodic fold bifurcation. Three long-term average relationships can exist between nutrition and workload in a preindustrial society with stable population[15]:

- (Z) Famine — zero nutrition, workload, and population.
- (P) Starvation — both birthrate and workload limited by poor nutrition.
- (G) High workload — good nutrition but the birthrate is limited by high workload.

For a level of nutrition initially some distance below the poor nutrition level, population decreases, nutrition improves, and the system returns to state P. If nutrition is too low, the population crashes, and the system moves to state Z.

For a workload outside of the G locus, fertility is decreased below the replacement rate, and the population decreases, improving productivity and reducing the workload required to support the population. For conditions between states P and G, fertility is greater than the replacement rate, and either the population increases and the system converges to state P, or the

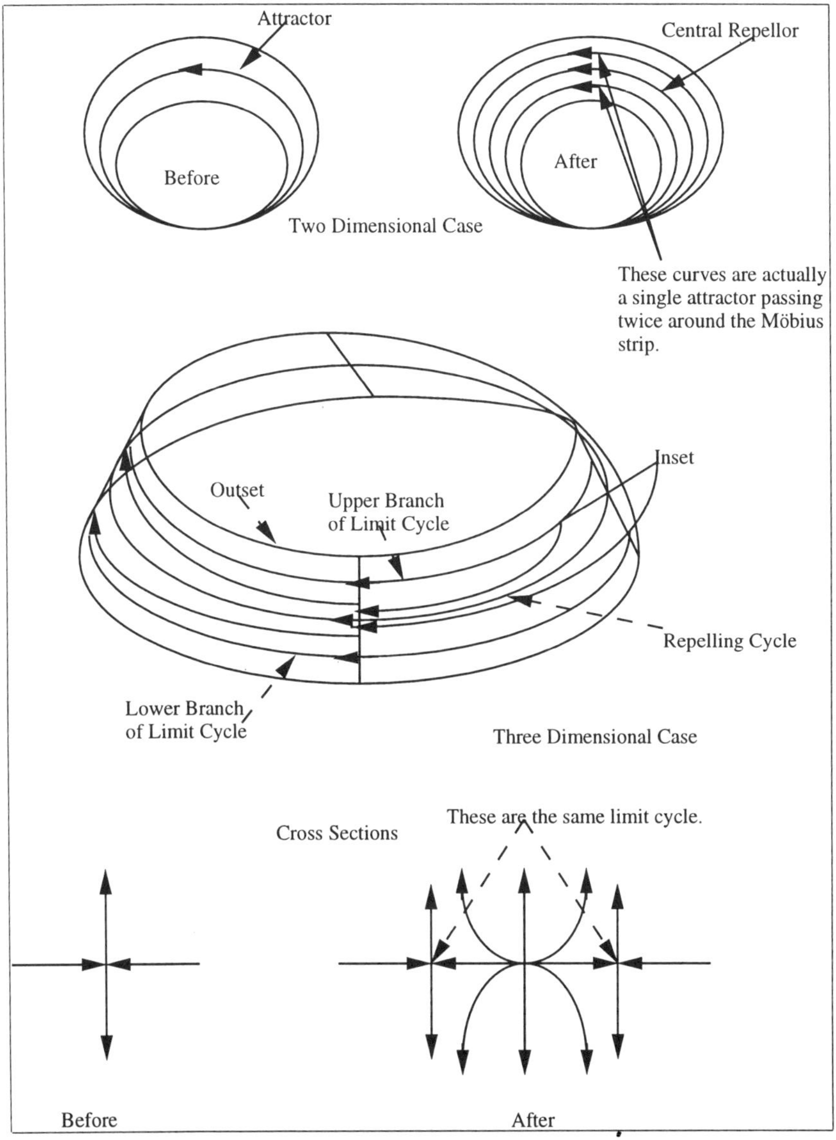

Figure 19: The period-doubling bifurcation

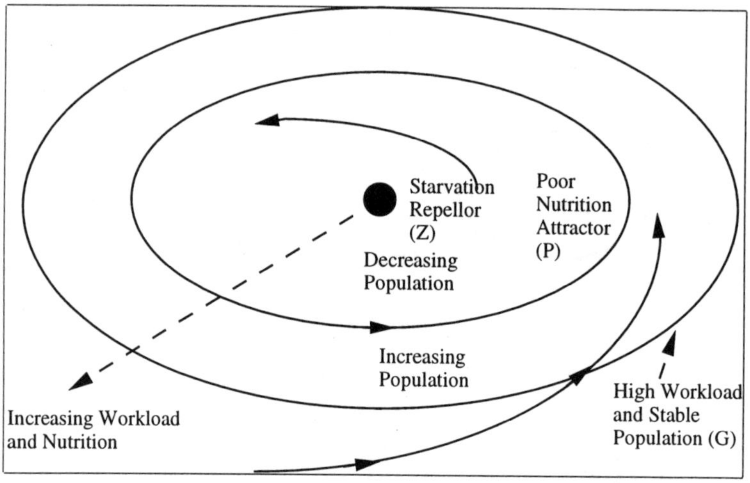

Figure 20: Emergence of agriculture

workload increases and the system converges to state G, the choice reflecting cultural factors.

The available subsistence strategies interact with the cycle of the seasons to define a dynamic flow. In preindustrial agricultural societies, state Z was a repeller, the poor nutrition case was an attractor, and the high workload case was unstable with the vector field crossing it to converge on the P locus. (See Figure 20.) The system was trapped in this state: slow increases in agricultural product resulted in increases in population but no permanent improvement in nutrition. A development policy that could increase the agricultural product per farmer was needed to allow the system to escape the trap and develop economically. That created a second repelling state, which then merged with the existing attractor to create a periodic fold bifurcation and free the system from the poor nutrition trap.

The system for hunter-gatherer populations was apparently reversed. State Z was an attractor — a hunter-gatherer group had to have a minimum average nutritional level to sustain its food collection activities. That minimum level defined a repelling surface; hunter-gatherer groups would migrate rather than remain near it. State P was near the repelling surface and vulnerable to famine, so hunter-gatherer groups apparently avoided it as well, resulting in a crossing vector field that converged on state G, which appears to have been stable (See Figure 21).

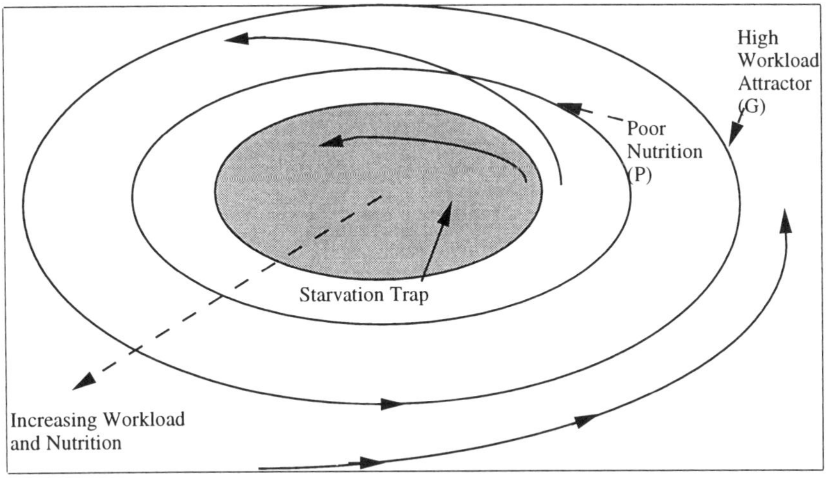

Figure 21: The hunter-gatherer system

The transition to agriculture can be modeled topologically as a switchover between the two cases. A change in climate or technology weakened the attractor at state Z, causing it to spawn a nearby attractive cycle associated with a slow decline in population but not a crash (a Hopf bifurcation — see Figure 22). Groups now had an alternative to moving out of a depleted area — they could follow a sedentary strategy of harvesting low-quality food sources to last out the famine. This modified the use of high workload for population control. Groups using the new sedentary strategy had to let their population recover during periods of high food availability, and so decreased their movements to reduce their workload.

As technology became more sophisticated, the sedentary strategy became more efficient and expanded towards state P, and a larger portion of the population relied on it. Population increased and the high workload component of the sedentary strategy also contracted towards state P as resources became scarcer. The poor nutrition repeller was eventually eliminated in a periodic fold catastrophe (Figure 23). Finally, after some population adjustment, the remaining attractor converged to state P and stabilized there.

4.2 Collapse of Complex Societies

S. E. van der Leeuw has commented: "urban cultures are remarkably unstable." [16] The collapse of an urban culture is an event more frequent in history than most observers realize[17,7,12,18], reflecting the resolution of short and long-

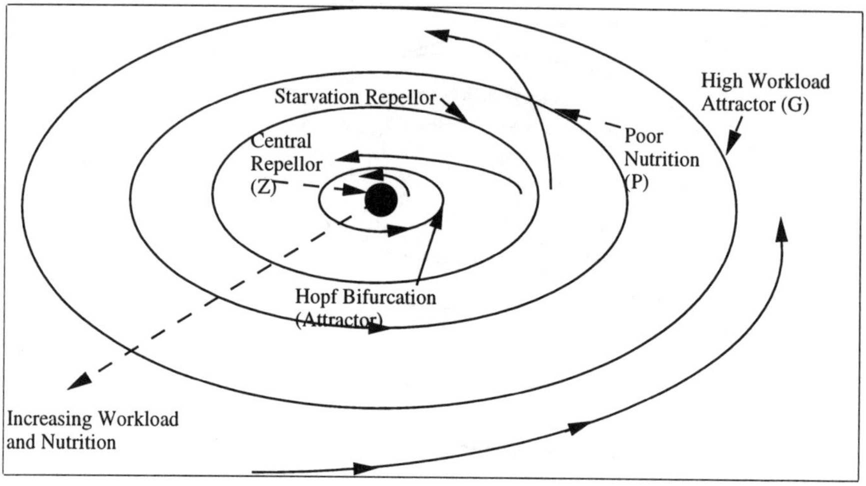

Figure 22: Early agriculturalists

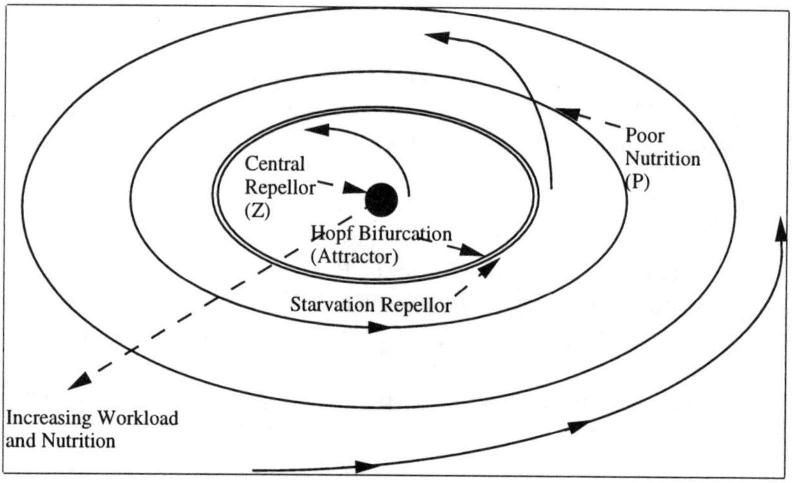

Figure 23: Late agriculturalists

term processes involving economics, demography, and social organization. Often, collapse has been well underway before societal elites and decision-makers became aware of it, leading to scenes of leaders responding retroactively and ineffectively as their society collapsed around them.

In *Economics of Feudalism*, Trout Rader develops general market equilibrium models of slave and feudal economies [19]. His use of the Poincarè-Bendixsen theorem is flawed where he attempts to prove that rural and urban populations can be expected to cycle, but his results become true with a few additional assumptions, and his model for the feudal economy can be extended be extended more generally to other systems. He assumes self-sufficient regional economies, extensive economic obligations to local elites, and free individual agriculture, limited by labor availability and the fertility of the land. He notes that the short-run interest of the elites is in maximizing the supply of elite goods, produced in 'urban' workshops. In Figure 24, that preference defines a relationship **S** between rural population and urban population that lies above the region where the population is stable or increasing. The locus **E** in the figure represents the urban and rural population combinations that result in a stable total population. The elite preference for elite goods, combined with the relative ease of collecting taxes and tribute from controlled urban areas, leads to short-run elite policy-making to favor rural-to-urban migration.

Rader notes that due to disease and crowding, the death rate in premodern urban populations usually exceeded the birth rate, resulting in a negative urban component in the vector of population change. On the other hand, rural populations below the malthusian limit saw a secular population increase. If sufficient rural-urban migration occured that the population was near the locus S, then the change in total population was negative, which led to a long-term drift towards zero total population.

Without enforced migration, the urban population was small (due to the high death rate) and the total population increased until the population vector approached a purely agricultural limit point at N'. Since the maximum population for a society with both rural and urban elements exceeded the total population at N', some natural rural-urban migration occured as the population approaches N' and the population vector then moved towards a limit at N" (Figure 25). As that drift began, the first formation of craft villages occured

Since elites were able to mobilize enough power to control the agricultural population once the craft villages grew large enough, it is clear why cyclical dynamics occured. This is shown in Figure 26: an agricultural population grew, and craft villages formed. Those craft villages provided a power base for village elites, and their policies encouraging urban good production led to a rapid

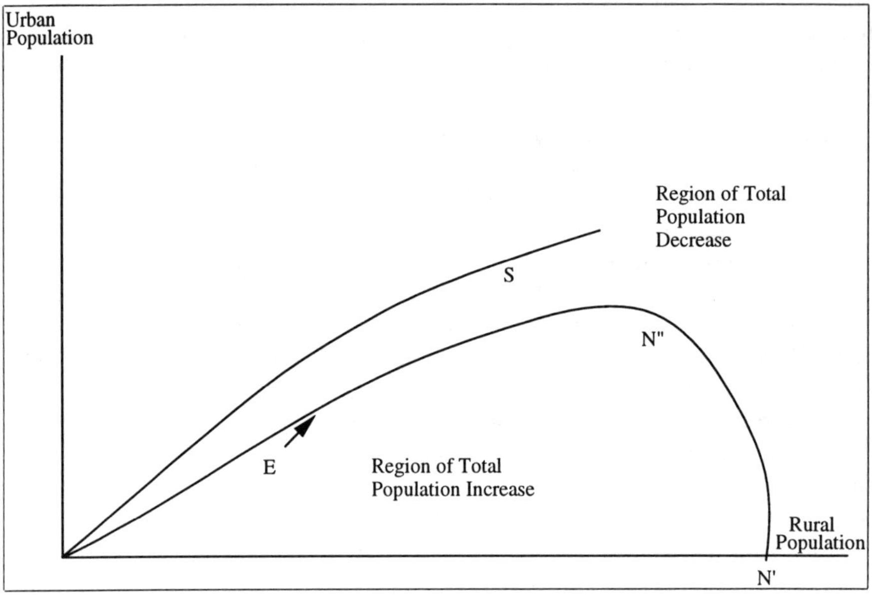

Figure 24: Elite preference locus

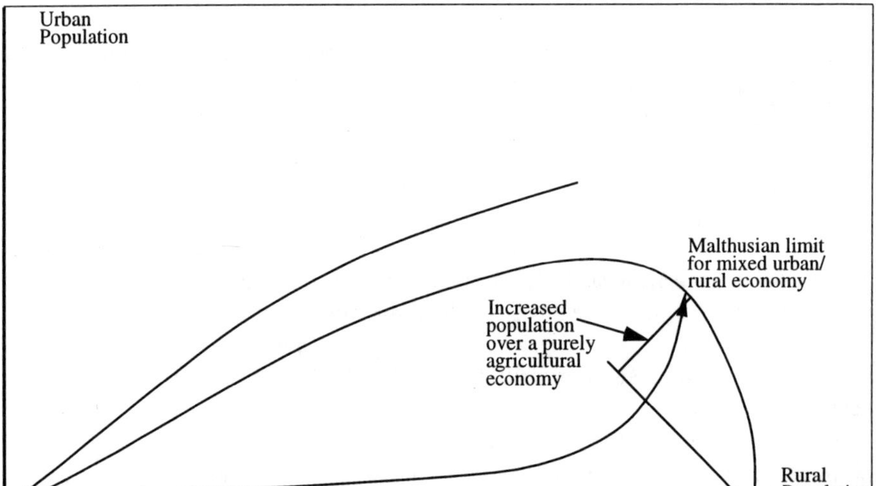

Figure 25: Unconstrained population dynamics

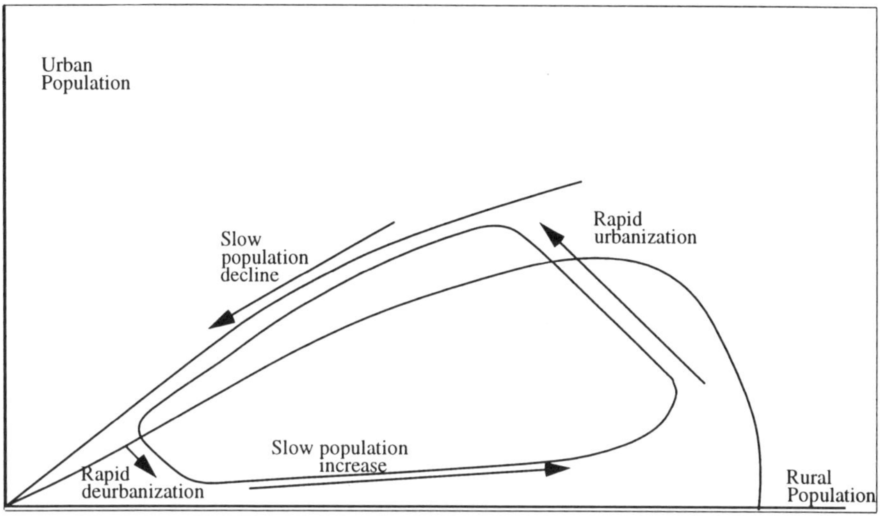

Figure 26: Cyclical population dynamics

population shift from the rural to the urban sector. That moved the community into a zone of net population decrease, but elite control over the increased urban population gave them enough power to continue to shift population out of the rural sector to maintain the urban sector. Regional population decrease continued until control of the urban sector was no longer enough to maintain control over the agricultural sector, and the population rebeled or emigrated to the countryside. System collapse took place, the urban centers rapidly disappeared, and the system returned to an agricultural economy (the first stage of the cycle).

Non-linear dependence of political power on urbanization is necessary to create these dynamics, especially the rapidity of deurbanization. Were political power linearly dependent on urbanization percentage, stability would be associated with the ratio of urban to rural population, not with the population size in either category. Given that, as an urbanized society lost population, the elites could (and would want to) enforce an evolutionary trajectory that eventually arrived at zero population. The historical pattern of system collapse at a non-zero regional population would not occur [17]. The dynamics seen in the historical record appear to be a "blue loop" bifurcation, analogous to a Van der Pol system:

$$\left(\frac{d^2x}{dt^2} + \alpha f(x)\frac{dx}{dt} + x = \beta p(t)\right)$$

with large α and zero forcing, β. This has an interesting implication — if a sufficiently large forcing function is applied to this type of system (representing large-scale migrations generated by the population dynamics of neighboring polities), chaotic dynamics are generated[20]. Those dynamics would have made long-term planning impossible for the elites and would have disrupted social and political relationships. That implies cooperative control over population movements should be observed as a primary concern of elites.

There is some evidence this process occurred in the Eastern Mediterranean basin between 800 BCE and 700 CE. Although classical civilization survived with many of its institutions intact over most of this period, Russell[21] shows that there was a five century population decline in Europe from 200 to 700 CE, and in the Near East the decline may have continued for another 300–500 years. Evidently, although the cities survived to the end of this period, their rural hinterlands became depopulated. Hodges and Whitehouse[22] have evidence that regional populations recovered in Tunisia and Italy once the urban sites disappeared, strongly suggesting a causal connection between urbanization and rural depopulation in the polis during this period. These data suggest that Rader's non-linear models hold for at least this case.

5 Conclusions

Although there are significant difficulties facing the social system modeler, particularly in validating non-linear models, significant insight is still available using topological methods.

References

1. R. W. Hamming. *Numerical Methods for Scientists and Engineers.* McGraw-Hill, New York, 1962.
2. R. M. May. Chaos and the dynamics of biological populations. In M. V. Berry, I. C. Percival, and N. O. Weiss, editors, *Dynamic Chaos*, pages 27–42, Princeton, 1987. Princeton University Press.
3. P. Manneville and Y. Pomeau. Intermittency and the Lorenz model. *Physics Letters*, 75A:1, 1979.
4. S. Newhouse, D. Ruelle, and F. Takens. Occurrence of strange Axiom-A attractors near quasiperiodic flow on $T_m, m \leq 3$. *Communications Mathematical Physics*, 64:35–40, 1978.
5. V. I. Arnold. *Mathematical Methods of Classical Dynamics.* Springer-Verlag, Heidelberg-New York, 1978.
6. Bernardo A. Huberman. Personal communication, 1990.

7. Harry R. Erwin. The dynamics of peer polities. Presented at the Cambridge Conference on Dynamic Modelling and Human Systems, 1990.
8. Harry R. Erwin. The evolution of emergent structures. In G. E. Lasker and G. L. Farré, editors, *Advances in Synergetics*, volume 1. International Institute for Advanced Studies in Systems Research and Cybernetics, 1994. Presented at the Seventh International Conference on Systems Research, Informatics, and Cybernetics.
9. David Beeman. Speeding up GENESIS simulations. In *The Book of GENESIS*, chapter 19. Springer-Verlag and Telos, New York, 1995.
10. Michael V. Mascagni. Numerical methods for neuronal modeling. In Christof Koch and Idan Segev, editors, *Methods in Neuronal Modeling*, chapter 13. MIT Press, Cambridge, Massachusetts, 1989.
11. Harry R. Erwin. Social system dynamics: Memes, genes, and neural networks. Presented at the First Interdisciplinary Conference on Evolutionary Systems, Georgetown University, 1993.
12. Harry R. Erwin. Dynamic models of social collapse. Presented at the Second Interdisciplinary Conference on Evolutionary Systems, 1994.
13. Ralph H. Abraham and Christopher Shaw. *Dynamics—the Geometry of Behavior.* Aerial Press, Santa Cruz, California, 1982–1988. Part One: Periodic Behavior, Part Two: Chaotic Behavior, Part Three: Global Behavior, and Part Four: Bifurcation Behavior.
14. Frederick David Abraham, Ralph H. Abraham, and Christopher Shaw. *A Visual Introduction to Dynamical Systems Theory for Psychology.* Aerial Press, Santa Cruz, California, 1990.
15. M. Harris and E. B. Ross. *Death, Sex, and Fertility.* Columbia University Press, New York, 1987.
16. S. E. van der Leeuw, 1992. Personal communication.
17. J. F. Cherry and Colin Renfrew. Epilogue and prospect. In Colin Renfrew and J. F. Cherry, editors, *Peer Polity Interaction and Socio-Political Change*, chapter 12. Cambridge University Press, Cambridge, England, 1986.
18. Joseph Tainter. *The Collapse of Complex Societies.* Cambridge University Press, Cambridge, 1988.
19. Trout Rader. *Economics of Feudalism.* Gordon and Breach, New York, 1971.
20. J. Guckenheimer and P. Holmes. *Nonlinear Oscillations, Dynamical Systems, and Bifurcations of Vector Fields.* Springer-Verlag, New York, 1983.
21. J. C. Russell. *Medieval Demography.* AMS Press, New York, 1987.

22. R. Hodges and D. Whitehouse. *Mohammed, Charlemagne and the Origins of Europe*. Duckworth, London, 1983.

Part III

CLINICAL STUDIES

LOVE IS A STRANGE ATTRACTOR: THERAPY AT THE EDGE OF CHAOS

ALAN PARRY

Family Therapy Program, Department of Psychiatry
The University of Calgary, Calgary, Alberta, Canada, T2N 4N1

Chaos and complexity have much to offer fields such as family therapy which, in a postmodern world, increasingly valorize disorder, uncertainty and randomness. The symptomatic problem that brings a family to therapy may be understood as a strange attractor toward which the system moves inexorably as if drawn. As such, the "noise" of the problem represents a cry for love and a "demand" that the system self-organize in more complex ways.

A. A violent order is disorder; and
B. A great disorder is an order.
These Two things are one.
Wallace Stevens,
"Connoisseur of Chaos"

"Don't just do something," the Buddha said,
"Stand there."
Father Daniel Berrigan, circa 1970

Chaos and its less unruly sister, complexity, have now begun to spread far beyond their origins in mathematics, meteorology, physics and chemistry into not only biology, but psychology and many of the social sciences. It has also penetrated, albeit from a different direction, into literature and literary criticism, particularly through the practice of deconstruction, which challenges every traditional and modernist assumption concerning writing as an orderly process. Those fields outside the conventional sciences that are beginning to valorize disorder, uncertainty and chance as vehicles by which systemic self-organization takes place in response to large increases in information and complexity find the highly evocative images of chaos and the edge of chaos to be very congenial. No term seems more descriptive of the rapidity and disorienting quality of post-, or what I prefer to call, hypermodern life. A consideration of the deep structures of order that appear to lie within chaos itself, and the self-organizing possibilities that apparently emerge best at the edge of chaos, might offer therapists and families alike a reassuring way beyond the assumptions which have accompanied modernity's technological conquest of the world, namely that a comfortable and contented order is the norm, that disorder, uncertainty and too much complexity are abnormal, even pathological, and must be returned to an order brought about by technologically derived improvements on human nature. Since, in addition, we live in what seem to many to be chaotic times, it might be wise to begin to consider some strategies for making the most of our circumstances rather than going along with much of the contemporary world's longings for yet another imposed return to a simpler order, which its own pace of change and increasing complexity renders impossible.

1 "A Great Disorder is an order."

The exciting and suggestive research discoveries of the first "chaoticians" [1,2], as well as related implications from information theory [3], and the emergence of complexity theory [4,5] especially in the life and social sciences contain exciting implications for therapy in general and family therapy in particular. With Edward Lorenz's now famous, fittingly accidental discovery of what has come to be known as the butterfly effect, chaos theory was born. Its discovery that chaos is not merely "chaotic," but generative of an order of its own has been demonstrated in a host of arresting ways, from the fractal universe of Benoit Mandelbrot to Robert May's intermittencies arising in the midst of the "period doubling route to chaos" to Mitchell Feigenbaum's discovery that when at least some dynamical systems enter chaos they moves toward their "strange attractor" at a universal rate. Chaoticians also took note of such physical phenomena as solitons, massive unities that, once connected by feedback exchanges, seemed to self- organize rather than randomly dissipate as might otherwise be expected. They also took note of the greater value of background turbulence over rigid orderliness within biological systems such as the heart which facilitate the degree of flexibility necessary for them to function optimally.

Strange attractors, which may be likened to "basins" toward which nonlinear systems move as if drawn or attracted, abound as turbulence wherever a system is changing from one form to another, whenever disequilibrium is introduced into a hitherto more or less stable order. A striking example of this for family therapists may be found in Karl Tomm's [6] pathologizing interactional patterns (PIPs) wherein family members find themselves caught in a situation in which each person is reacting to what the other is doing: one might be pursuing greater intimacy because the other is avoiding intimacy and that person is avoiding because the other is pursuing. The current preference for solution-focus in systemic family therapy seems to favor identifying PIPs like these solely in order to identify such "unique outcomes" or "exceptions to the rule" as might enable the family to amplify them, so that a healing interactional pattern (HIP) can more quickly replace their problem-focus and a more enduring wellness pattern (WIP) be established. Systemic and strategic family therapists were not always so "helpful" or determined that order be quickly restored. Inspired by those maverick geniuses, Milton Erickson and Gregory Bateson, such predicaments were often either actually prescribed or positively connotated in the unspoken hope that the family would stop fighting the problem, thereby freeing the system to self-organize to a more complex order. They were invited to embrace rather than flee their own turbulence. As such, family therapists of a decade or so ago might be said to have been operating within assumptions resembling chaos theory if from a slightly different perspective.

Ilya Prigogine[7] won the Nobel Prize in chemistry in 1977 for his work on dissipative structures, systems which, when far from equilibrium, began spontaneously to reorganize themselves into a more richly informed and complex order. In a like manner, a couple or a family caught within the self-referential paradox of one of Tomm's PIPs might be invited to remain there awhile longer, while reflecting upon the "advantages" to be found in the chaos of their particular impasse until the stark awareness of its futility drives them toward the self-organization of a new, more complex way of ordering their relationships.

I have already mentioned the soliton as one of the most intriguing and formidable natural instances of systemic self- organization. When waves that would otherwise either break up in turbulence or dissipate instead self-organize into the majestic power of a soliton, it is the *connectedness* that arises through a weaving of feedback loops that unite them. Smaller waves are coupled into one, apparently communicating with one another, even displaying a kind of organizational memory as this single wave moves on sometimes over hundreds of miles, remaining intact even as it passes through other solitons. In this instance of chaos giving rise to order, the question arises as to what happens at that critical juncture that determines whether a chaotic system self-organizes, or whether it dissipates into a thousand wavelets. According to Prigogine systems that exist in a state far from equilibrium are capable of maintaining themselves only by continuing to remain open to the flux and flow of their surroundings. The determination as to whether a system stays together or dissipates occurs at *bifurcation points* which, given their sensitive dependence on initial conditions, are highly subject to slight random events. At such times they may either cascade into a procession as period-doubling leads it to fragmentation, or stabilize through feedback looping that couples elements within the system, as well as to new changes in the environment. Once stabilized it may remain robust enough to resist further changes for extended periods of time–within the evolutionary process, for instance, sometimes for millions of years at a time.

The operative theme is the "weaving of feedback loops" which "is what Prigogine means by 'communication'."[2] The difference that makes the difference, then, as to whether any turbulent system at a point of bifurcation breaks apart or self- organizes is the presence of a force that keeps its parts "in touch" with each other. In families there is an equivalently powerful force at work that keeps its members connected. It is provided by an *activated* feeling of love or sense of compassion that is either still operative within the turbulence, or cries out to be awakened. Without its presence, further fragmentation in family relationships is the likelier consequence.

2 Chaos as Information

Information theory also belongs at the interface of chaos theory, on one hand, and poststructuralism, on the other. It emerged at the same time and place as cybernetics, in the late 1940s at M.I.T. as an outgrowth of the vital role that the coding and decoding of information played in World War II. Claude Shannon, a Bell Laboratories engineer was to information theory what Norbert Wiener was to cybernetics. Shannon demonstrated that in order for a message to become meaningful it had to contain approximately equal portions of redundancy, or repetition, and novelty, or surprise; too much redundancy and the listener stopped listening- -it was boring; too much novelty and the message was meaningless. It lacked enough familiar references points for the new information to have meaning. When a system, therefore, is confronted by an influx of new information it is being challenged at that very sensitive point of bifurcation; it is in turbulence. Thus it may be said that, for living systems, information replaces energy as the primary operating ingredient and becomes chaotic when it is inundated by more information than its customary ways of decoding are able to process. At the bifurcation point it must 'choose'. For humans the question becomes: Will the choice be to disregard the new information because it is incompatible with or does not fit the familiar parameters, or to expand those parameters to integrate the new demands?

The answer to this question is complicated by the pervasiveness of language and its capacity to construct a world which then tends to retain a high and often rigid, degree of stability over time. That world-construct gets reinforced by the repeated narrativization of peopleUs lives: the oncoming events in their lives tend to be absorbed into a prevailing narrative which is to say, the meaning given the events of a life as read retrospectively. In other words, although a story of a life is lived forward, it is narrated backward. Ironically, then, it seems to be the human propensity for developing a model of the world, a world-narrative in the first years of life, that gets in the way of receiving and understanding new information or its significance, particularly information that does not fit into the prevailing narrative. This is all complicated by a further propensity to equate the model with reality-itself. On account of this, new information is received instead as unwelcome noise to be rejected or excluded as incompatible with what we assume and mistake for reality itself. As a consequence, adherence to the model has the effect of containing, when not actually eliminating, feedback circuitry, our primary resource for effective interaction with others and the environment, in effect turning a nonlinear system in the direction of linearity.

The same language that enables us to compose a world-narrative with which to orient ourselves in the world and provide a familiar framework into which to fit new information also allows us to use feedback with a degree of complexity unavailable to any other living system. As a strange consequence of all this, language gives us the capacity to control the flow of information in the form of feedback as well as to lie and deceive not only others, but ourselves. Why do we engage in what would appear to be such self-limiting and self-defeating behavior when full openness to feedback creates so many possibilities for us? To live willingly in the uncertainty of a nonlinear state is to live with unpredictability. But we are beings who know that death awaits us all. We have the capacity to anticipate the future, probably shared in one form or another by all living things, but to an extent unavailable to any other. We also have the capacity to alter our surroundings to fit our model of the world and to make them safer, more comfortable and secure. In short, we are unique in combining anxiety about the uncertain future with the capacity to alleviate it by imposing order upon the world. It is the source of what Pascal called "the grandeur and the misery of [humanity]".

When we exercise our capacity to select certain information out of the whole and ignore the rest we tend to select the most tangible and regularly recurring features of a situation–its redundancies–disposing of the rest as "noise" and "the random". We tend strongly to assume that the features we have selected are the whole picture, the "truth", reality. Salman Rushdie says in his novel *Shame* [8], "Every story is a kind of censorship. It prevents the telling of many other stories". The experiences we tend to assimilate are those that fit into our world-narrative. Those that do not fit in are the "censored" stories which are, accordingly, dismissed as of little or no significance, random events, luck, exceptions to the rule which can be safely ignored.

There seems to be a contradiction at the very heart of human development. The same world-narrative that is able to serve as a familiar receptacle for providing the redundancy necessary to provide meaning for the complexities, the noise and the surprises of life tends, when a person is anxious, to serve less as a receptacle and more as a fortress. It becomes a bulwark against the very change that it is otherwise so well-equipped for absorbing and which enables it to be continually open to revision. The latter is essential for continuing adaptation and development. Its original assumptions were made, however, when, as children, we were open and resilient, but also helpless and vulnerable. Nevertheless, what was good for childhood invariably proves woefully insufficient for the demands the rest of a person's life imposes.

Tribal peoples the world over have addressed this problem by developing dramatic and sometimes harsh rituals to simulate a death-rebirth experience so that the child might "put away childish things and become a man or woman, responding henceforth to the world with appropriate adult resources." We are left with, at best, the sacrament of Confirmation in some churches wherein, in the Roman Catholic tradition, the bishop applies a gentle tap on the cheek, a pale reminder of the undoubtedly vigorous slap that, in turn, is itself a mere shadow of those rituals where the initiate had, for instance, a tooth knocked out or a penis sliced open.

The stories we live represent ways of ordering the world to make it more meaningful and to enable us to anticipate the uncertainties of the future. The whole of life itself is the chaos of new information that rushes in, overriding our desperate orderings, whenever our familiar stories are found to be insufficient to deal with new challenges or circumstances. A family in such a position is apt to experience this rush of new information as chaos indeed, but of the sort that threatens destruction rather than a new creation. If they come to family therapy to find out what they should do to re-establish their old familiar order, they will if they are lucky, find themselves instead with a therapist who enables them to understand their chaos as opportunity, new information rather than noise to be silenced to keep their old story intact.

A reversal from serving as agents of re-established order to the amplification of chaos would represent a severe challenge to a field which has barely embraced the paradigm shift implied by quantum field theory. Despite its dalliance with cybernetic theory and its recent embrace of all things postmodern, family therapy in many ways still languishes in a largely Newtonian universe where order is the norm and chaos is unwanted, a noisy aberration to be silenced so that order can be restored, in a place where time is reversible and an old order can be returned once the right technique has been applied.

It is, rather, the old order that has become the problem against which the disorder of the identified problem may be understood as a strange attractor, that region of the phase space to which the particular dynamics of that family system pulls it. Chaos rushes in behind the problem as attractor in the form of new information which the family experiences as a disturbing noise, interfering with all the assumptions and structures of the old order. Should the therapist ally with the parents and the old order, she/he unwittingly conspires to withhold the feedback possibilities contained in the rush of new information. The problem brought by the family invariably involves the behavior of one of the children. It represents disorder or noise. *But what if this "noise" is a message that simply has not yet been understood?*

Since a family is a nonlinear system in that it possesses the capacity to respond to random feedback no matter how distressing, it benefits best from the power of information to give it the opportunity to allow a more complex order to organize itself out of the inrushing chaos of new and challenging information. The crucial issue for the family will be what takes place at the bifurcation points where, in the absence of conscious choices or decisions, the family system is highly sensitive to initial conditions and, therefore, subject to random influences that can send it, willy-nilly, in either direction: toward greater fragmentation or more complex self-organization. If the family chaos continues to be interpreted negatively as a threat to a preferred order, further fragmentation is likely. If, however, it is interpreted as a call toward greater love, then the possibility is increased that the chaos of new information and complexity (which the therapist further maximizes by completing more and more feedback loops) will not only become available to the family, but will put such a strain on the old stories that they will have to expand and self-organize in the direction of greater complexity. Love might then replace chaos as the family's strange attractor, the new region of phase space toward which the challenge of the new information guides them, the new relationship complexities they are being called upon to seek out.

In order to facilitate the latter, the therapist who would embrace chaos must keep in mind, not only the importance of a compassionate interpretation or misreading[9], of the family's chaos, but the importance of resisting the almost visceral urge to help the family achieve closure and good feelings as each session ends. Therapy that is willing to work with, rather than against, the family's chaos must be prepared to tolerate discomfort involved in pondering the fuller implications of their choices in order to allow chaos the opportunity to self-organize.

3 From Chaos to Complexity: The Edge of Chaos

Chaos theory for the most part is of value for therapists for what it offers in the way of richly suggestive analogies, connected primarily by the operation of feedback loops and information-exchange in nonliving as well as living systems. More immediate implications show up quite persuasively though in the not quite so unruly sibling of chaos theory known as complexity, or, more evocatively, the *edge of chaos*. Thus, although perhaps not quite as startlingly provocative as chaos theory, complexity probably speaks more to family therapists and other practitioners of the life sciences in that most of the work in this area is being done in efforts to understand *living* systems of every sort, those studied in everything from evolutionary biology to economics in all their complexity. Chaos theory, while very pertinent in that people are apt to seek therapy when their lives feel as if they are in chaos, has tended to have cut its teeth on mathematical and physical systems. As such, too ready an application of its findings to human systems might lend itself to the criticism that such comparisons are premature and even misbegotten.

Just as chaos theory did not so much begin with Edward Lorenz's accidental discovery of the butterfly effect as vividly exemplify rich and unexpected possibilities for change where it was least expected, the edge of chaos was similarly discovered by accident. Christopher Langton, one of the major scientists working in the area of artificial intelligence, discovered, initially by accident, that when you leave ordered territory and enter the region of chaos you cross an area of maximum information exchange which he called "the onset of chaos," but which was later given the more evocative term "the edge of chaos" by Norman Packard, one of the legendary members of the Dynamical Systems Collective that had sprouted up at the University of California, Santa Cruz in the 1970s.[4] As Lewin points out, this region is the place in the physical world at which energy exchange gives way to information processing and the dynamics of systems become most interesting.

Such changes are not only more interesting than what happens in the realm of chaos, but appear to be capable of generating spontaneous self-organization. Once a complex system enters the realm of chaos it is so acutely sensitive to its chaotic conditions that, although it remains subject to deep structures of order implicit in the emergence of bifurcation points, everything is operating so randomly that chance overrides the capacity of the system to optimalize and regulate its own changes. This is so of families in chaos as well. Indeed, intended change at such times is apt to be even more elusive for the sense of urgency and anxiety that chaos engenders in humans tends to give rise to desperate attempts to impose order in precisely those ways that may have brought the family to its present chaos in the first place. At the edge of chaos, however, the capacity for intelligible if unpredictable change is optimal. The closer the system is to stability, on the other hand, the less likelihood there is for it to move in the direction of change by self-organizing differently. It is not being sufficiently challenged by an exchange of new information from other surrounding systems.

Changes at the edge of chaos have as much to do with what systems adjacent to it are doing, the changes they are undergoing, as what it is doing within itself. Changes in natural systems, in other words, are coevolutionary. Each is evolving to the extent that it is dealing with what the other is doing. Doyne Farmer, one of the pioneers in the science of complexity, is quoted by Mitchell Waldrop as proposing that "learning and evolution don't just pull agents to the edge of chaos; slowly, haltingly, but inexorably, learning and evolution move agents *along* the edge of chaos in the direction of greater and greater complexity".[5] In other words, it is at the edge of chaos and not amidst order, predictability and certainty that change most readily occurs and complex systems self-organize.

Jack Cohen and Ian Stewart, the former a reproductive biologist, the latter a mathematician, have joined forces in an effort to join chaos and complexity theory together and, at the same time, to place them within and not without the traditions of modern science [10]. They suggest that the universe operates according to the interaction between two meta-rules which they call *simplexity* and *complicity*. They conclude their ambitious enterprise with this arresting passage:

> Reductionism seeks to explain all patterns in nature, obvious or hidden, as simplexities arising from underlying internal simplicities. We think that many patterns do not fit this description at all; they are complicities, arising from internal complexities and simplicities under the influence of external complexities and simplicities. Because our brains themselves evolved through complicity between their internal representation of reality and the external reality itself-between their content and their context–they can recognize features, analogies, and metaphors, and see patterns in them. Our complicit brains are both made with and aware of genuine complicities in the universe and the ringmaster in our head tells himself stories about them (p. 435).

Per Bak, another of the provocative pioneers of complexity proposes that all large complex or dynamical systems "naturally evolve toward a critical state".[4] For him*criticality* is that state. He agrees that his term is probably "the same kind of phenomenon" as the edge of chaos. Bak's suggestion of a natural evolution, or what evolutionary biologist Stuart Kauffman calls coevolution toward the edge of chaos or criticality as the optimal place for adaptive change to occur, implies an unprecedented challenge to the currently still firmly established worldwide cultural assumption of order as the safest and most desirable place to be. Ever since Darwin, according to Kauffman, biologists have regarded natural selection by the environment as the sole source of the order that emerges in the evolutionary process. Kauffman[11] proposes, however, that in that very process self-organization has evolved as a distinct property of complex systems. It is such self- organizing order that interacts with the selective constraints imposed by the environment that allows complex systems to evolve and adapt.

Kauffman believes that it is more appropriate to speak of coevolution rather than simply evolution because natural selection, in the manner thus described, involves complex living systems continually adapting in response to ongoing adaptations of adjacent complex dynamical systems. "The evolutionary success of one species may . . . be as much a function of what other species do as what the species itself does." He discusses such changes in terms of what he calls "fitness landscapes" or "rugged landscapes." Kauffman likens ecosystems to vast landscapes in which those most advantaged at particular moments loom much higher like lofty mountains and others are lower according to their adaptational fitness at the time.

"Imagine a fly," said Stu. "It has a fitness landscape. Now imagine a frog. It has a fitness landscape, too. But they're not independent. The frog shoots out its tongue, zap, the fly's gone. That's part of life. Now suppose the fly evolves slippery feet so that the frog's tongue doesn't stick. The frog goes without dinner, and its peak on the fitness landscape goes down: it's less fit. The fly is fitter, and so its peak rises. So the coupled landscapes change, each responding to the other." The next step in the story is that the frog evolves hairs on its tongue-or some such device–and is able to catch the fly again. Fitnesses change, landscapes change. "It's the classic biological arms race," explained Stu. "Predator and prey constantly trying to be one step ahead of the other".[4]

Leigh Van Valen, another biologist, has labelled a similar phenomenon the Red Queen effect after the character in *Through the Looking Glass* to whom Alice complained that she had to run as fast as she could just to stay in the same place. Something similar might be said to occur in families. Each member does as she/he does in response to what others are doing. A problem or symptom might be understood as a demand to be heard that has gained a jump on the family's current manner of ordering its relationships. To the extent that the symptom-bearer's "noise" continues, others in the family find themselves stymied, powerless to eliminate that noise by methods currently favored. They will have to "run twice as fast," as the Red Queen says, if they are to "get somewhere else," perhaps by learning how to decode the message in the noise that will not be silenced until it is heard, understood and answered. If and when that happens the family will have coevolved to a more inclusive, hence complex level of self-organization.

Whenever two struggle with each other for advantage on the rugged landscape of interaction each might be expected to call upon whatever resources are necessary to meet the challenge or noise of the other. Maybe chimpanzees do it too, but we humans have raised to uniquely calculated proportions ways of inviting or even being invited by a third party to bolster them against the other, particularly when one seems or feels at risk of losing the struggle. Out of this has arisen the remarkable phenomenon of victimhood as a self-conscious identity separate from the actual tragedy of victimization by misfortune. The first of these, victimhood, often confers on the "recipient" an exemption from all requirements of striving to surmount the rugged landscapes of their lives, rather that someone else "should" rescue them from the challenge of those demands. Many others, indeed, respond, some to the point of symbiosis, others by establishing rescuing as a virtual life-style that often gives a sense of meaning to their lives. The most questionable thing about this uniquely human phenomenon is that it tends to short circuit the very process through which life is enhanced, that is through getting itself to the edge of chaos where it is best positioned to self- organize toward a greater level of complexity and adaptability. Michel Serres [12] discusses this phenomenon imaginatively and at length in his concept of the parasite, the troublemaker, an unlikely image which might serve as a possible alternative for therapists to the otherwise ever present temptation of the rescuing role. Suffice it to say that, whereas two can usually work something out between them, the intervention of a third invariably means trouble–and lots of noise. Will it be a troublesome noise that *challenges* a settled but toxic arrangement, or one that regards the embattled other as either a victim to be rescued or a troublemaker to be punished, thereby consolidating her/him in one position or the other.

Jenny was a noisy, demanding 14 year old girl with a "major attitude problem." The noise she made left her parents in a completely exasperated state and they freely admitted that they could not stand her. Their own relationship as a couple was very comfortable, but felt that their daughter's demanding attitude had rendered them powerless as parents to influence her, hence they wanted to have nothing more to do with her. They feared a bad future for her, but wanted to be no part of that either. Her mother did acknowledge some responsibility for how Jenny had turned out. Jenny took much of her "attitude" to school, but was in no serious trouble anywhere in the community and gave no evidence that she was headed in the dire directions predicted by her exasperated parents. She was simply a noisemaker reacting against overt rejection and exclusion. She was saying by her actions "I won't get out of your face until you say 'I love you!'" even as the noise of her actions gravely lessened the likelihood of her ever hearing those words. We have all met too many of the Jennys of the world, in and out of clinical practice and, alas, frequently in the places her parents feared she was heading. The first step in preventing such outcomes should be in hearing their noise, however raucous, as a cry for love.

4 Love is a Strange Attractor

Its unwillingness to take the role of power in families with sufficient seriousness eventually threw into question the relevance of the purely systemic model in family therapy, particularly for dealing with families in which violence and abuse were salient issues. It failed, in fact, to give credible answers to the questions posed by the feminist and White-Epston critiques on the differential uses and abuses of power within families. Would a systemic approach strengthened by the addition of chaos and complexity theory do any better in addressing such issues?

I suggest that it would and that it is perhaps time to attempt a wedding of the lighter hand of therapists using a purely systemic understanding of families with the sociopolitical awareness of power found in social and narrative constructionist approaches. The emerging realization that there are deep structures of order at work within chaos and a perhaps universal capacity to self- organize toward greater complexity and connectedness at the edge of chaos offers hope and regeneration for a world in which rapid, destabilizing change has become the norm rather than an alarming exception. The world that is described by social constructionists as one in which people find themselves accepting a dominant story which legitimizes their oppression is the same violent order which Wallace Stevens equates with disorder. As such it presents an irresistible invitation to challenge it in the name of the self-organizing possibilities for a richer and more complex order that exist when things are stirred up and even pushed to the edge of chaos. Chaos and complexity, therefore, offer a way of looking again at systemic understanding that allows us to view the problems families bring to therapy, not as disorders to be restored to order, but as attractors that are calling the family to be new to themselves. The therapeutic question concerns one of facilitating an openness in the family that will enable it to regard the problem as an opportunity to be understood rather than as a disturbance to be eliminated by strengthening an old order with its tried and true solutions.

Something not yet understood has burst into the open upon a family. Briggs and Peat propose that: "Instead of thinking of the whole as the sum of all parts, think of it as what rushes in under the guise of chaos whenever

[we] try to measure dynamical systems as if they were composed of parts".[2] Or, whenever we try to treat a family system as if it was composed only of individual members. "Missing information" is rushing in as chaos, throwing a hitherto more or less stable system into disharmony. Unwanted information gives rise to a symptom or problem which is serving as the family's strange attractor, its "shape of uncertainty", the region of "phase space" toward which its energies are drawn. To the extent that family members follow the strange attraction of the problem but, in their rage for order interpret it as residing in one who "has the problem", the overlooked whole will continue to rush in as chaos demanding to be followed. But where? Precisely where their fear attempts to bar the door, but where the therapist must encourage them to go.

If a family problem were understood as a noisy protest that is seeking to follow the pull of love as a strange attractor then it must be listened to rather than silenced, for if silenced the family will be drawn irrevocably beyond the edge into chaos itself where resolution or dissolution are each as apt to take place as the other in that realm where the random is supreme. May we not go further, in fact, and affirm that love is indeed the strange attractor of human life and if it is not followed chaos will follow? Is it too much to suggest further that the family has evolved in human affairs as the primary vehicle of its operation? As such, perhaps we can regard the institution of the family as the foremost among all human developments to offset that other strange attractor, death. Indeed, the very capacity of life to propagate itself represents its victory over that other attractor.

Cohen and Stewart [10] state that: "An attractor, in general, is a region of phase space that 'attracts' all nearby points as time passes. . . If you watch a typical dynamical system and wait for a while, it ends up on an attractor" (p. 205). They add: "Attractors are the things that the dynamics converge toward if you wait long enough; but once they reach the attractors, they promptly diverge again–and drastically" (p. 206). Stable systems follow point attractors according to fairly discrete and predictable patterns. "Chaotic systems," on the other hand, "wander irregularly over fractal attractors" along elegant but unpredictable paths (p, 205). "It is only possible to know what the attractor looks like by watching where the initial states go" (p. 207). Nor do systems somehow "know" in advance where they are going, rather they seem to be pushed around in accordance with certain rules. "The existence of attractors," Cohen and Stewart conclude therefore, "does not imply that dynamical systems are goal-seekers: on the contrary, they are goal-*finders*, which only recognize what the "goal" is when they have found it" (p. 431).

Toward which attractors, then, are human systems invariably moving? As the only animal that knows that its end is death and whose life, therefore, consists in a significant way of ignoring, avoiding or postponing its inevitable arrival, death may be said to be even more of an attractor or basin toward which all life is irrevocably moving. Until and against that day, however, we seem to be drawn with at least equal inexorability toward love of, and by others– and failing love, or love failing, toward power. The institution of the family may be described as a way of defying death and making ourselves immortal by passing on a legacy according to which we may not only be remembered but continue to influence others in, it might be hoped, life-enhancing ways. It is in the family that we first of all learn about love by receiving it and how to love in return. At least that is how it is "supposed" to be, what perhaps we are even genetically programmed to expect. Of course, when we consider the family it is not the modern nuclear family that is meant. The latter is an historical aberration, an accommodation to the primacy of economics of the kin-based system of relationships known as the extended family. Be that as it may, the family, as Glantz and Pearce point out, "is a kin-based system that once prepared children for a world based on reciprocity through exchange". The exchange of love is probably the child's first experience of the reciprocity that Glantz and Pearce suggest to be fundamental to the centrality of interaction for personal and social well-being in the human species. Thus, whether experienced or not, I would suggest that love is followed as a strange attractor wherever human life is found, indeed where, in some equivalent form, perhaps where any life is found.

Where, however, love is wanting or feared and life begins to follow the inescapable strange attractor of death, the illusion looms that it can be postponed or even denied through the accumulation of power over people or things. I say illusion because such power is but death in disguise. Ernest Becker, in fact, in his dark sequel to his uplifting classic *Denial of Death* (1972), *Escape from Evil* (1975), proposes that humanity's propensity for mayhem and monument-building alike proceed from the determination to use power instead of love to keep death at bay and make ourselves immortal.

Sex and procreation, at the biological level, have always represented an even more basic answer to the strange attractor of death. If power over others has seemed like the promise of immortality without the risks attending love, love for and by the other represents a different kind of defiance of death, the risk of suffering through the possibility of loss. By allowing ourselves to be led by the strange attractor of love, with all its attendant risks of pain and hurt, we are choosing to live on the edge of chaos. When we take an even greater risk, a uniquely human one as far as we know, and give love, we are truly at the edge of chaos. On the other hand, the temptations of power and certainty, to use Amundson, Stewart and Valentine's apposite phrase [13], represent a retreat from that life-enhancing edge back into the illusions of an imposed order-remembering once again Wallace Steven's line that serves as part of the apothegm for this paper: "A violent order is disorder . . "

According to Kauffman, the success of the evolutionary process itself is an exemplification of what he refers to as "order for free." Lewin goes on to quote Kauffman: "A coevolving system," he says, "gets itself to the edge of chaos" which is "a *favored place to be*" because "that's where computation is maximized" and because "the system optimizes sustained fitness there" (p. 186). The edge of chaos may be scary where the risk of letting ourselves be "out of control",[14] where death itself may be our companion, but its very challenges apparently bring forth optimal efficiency and adaptability. If this be the case and both love and death are to be found there, as the place where coevolving systems make sure they get themselves, the edge of chaos is, in fact, life's constant attractor.

If the love that transcends death at the edge of chaos is humankind's strange attractor, the control which the imposition of order promises is our great temptation. Gregory Bateson[15] equated these with what he called "conscious purpose" which, he insisted, interfered with the self-correcting feedback loops that otherwise spontaneously connect us to the larger social milieu even when they are outside the necessarily limited range of our conscious purposes. Things go awry for human beings once we begin to confuse "the unconsciously edited version of a small percentage" of what we see with "the larger interactive system" of which each of us, or our particular family or tribe are only a part. Bateson's self-correcting feedback loops seem virtually identical to what Kauffman has in mind with his concept of self-organization in coevolving systems at the edge of chaos. Both versions, involve an attitude of trust, a willingness to yield to the self-organizing "wisdom" (which Bateson termed the recognition of the operation of circuitry) that lies beyond the control to which the limited and self-interested perceptions of conscious purpose is so readily drawn. A crucial human conflict takes place between the attraction to the order and illusion of power that conscious purpose seems to promise us, on one hand, and the edge of chaos where fitness is actually optimized by the total organism and its connections with its coevolving ecosystem, on the other.

The yielding that I suggest above implies a kind of surrender to the unknown possibilities of an order not yet seen that can be trusted to take place, not only at the place where we feel least in control, but where, as a consequence, we are apt to feel most anxious: the very edge of chaos. Indeed, Kauffman's "order for free" sounds suspiciously and wonderfully like the religious word grace only that it refers to processes imminent in the order of things themselves. In psychological terms some of what I am describing might seem merely like one more version of the interplay between the more purposive "conscious mind" of every dynamic psychology since Freud and the mysteriously active ever- connecting "unconscious mind," more, however, in the creative, healing potentialities emphasized by C.G. Jung and Milton Erickson than the realm of repressed, mostly antisocial urges proposed by Freud. I am even more struck, however, by the resemblance to the distinction family therapists invariably encounter between the illusion of individually autonomous perception that tends to be enhanced by purposive consciousness, on one hand, and the strong connectedness that spontaneously emerges and continues to operate between family members and in other important relationships. Such bonds as these, together with the propensity complexity theorists are discovering for an inherent movement toward, rather than away from, the edge of chaos where more complex self-organizing is able to take place, provides the basis for a massive reappraisal of therapeutic practices.

5 Doing Therapy at the Edge of Chaos

Judith, a 45 year old woman, and her 15-year old daughter, Helen entered family therapy in response to a crisis between the two of them that had resulted in Helen leaving home angrily and going to live in the home of her best friend. Judith also had two older sons who refused to participate in therapy because they were angry with Helen for the distress her actions were having on their mother. Helen's father had died ten years earlier of cancer. His death, when she was only five years old, had long been felt by her as an exclusion of her from the rest of the family in that she alone had very few memories of the man who had been such a loving husband and father in this family.

The relationship between Judith and Helen had been close, but often tumultuous. Helen returned home after having stayed with her friend for six weeks. A honeymoon period ensued in the midst of which I saw mother and daughter who wanted to understand better what had happened in order to prevent such a thing from occurring again. In the absence of any sense of urgency I had scheduled another appointment a month hence. I received an anxious call from Judith three weeks after that session, informing me that Helen had taken an overdose of tylenol tablets. The session that was immediately scheduled was extremely intense and grim. Try as I might, I was unable to construe any middle ground scenario that could bridge Judith's intense pursuit of reassurance regarding her daughter's suicide attempt and her own mothering, and Helen's ambivalent but insistent continuing withdrawal from her mother. The session ended by the necessity imposed by time before I could propose an intervention that might afford them some sense of hopeful closure. Both left angry with the other and very discouraged. I felt as though I had left them quite literally dangling at the edge of chaos.

I wondered and worried about them, the more so due to Helen's recent suicidal disposition, reassured only by Helen's readily given agreement not to attempt suicide before, at least, the next appointment. Yet, when I went to greet Judith and Helen in the waiting room to begin the session it was immediately evident that something quite remarkable had happened. They were sitting together, faces relaxed but animated by smiles, chatting amiably. In the session they explained to me that they had indeed left the session in a state close to hopelessness. Matters even continued to deteriorate further until the weekend. Judith decided that, as much as she wanted Helen back home and a happy and willing part of the family, she simply could no longer endure the pain involved in trying to bring this about and having her daughter react as though she was imposing something unwanted on her. She would make what arrangements she could, including the possibility of Helen living permanently with her friend outside the family. She phoned Helen to inform her of her decision. Helen asked her mother if she really meant that she wanted her to live elsewhere and, more importantly, did this mean that she had given up on working things out between them. Judith was nonplussed by Helen's questions. No, she did not *want* her daughter to live elsewhere and, *certainly*, she wanted, more than anything in the world to work things out between them.

Her mother's action, followed by her clarification of her heartfelt desire for reconciliation, was received by Helen as a new possibility, namely that her mother was no longer insisting, as she had believed, that their struggle had to be resolved with her still in control, or not at all. This apparently meant to her that her mother's imposition of *her* order did not have to be the only way open to her as the price of reconciliation. She no longer had to resist and remain on guard. She and her mother could now talk and negotiate person-to-person rather than from the parent-to-child position from which Helen had felt at an increasing disadvantage.

I learned from Judith and Helen the importance of leaving even a family on the edge of chaos, the time and the space in which to make their own choices, out of my control. That choice, however, had best be made within an awareness that is made clear to them of the likely consequences of a reactive rather than a reflected choice. When a family is on the edge of chaos, the choice made at a bifurcation point seems more apt to be one that will provoke self-organization, leading to a higher level of family complexity to the degree that its members are made aware of the effects of their actions on the others involved and how others in the family are interpreting one's own actions. Equipped, then, with as much information as possible in the form of feedback, they can be asked to make their choices on the basis of both the unwanted consequences of continuing on the present problem-saturated course, and of misreading, instead, each other's motives lovingly. All things being equal, a family at the edge of chaos is more likely to make choices at critical points that further fragment them *to the degree that its members strive to impose their order on others while misreading their motives negatively*. All interpretations are misreadings. It is difficult enough for a human being to know oneself truly. It is impossible to know the motives of another. Hence we must choose our inevitable misreading with care.

A family in which the choice is made of a loving misreading is freed thereby to live with the uncertainties introduced by the bombardment of new information and the degree of complexity that a problematic situation is calling for as the next step in a family's development. It is love that provides the viscosity necessary to allow the individual members to reform in the more complex way that new circumstances are imposing on them. Where anger, anxiety, and hurt elicited by an escalating problem tends to provoke fragmentation out of fear and antipathy toward differences, letting themselves be drawn by love as a strange attractor allows the family to come together in appreciation of the complexities involved in living with each other's differences. It alone has the power to replace chaos as the attractor that is pulling the family's turbulence into a new form.

6 When the Pull to the Edge of Chaos is Resisted

While an idea such as self-organization within the context of the evolutionary process may be very inviting it, in turn, invites the question: What is it that organizes itself in families? Perhaps even more to the point, what *prevents* self- organization from taking place within persons, families and larger social systems? I will, first of all, proceed on the postulate that love *is* a strange attractor in human life, that is, all things being equal, human beings are not indifferent to one another, rather are drawn toward each other, not only in a simply affiliative sense, but according to a fundamental yearning to recognize and be recognized by an other as well as to cherish and be cherished. I will assume, in addition, that the institution of the family has emerged in human relations as the most likely place for this yearning to be born. In the absence of the kind of love participation in a family promises, the individual human being is rendered dangerously vulnerable to the inescapable biological attractor of death–not, alas, only in a surrender to her/his own death, but through indifference and violence to the other.

The family in the example given above had entered a developmental crisis. This family was able to be pulled along the edge of chaos because its members were struggling to learn anew that their only hope lay in their willingness to trust in the power of the love that had been established between them. They discovered, especially through Helen's suicide attempt, that attempts at trying to impose a familiar order, based on an earlier successful phase of family development, upon the errant member were not only counterproductive, but in this case, could have proven even fatal. Once this message was received Judith, in particular, chose to wait out the uncertainty involved, then made an offer which involved letting go of her own wishes in the hope of at least eventually restoring her relationship with her daughter. Her offer freed Helen to make it clear that her own wish was to maintain that loving bond.

What would have prevented the happy outcome which occurred? In a word, had they continued each trying to overpower the other with the rightness of her own position: the mother insisting that it was *right* for her daughter to return home immediately, and the daughter that it was *right* for her mother to let her exercise her freedom to choose where she lived. Ironically instead, once Judith surrendered the control she never had and gave Helen that choice, Helen chose to return home.

It would appear that the unilateral imposition of one person or group's order upon another in the absence of their consent (whether implicit or explicit) is a violent order which gives rise to disorder. On the other hand, a willingness to risk disorder in the name of love is likely to give rise to a new order. The culprit in the imposition of order appears to be the propensity of purposive consciousness to "do something" when its own preferred order is challenged. What it wants to do at such times tends to involve the restoration of that prior order. Since the threatened disorder is apt to involve the preference of another for a different order the resulting clash of "orders" makes the violent order which is disorder all but inevitable. It is a willingness to risk and even tolerate disorder by a loving consideration of the preferences of the other that makes a new order possible based upon a *coevolved* self- organization.

Purposive consciousness is a remarkable instrument of human accomplishment. Its fateful defect is that it is a property unique to the individual person who is condemned to follow the biological attractor of death. David Bakan [16] has proposed that human life involves a dialectic process between what he called a mode of *communion*, which seeks participation of the individual in a larger body, and a mode of *agency*, in which the individual seeks his/her own advantage in acts of self- assertion. He further suggests that agency unmitigated by communion is nothing other than a demonic force in human life: what Freud was groping for with his controversial concept of the death instinct, what the great monotheistic religions mean by Satan; it is Goethe's Mephistopheles and, as such, the driving force behind modernity itself, including capitalism, science and individualism. Unmitigated agency, in short, is the impulse associated with the attempt to dominate the world and to control others rather than, as in communion, to join with them. It is interesting then that many of the dominant approaches in family therapy today, the narrative and the solution-focused in particular, place a heavy and assumedly liberating emphasis on the element of personal agency, of dealing with a problem by "doing something", with therapy itself an exercise in how to do things differently.

Finally, then, in answer to the questions, "What organizes itself in families and other groups?" and "What interferes with such self-organization?", I propose that love coevolves the family to increasingly more complex levels of organization. This is evidenced by the willingness of family or group members to listen to one another's perspectives and to be prepared to moderate one's own accordingly by increasingly involving the meeting and melding of individual perspectives. This is an ever- complexifying process as the opinions and perspectives of younger members are more and more included as their maturity is demonstrated and acknowledged. The greatest obstacle to a coevolved self-organization within families is the agentic activity of purposive consciousness–the wild card of evolution, "the old Adam," socially constructed as the ego and easily aroused by anxiety at virtually any prospect of losing control- confusing its necessarily limited and self-justifying perspectives with reality itself.

7 "Just stand there."

Perhaps, then, therapists might be advised to respond less to our own and the family's anxieties to "do something" and more to the formidable challenge to "just stand there" in order that the family's systemic capacity for self-organization may be liberated from a constraining orderly disorder to follow the strange attractor of love through that disorder to a new, more complex order. How does a therapist "just stand there", much less an anxious and strife-ridden family? By remembering, in the first place, that the urgency to do something is almost always a reaction to a strong negative emotion such as anxiety. Such emotions tend to have evoked scenarios of old stories with unhappy endings which those involved have been seeking to avoid at all costs. Such scenarios are inherently self-fulfilling. To offset an urgency to do something based on the past and to offer a way of regarding the present with more hope, the joint creation of a "re-vision" of the family's stories[9] might be advised. Such a re-vision would give meaning to the family's crisis in terms of the age-old human experience that all the most enduring and absorbing stories are about life where it is most alive, at the edge of chaos, the role that myths and tales of old once had before they lost their canonical status. A reversal of fortune takes place: a once-tranquil order is rudely shattered, chaos looms, but by struggling with its trials and tribulations a new and happier order of things eventually emerges.

References

1. J. Gleick, *Chaos: Making a new science*, (Viking, New York, 1987).
2. J. Briggs and F.D.Peat, *The turbulent mirror: An illustrated guide to chaos theory and the science of wholeness*, (Harper and Row, New York, 1989).
3. J. Campbell, *Grammatical man: Information, entropy, language and life*, (Simon and Schuster, New York, 1982).
4. R. Lewin, *Complexity: Life at the edge of chaos*, (Collier Books, New York, 1992).
5. M. Waldrop, *Complexity: The emerging science at the edge of order and chaos*, (Simon and Schuster, 1992).
6. K. Tomm, *The Calgary Participator* **1**, 21-24, (1991).
7. I. Prigogine and I. Stengers, *Order out of chaos: Man's new dialogue with nature*, (Bantam, New York, 1984).
8. S. Rushdie, *Shame*, (Viking Press, New York, 1983).
9. A. Parry and R. Doan, *Story re-visions: Narrative therapy in the post-modern world*, (Guilford Press, New York, 1994).
10. J. Cohen and I. Stewart, *The Collapse of Chaos*, (Viking Press, New York, 1994).
11. S. Kauffman, *The origins of order: Self- organization and selection in evolution*, Oxford University press, New York, 1993).
12. M. Serres, *The parasite*, tr. L.R. Schehr (Johns Hopkins Press, Baltimore, 1982).
13. J. Amundson, K. Stewart and L. Valentine, *J. Marital Family Therapy*, **19**, 2, 111-123, (1993).
14. K. Kelley, *Out of Control*, (Addison-Wesley, Reading, MA, 1994).
15. G. Bateson, *Steps to an ecology of mind*, (Ballantine Books, New York, 1972).
16. D. Bakan, *The duality of human existence: Isolation and communion in Western man*, (Beacon Press, Boston, 1966).

THE SELF-ORGANIZING PSYCHE: NONLINEAR AND NEUROBIOLOGICAL CONTRIBUTIONS TO PSYCHOANALYSIS

A. H. STEIN

200 West 70th St., Suite 9K, New York, NY, 10023, USA

Abstract

Sigmund Freud attempted to align nineteenth century biology (and the dynamically conservative, continuous, Newtonian mechanics that underlie it) with discontinuous conscious experience. His tactics both set the future course for psychoanalytic development and introduced seemingly intractable complications into its metapsychology. In large part, these arose from what we now recognize were biological errors and dynamical oversimplifications amid his physical assumptions. Their correction, brought about by integrating nonlinear dynamics and neurobiological research findings with W. Bion's reading of metapsychology, fundamentally supports a psychoanalysis based upon D. W. Winnicott's ideas surrounding play within transitional space.

1 Physical and Metaphysical Foundations of Classical Psychoanalysis.

1.1 Closed, Linear, Conservative Systems and Dualism

The mechanical and mathematical underpinings of Freud's strategies for aligning nineteenth century biology with discontinuous conscious experience emerged during the Sixteenth century when Newton extended Galileo's celestial mechanics to earthbound pursuits using the newly discovered calculus. He developed sets of partial differential equations that described the general motions of planets revolving around the sun. The mathematical solutions to particular problems involving more than two bodies interacting under gravity, however, proved unmanageable; they were only resolved in 1963 with the advent of the computer and the notion of "chaos," a mathematical concept unknown to science until quite recently.

Though Newton never intended his mathematical efforts to address causal issues, they became widely used in this regard because, in addition to providing fair approximations to various many-bodied problems within closed systems, they asserted our analytical mastery, i.e., that problems of almost any kind could be solved by differential equations. His efforts were also readily integrated with acquired philosophical notions that inferred the necessity of transcendental *eidos* for organizing inherently "dumb" matter.

Linearity, Independence and Predictability

This "world hypothesis" [1] presumes that simpler elements make up more complex systems and carry on independently of each other, i.e., they do not feedback to effect one another. Their contributions are additive, i.e., linear, and their trajectories are differentiable with respect to time, i.e., they are "smooth" and continuous, and they do not "double back" upon themselves. They can be described by differential equations of the form $\frac{dX}{dt} = a(\lambda)$, where, a, is a differentiable function and λ a set of "tunable" parameters, including time, t.

Figure 1 depicts a graph of the linear function, $X = a\lambda + b$. Any two points

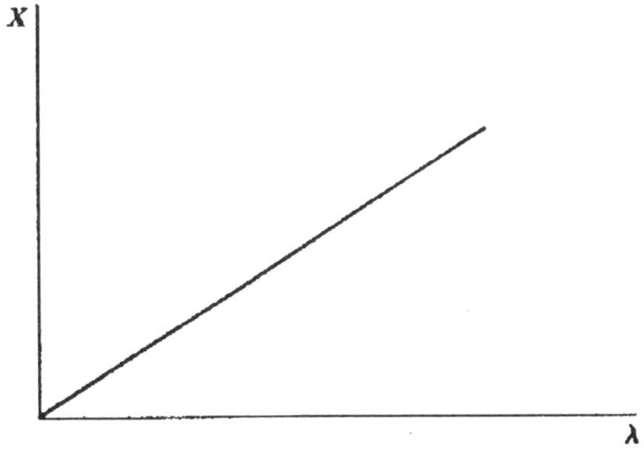

Figure 1: The linear function $X = a(\lambda) + b$.

on the line, or one point and the slope, determine the line exactly. Also, the addition of any two values on the horizontal axis (the independent variable) gives rise to unique points that lie upon that continuous trajectory. This implies that there is one and only one possible value on the vertical axis (the dependent variable) that corresponds to a value of the independent variable; it is analogous to the solutions of a linear differential equation, though "linear" does not have the same meaning in each.

It is not only straight lines that behave so nicely. Quadratic functions of the form $X = a\lambda^2 + b\lambda + c$, where a, b and c are known constants, have the property that knowing any three points on it determines it completely. On the other hand, if all we know about a function is that it is continuous, no finite number of points on the graph will let us determine it exactly. This is rather like the case in solutions of nonlinear equations. See Figure 5 and discussion, below.

In physics, the world is modelled *via* differential equations. In order for such equations to be tractable, that is, admit exact solutions, it is often necessary to make idealizing assumptions which ignore some features of the "real world." Forces such as friction, which dampen ideal dynamics, properly belong in equations that describe such a world; they don't exist in celestial mechanics, however, and, for simplicity are generally omitted from earthbound ones. This provides us with equations that, while only approximate, are fairly easy to handle.

A time series of such an ideal system (in this case representing a pendulum) would look like Figure 2, where, θ, indicates position. Figure 3 removes time

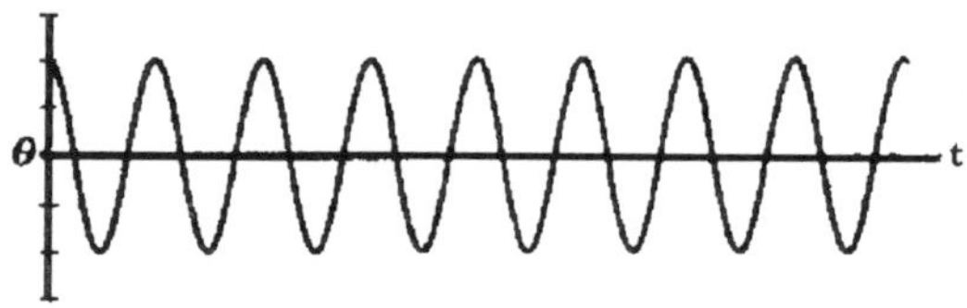

Figure 2: A time series for an ideal pendulum's position as a function of time, t, $\theta = \sin t$.

as a dimension and plots only the pendulum's velocity ($\frac{d\theta}{dt}$) against its position (θ); such a representation describes the entire system at a given moment and is called a *phase space* diagram. For the ideal pendulum, in dynamic equilibrium, its trajectory is symmetrical, in this case circular, and it persists forever since there is nothing to move it, slow it down or stop it.

Ideal systems are *conservative* which simply means that, as shown in Figure 3, the area within the circle is conserved over time ($A_1 \equiv A_2$)—it does not increase or decrease. Time, in such conservative systems, is reversible in the sense that one can reverse direction and as confidently move from later to earlier as from earlier to later, assured that the same pairs of independent and dependent variables will be retained—the system is ahistorical. The X value at any given time is given by $X_i = a(\lambda_i)$, where X_i indicates a particular value for λ_i, at time, i, and a does not evolve.

Such systems are *closed*, which means that they do not exchange energy with what lies beyond their defined boundaries. Since, over time, no additional energy can be supplied to closed systems, they spontaneously equilibrate to states of highest entropy. This final "teleonomic" fact became joined to Plato's and Aristotle's conviction that "raw matter lacks the capacity to develop...complex form...."[2] and its corollary that organization requires activity by some external, metaphysically locatable, extramechanical *eidos*; this,

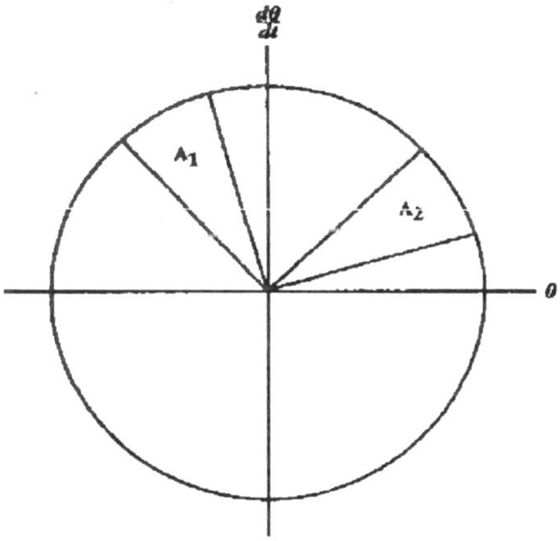

Figure 3: A phase space diagram of the ideal pendulum whose time series is depicted in Figure 2.

in turn, gave rise to "teleological" or "equilibrium seeking" interpretations[3] of organization and order.

All these ideas were sufficiently familiar to the Nineteenth century European intelligentsia that in "The Future of an Illusion" Freud[4] could argue for psychoanalysis' "impartiality" as an observational instrument, by comparing it to the "infinitesimal calculus," who's premises it largely shares. Yet, while he hoped to argue that, like the calculus' differential equations, psychoanalysis makes no *a priori* hypotheses about what it observes and understands, *both do hypothesize!* And, since holding too closely to any hypothesis, including these, without appreciating the limits of its applicability, constrains our capacity to candidly understand the objects of our study without distortion, I will consider their limitations in light of their suppositions. Only the questionable assumption that time is always reversible, for example, supports the psychoanalytic premise that regression can give up personal developmental history and the possibility of localizing early trauma. A nominal step then expects that interpretation, which can link personal and presumed genetic events (see below), can lead, thereby, to cure.

Additivity of independent factors supports the patient-on-the-couch producing material independent of the environment, including the analyst-in-the-chair. The ostensible equilibrium-seeking behavior of closed systems supports

Freud's idea that the psyche seeks equilibrium as a normative state and that disruption of equilibrium leads to pain.[5,6] Any energy arriving from outside the system which could penetrate its boundary and compromise its "closedness" must, then, be either immediately discharged without effecting the system's history and state or elements within the closed system must operate upon it in some way so as to neutralize its potentially disruptive influences; these mechanisms support *eidos*, in the form of race, culture, parent, analyst or what would later become ego to organize id, as "dumb matter," unable to organize itself.

1.2 Freud's Search for eidos and the Origins of Psychoanalysis Proper

Freud began his search for *eidos*' representatives pertinent to psychic functioning and their mechanisms of operation in his "Project for a Scientific Psychology,"[7] wherein he proposed "attention" and "biological experience" as organizers. In "Totem and Taboo"[8] he proposed the Oedipus complex as an organizer; he generally fleshed this idea out in the "Ego and the Id."[9]

In the "Project," Freud sought to explain the operation of the psyche in terms of nineteenth century biology. It failed, as Sulloway[10] points out, when it became obvious that to succeed, it required metaphysical elements. Consider Figure 4, adapted from Freud's diagram,[11] in which he attempts to depict the process of defense against unpleasure, i.e., repression.

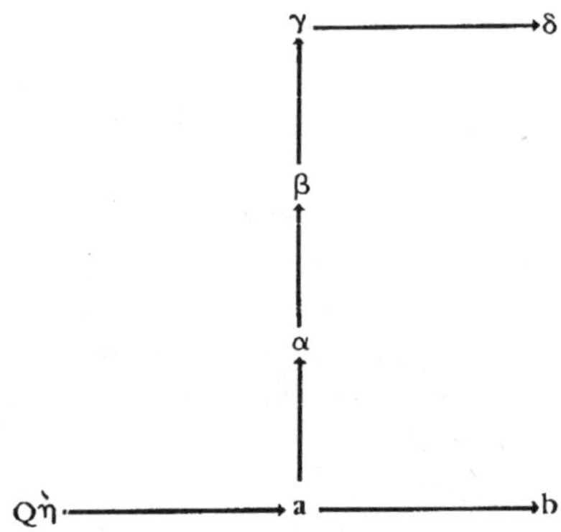

Figure 4: An abstraction of Freud's diagram describing repression.

Arriving from the left, a, is a "$Q\grave{\eta}$"-loaded memory trace of positive or negative valence. If it is negative, it must be repressed by some mechanism before it arrives at b, a "key neurone" and potential source of unpleasure, that could disturb equilibrium. A non-negative trace, on the other hand, can proceed directly to b and, perhaps, even reach consciousness, unmolested. Freud posited that "attention" and "biological experience" could account for repression; together, they could somehow "look ahead," in the manner of Maxwell's demon, and divine the valence of the memory trace before it reached its functional evaluation site, also at b, so that only images of non-negative valence could pass on to b, unmolested, while those of negative valence would be "laterally cathected" along the path $\alpha, \beta, \gamma, \delta$, etc., and, thus, prevented from reaching b or consciousness where they would disrupt the system. Even Freud, however, realized that, like Maxwell's demon, "attention," and similar *eidos* forms, violated physics' conservation laws and had to be abandoned. It was at this point, according to Sulloway, that Freud forsook the "Project."

Beginning in "Totem and Taboo," Freud developed and extended the idea of "biological experience" by proposing that the Oedipus myth, originally understood as a mythic/symbolic tale, be understood, instead, as an inherited, universal, biological experience that operates intrapsychically; dynamically, it functions as an internally consistent symbolic calculus. When coupled to the differentiable dynamics of the system's symmetrical biology, the two become a single continuous system that "biological experience" entrains[12], driving the latter and causing it to bifurcate into momentary asymmetrical "stable states" which correspond to our discrete contents of consciousness.[13,14]

This argument reduced the importance of the manifest content of psychic experience in favor of "biological" causes within the mythically ordered, dynamic unconscious and gave rise to what has become the "main trunk" of psychoanalysis' development—a competition among various mythic structures, devised by its founders, Freud, Jung, Klein, Adler, etc., and their disciples, for which unconscious model best drives the system's continuous biology, i.e., most closely gives rise to observable behaviors including the experienced discrete contents of consciousness.

Such coupling became and remains necessary only as a consequence of Freud's premise that biological systems, including human ones, are closed, differentiable and conservative. Yet, evidence that this is neither the case for most natural systems nor for all biological ones abounds. I will presently consider the biological data developed by neurobiologists Wolf Singer (on the cat visual cortex) and Walter J. Freeman (on the rabbit olfactory system) who convincingly demonstrate that biological systems are, instead *open, nonlinear* and *dissipative* and the consequences of this for psychoanalysis. But first I will review the nature of open, nonlinear and dissipative systems.

2 Alternate Physical and Metaphysical Foundations for Psychoanalysis

2.1 Open, Nonlinear, Dissipative Systems, Chaos and Self-organization

Unlike systems we have considered so far, open, nonlinear, dissipative ones do exchange energy with their environment, they all potentially have multiple dependent values associated with given independent ones—their actual values appear only in the moment, as a result of their particular histories and how we choose to examine them. Unlike pendula in frictionless environments that go on forever without energy supplied from outside, they stop under the effects of gravity and friction and, *when they are far-from-equilibrium, they self-organize*, making the imposition of mythic or physically impossible mechanisms to drive them unnecessary and alien.

Figure 5 illustrates the fixed points of a nonlinear dynamical system as a function of the parameter λ.

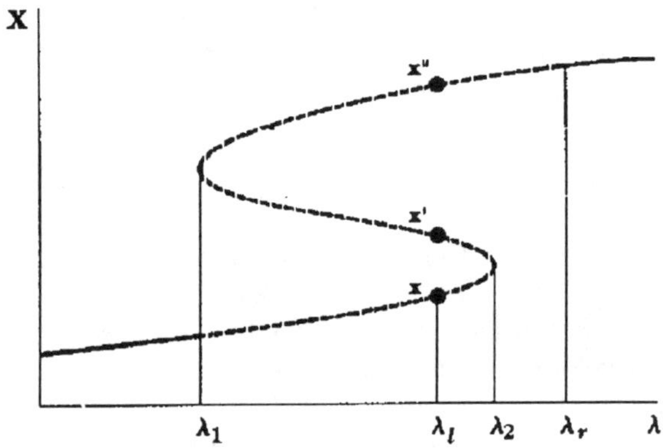

Figure 5: The fixed points of a nonlinear dynamical system as a function of the parameter, λ.

Assume that the parameter λ is varied slowly so that the trajectory of the system remains close to the associated fixed point. Assume also that the system is on the lower most portion of the curve and that the parameter λ is increasing. The value, λ_l, is allowed to increase to λ_2 and then slowly decreased back to λ_1. What happens? As λ_l decreases, the system can "choose" among an infinite number of fixed points, three possible ones being, x, x' and x''. Sometimes it will return to x, retracing its original trajectory. Sometimes, however, it will move to x' or x'', differing markedly from its original course.

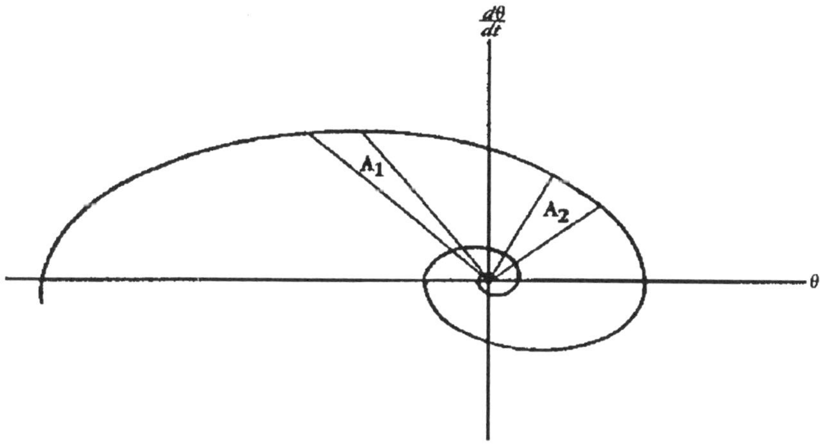

Figure 6: A state space diagram of a dissipative system appears not as a circle, but in this case, as a spiral that eventually approaches a fixed-point attractor at the origin.

If λ is interpreted as time, then the trajectory from λ_1 to λ_2 is not the same as from λ_2 to λ_1. In such nonlinear systems, time is not reversible and history is not preserved; regression will provide us neither with "lived" nor objective, historical experience. This raises significant questions regarding the value of regression and genetic interpretations.

Dissipative systems do not remain within specific equilibrium conditions since they seek no single equilibrium condition. In natural environments, unless they are open and therefore supplied with energy from without, pendula do not remain in continuous dynamic equilibrium but eventually come to rest. The state space diagram of such a system (Figure 6) appears not as a circle, but as a spiral that eventually approaches the origin. Unlike the case of the frictionless pendulum, above, the area of the original state space is not conserved, but "dissipates" as it approaches its origin, $(A_1 > A_2)$. Here, the origin serves as, and is therefore called, an "attractor" because it seems to attract viable elements of the pendulum's motion (constituting the pendulum's dynamic *basin*) to itself.[15]

Attractors are characteristic of dissipative systems. In the case of the dampened pendulum, the attractor is a single point. There are other kinds of attractors: a limit cycle attractor which is depicted two-dimensionally in Figure 7, a two-dimensional torus (and its non-orientation preserving counterpart, the Möbius strip, which was of particular interest of Lacan[16,17] as a model for

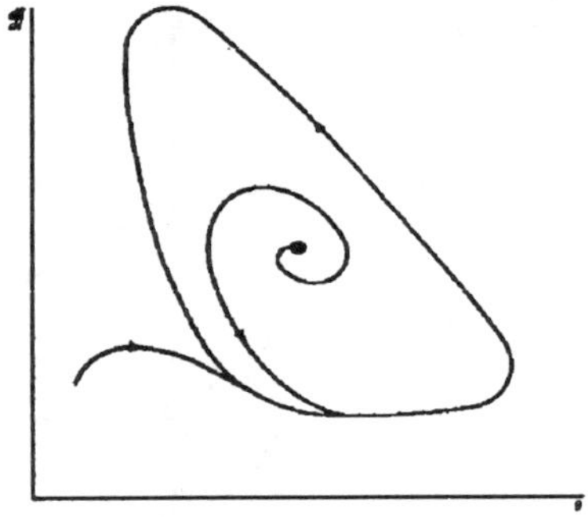

Figure 7: A 2-dimensional limit cycle attractor.

psychic structure) and a chaotic attractor (Figure 9), which exists only for basins of dimension three or greater. Attractors persist at lower dimensions than the basins they attract making the system more ordered.

Equations describing nonlinear systems differ from linear, differentiable ones in that solutions are not additive. The whole is not the sum of the parts. One consequence is that any interaction, no matter how small its contribution may appear to be, may have a significant impact on the behavior of the system. Therapeutic neutrality becomes another myth.

Nonlinear dynamics have proven valuable as a tool for describing the behavior of various natural systems ranging from the stock market and economies to the behavior of ant colonies and natural selection.[18] Yet, in the case of natural selection at least, Freud rejected them in favor of simpler, intentionally causal Lamarckian inheritance based upon biological need or desire[19,20]. This is particularly intriguing in terms of psychoanalysis' fundamentals because, as noted, when they are "far-from-equilibrium," open, dissipative systems *organize themselves* rather than requiring external organization by *eidos* or its surrogates.

Figure 8, a three-dimensional version of Figure 5, adopted from Zeeman's[21] classic work on catastrophe theory demonstrates how, if we consider the projection upon the horizontal plane of the diagram, we note that nonlinear systems can describe our discontinuous phenomenological experience of psychic reality more accurately than linear ones.

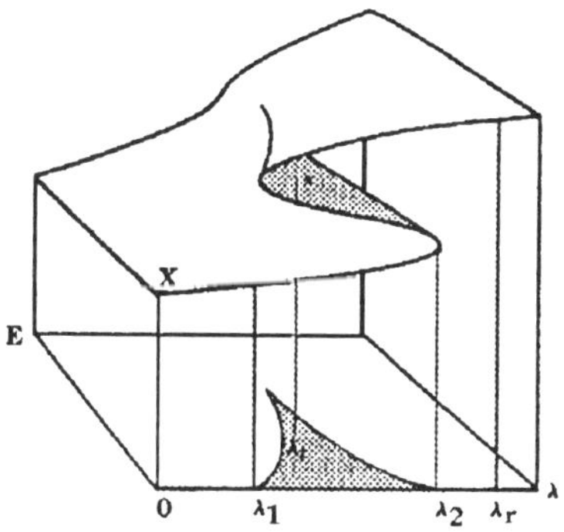

Figure 8: A 3-dimensional version of Figure 5 indicating the apparent discontinuous behavior of a nonlinear dynamical system when projected onto a plane.

2.2 Some Nonlinearities within Biology

Wolf Singer

Wolf Singer[22,23,24] has researched biological mechanisms for learning the contents of visual experience in cats. The relevant anatomy includes four segments: 1., independent parts of the retina which are sensitive to particular aspects of external reality including changes in contrast, horizontality, etc., 2., a way station, the lateral geniculate nucleus, in the mid-brain, that relays the retinal signal to 3., the primary visual cortex, in the rear of the brain, which separates elements of the incoming patterns and sometimes stores them temporarily as components that, when "synchronously bound,"[25] i.e., constructively or reconstructively assembled, seemingly, at least, reflect something reminiscent of an "original scene," 4., finally, the mesencephalic reticular formation (MRF) which resides in the brain stem as a neural net: it is involved with attention, not the "attention" of Freud's "Project," but the attention that separates being asleep from being awake.

When the MRF sends signals to the lateral geniculate nucleus while information is arriving there from the retina, a "now print" signal[26] becomes established that sets perception into memory. If this channel is severed, memory is not set down; if it is severed and its function is replaced with externally generated signals, memories are set into place.[27] Since the operation of the MRF is nonlinear, the setting down of memories must be nonlinear, as well: they must depend upon nonlinear[28] signals originating from the deepest and most evolutionarily primitive parts of the brain to which neither conscious nor unconscious contents (as internal objects, i.e., thoughts, ideas or intentions) can penetrate or influence. Rather than being an objective record of the environment, as it was presented to the organism, memory is nonlinearly and jointly *selected* both from the organism's biology by the environment (through "binding by synchrony"[25]) and from the environment by the organism's biology, *in toto*. For higher organisms, the environment includes, but is not limited to, familial authority.

Walter J. Freeman

Walter J. Freeman[29,30,31,32] has studied the rabbit olfactory system. His experimental paradigm called for inserting fine platinum electrodes (generally 64 of them in 8 by 8 matrices) into the olfactory cortex of the rabbit. He discovered that when sets of different smells were presented to rabbits as they inhaled and the responses of their cortices were measured, different rabbits recorded the same smalls differently in phase space and that in a set of any six odorants, presented to the same animal, the second odorant gave rise to a different pattern than the first and the third different from the second, etc. These findings, in themselves, were not surprising. What was surprising. however, was that when the first odorant was again presented to the same animal after the sixth odorant, its pattern became different form its pattern the first time. The animal's experience with odorants 2-5 had modified its system sufficiently that the first odorant was represented the second time around as a different pattern from its pattern the first time. Like the case of the nonlinear function above, returning along a trajectory from later to earlier does not give rise to the same history laid down from earlier to later.[a]

When Freeman plotted a phase space diagram for the smells, he discovered that each took the form of a high amplitude, asymmetric, limit cycle attractor that persisted for upwards of 50-150 ms. Each was preceded and followed by a different low-level, symmetric, chaotic attractor; the previous chaotic attractor had bifurcated as a result of a coupling of input and feedback gain that arise from a particular odorant in the environment thereby giving rise to its corresponding limit cycle attractor. The subsequent chaotic attractor included the previous odorant in its geometry.[34]

[a] Arnold Modell[33] argues that Freud was referring to this idea by the concept "*Nachträglichkeit*" or "delayed action."

2.3 Chaos

Figure 9 represents a chaotic attractor which, in this case, is a relatively low dimensional geometric structure upon which basin elements converge. Unlike

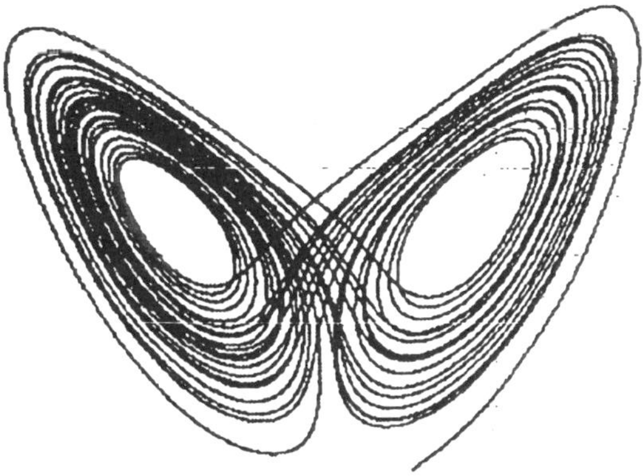

Figure 9: A chaotic attractor which displays "unstable aperiodic motion" and generally cannot overcome environmental perturbation.

other attractors, however, it arises from systems that, like the MRF, display "unstable aperiodic motion," i.e., which generally cannot overcome perturbation by their environment.

According to Kellert[35] the activity of chaotic systems appears to be random but is actually deterministic and reproducible if the input and initial conditions can be replicated. This is not a practical goal since they exhibit *sensitive dependence on initial conditions.* This means that trajectories that may have initially been infinitely close together diverge rapidly and points of infinite closeness can unpredictably shift lobes. While all points on the attractor's surface are required to have nearby points that diverge exponentially, the topological facts of a two-dimensional surface dictate that this would lead to trajectories that cross. Yet, they must not cross since that would contradict the deterministic nature of the system of differential equations: intersecting trajectories would mean that the system would have a "choice" of paths when it reached the intersection. The attractor's dimensionality, then, must be fractal—in this case greater than two but less than three.

A set of first order differential equations describing a chaotic dynamic system must have at least three independent dynamical variables and the equations of motion *must contain a nonlinear term that couples several of the variables.*[36] Note how this coupling differs from the additivity among independent members of differential equation sets, above, whose predictable outcomes were part of Freud's premises.

The main geometric features of chaotic attractors arise from a combination of "stretching" and "folding" wherein nearby points are stretched apart in one direction, thus creating the local divergence responsible for unpredictability, while the system "folds" points together that are at some distance, causing a convergence of trajectories in a different direction.[b]

3 Constructing Psychoanalysis Upon an Alternate Physical and Metaphysical Foundation

3.1 Wilfred Bion's Metapsychology

The applicability of Singer's and Freeman's research to psychoanalysis readily appears when we compare it to Wilfred Bion's[37] model for thinking, $Ps \longleftrightarrow D$, which I previously[38] summarized as ideas (archetypal "thoughts without a thinker," whose phylogenetic origins predate the ego) being presented to consciousness sequentially. Structurally, these ideas are members of a class, D.[c]

[b]For me, this questions how best to envision the goals and processes of psychoanalysis, if not of life, itself. To the extent that we take seriously Freud's therapeutic intention, "where id was, there ego shall be," as an accomplishable goal, we may, for example, imagine a three-dimensional linear graph, similar to Figure 1, that contains a trajectory connecting points (time=0, ego=0, id=1) and (time=1, ego=1, id=0). Our task, as analysts, then, is to note when our patients' actual path diverges from this theoretical one and to supply interpretations that rectify this situation by addressing their "resistances," "transference constellations" or our own "countertransferences" that contribute to the divergence.

If, however, we appreciate, instead, that the unfolding of psychic reality, like all biological unfolding, though deterministic, is neither linear, predictable not goal seeking, but takes unique self-organizing geometrical forms for each of us, constrained by stretching and folding, and ultimately based upon our histories and initial conditions, then our picture of psychic unfolding and out understanding of interpretation must necessarily change. It must become one of a self-constructing object whose internal experiential points emerge only in the moment, as realities, perhaps following Schrödinger's uncertainty principles; interpretation becomes a "playful" response to another rather than a corrective response to another's goals which, even if they did exist, intentionally, we could never trust or understand, biologically. See the discussion of play, below.

[c]Relatively unique among psychoanalysists, Wilfred Bion was both a physicist and a metaphysicist. His training in physics prepared him to reckon systems formally. As an example, he employed a matrix, which he called "the grid," for reliably locating psychoanalytic ideas in relation to each other, aspired to devise an "algebraic calculus," within which to rationally define the dynamics of "psychic reality" and the dynamics of psychoanalytic transformation, and to align psychoanalytic ideas with each other, and, from within which to map psychoanalytic ideas onto biological ones. The latter was an attempt to repair a situation that continues to exist as a result of Freud's decision to sever the dynamics of psychoanalytic metapsychology from those of physical science: the separation of thinking in

Bion derives his sign, D, from Melanie Klein's[39] formulation of the depressive position, a developmental milestone whereby previously "apparently unrelated elements, associated with feelings of persecution (the β-elements), come together as an integrated whole associated with feelings of depression."[40] This "integrated whole" is conceived as a bounded, relatively stable, "internal object" (an asymmetric limit cycle attractor in Freeman's language), which, for Bion, contains within itself various conceptual polar opposites. As β-elements, these might previously have been pathologically split and projected outwards into an alter and experienced by the ego as arising there.

D, consists of α-elements, the products of α-function, an undefined but probably continuously developing function of the psyche which transforms β-elements into α-elements.[41] α-elements can be used for thinking, for forming symbolic statements, for remembering and for dreaming.[42]

Between the experiences of sequential D's lies the more developmentally primitive experience, Ps. Ps is a symbol used by Bion to refer to M. Klein's formulation of the "paranoid-schizoid position,[43] a psychic state in which "objects" are β-elements (sense impressions, primitive affects, "things-in-themselves"). Ps takes the form of an apparently disordered field and, though as an "object" it may be useful for projective identification or control of non-sentient objects, it is symbolically and semiotically empty[44]. It is not useful for thinking, dreaming, symbolizing, etc.

For Bion, the state, Ps, can result from two possible sources. A process associated with the primitive ego can reverse α-function,[45] "cannibalizing" previously constituted α-elements (D's) by splitting them and, hence, transforming them into "bizarre objects" (rudimentary ego and superego fragments combined with the β-elements from which they came) or Ps can ensue as a normal consequence of an immature ego, i.e., one that is either completely devoid of, or possesses only rudimentary α-function. Bion refers to the oscillating experiences of Ps and D by the statement, $Ps \longleftrightarrow D$;the double headed arrow indicates that each state leads to and arises from the other.

---

terms of "extra-sensual" objects, more generally associated with the physical sciences, from thinking in terms of "sensual" ones, associated with psychoanalysis, as Bion understood it. One of Bion's methods of repair was to escew the sensual- extra-sensual demarcation by seeking a common formal structure for both.

Bion, then, spoke in quasi-mathematical and physically dynamical language about metaphysical and psychoanalytic entities: these included Kantian β-elements, psychoanalytically defined depressive and paranoid-schizoid positions (D and Ps), undefined α-functions and "selected facts," given up by transcendental "O," the Kantian *ding-an-sich*, through Faith.

In spite of these seeming confusions, however, Bion generally separated the physical from the metaphysical aspects of his own theory making it possible for us, today, to analyze the latter, at least functionally, to determine if and how they can be explained in terms of the former, given our new tools provided by nonlinear dynamic systems theories.

The Selected Fact

More comprehensively, Bion posits that D arises from Ps as α-function operates on β-elements in the presence of the "selected fact."[46] The selected fact is a β-element or a previously constructed α-element that functions as a seed might function in a supersaturated solution, allowing disparate elements (β-elements) floating, as it were, in psychic suspension, to agglutinate upon it so that a well defined precipitate (an α-element) can "fall out" by its presence. The precipitate corresponds to and provides a novel vertex or perspective.

The selected fact appears somewhat mystically for Bion. Though it can be hoped for, it cannot be predicted. One can only have "Faith in O,"[47] "ultimate reality...absolute truth, the godhead, the infinite, the [unspeakable] thing-in-itself,"[48] that ultimately the selected fact will be provided, and hence D will reoccur—that the world will again make sense. Bion proposed that the origin of the selected fact lies ultimately *in the environment*; access to its contents require the subject to allow "linkages" to form. Bion[49] defines three modes of linking: Love (L), Hate (H) and Knowledge (K).

If we call α-function, α, and the selected fact s, we can restate Bion's proposition by the statement $D_i = \alpha_j(Ps, s)_{Faith\ in\ \text{“}O\text{”}}$. Note that unlike α, in the equations, above, here, α has a subscript, j, because it is not constant over time, but evolves. When we further recognize that in speaking of a "developing function of the psyche," Bion probably intends, α, to evolve continuously, i.e., according to differentiable evolution laws, and that s, by virtue of being given discontinuously and unpredictably by "O," representing the godhead, is non-differentiable, we can separate the continuous (differentiable) from the discontinuous (non-differentiable) components of the statement and rewrite it as $D_i = \alpha_j(Ps) + s_{\text{“}Faith\ in\ O\text{”}}$. Newton's partial differential equations are inadequate for constructing a psychoanalysis that is reflective of, and responsive to, non-differentiable biological dynamics, but so, too, are *universal* symbolic systems.

Self-organization in lieu of the Selected Fact

Dynamically, this situation is identical to Freeman's description of the oscillation between symmetric, chaotic attractors (Bion's Ps or Singer's "separated visual-elements-awaiting-MRF-input") that punctuate asymmetric limit cycle attractors which, themselves, correspond to individual smells in the rabbit olfactory cortex (Bion's D), i.e., $\{\text{chaotic attractor}\} \longleftrightarrow \{\text{limit cycle attractor}\}$. Yet, for Freeman, limit cycle attractors do not arrive as functions of unpredictable, transcendental "O," operating "through Faith," as D does for Bion. For Freeman, the limit cycle attractor arrives as an "ordinary" nonlinear event, a joint product of the organism's instantaneous biological reality (based upon its unique initial conditions and history) and the particular environment in which it finds itself at a given moment. Indeed, it is the instantaneous environment which provides "the input surge and the attendant increase in gain [that] may cause bifurcation to a limit cycle attractor."[d]

Freud, then, was correct to assume that physical and biological realities set necessary parameters within which psychic reality, whatever its nature, must operate. Though nineteenth century biology did not provide a sufficiently robust foundation for his purposes, Singer's and Freeman's biology does for ours. Psychoanalysis must heed these if it is to address mind without a dependence upon dualism. We must take into account, for example, that memory is neither static, stable nor recoverable but, like memories of odors in the rabbit are constantly updated and literally created as we live out individual lives. As Freeman visualizes it:

> In brief, we conceive that the olfactory system maintains a global chaotic attractor with multiple wings or side lobes, one for each odor that a subject has learned to discriminate. Each lobe is formed by a bifurcation during learning, which changes the entire structure of the attractor, including the pre-existing lobes and their modes of access through basins of attraction. During an act of perception the act of sampling a stimulus destabilizes the olfactory bulbar mechanism, drives it from the core of its basal chaotic attractor by a state transition, and constrains it into a lobe that is selected by the stimulus for as long as the stimulus lasts.... The activation of the lobe is expressed by the spatial pattern of bulbar activity, which on transmission to the targets of the bulb comparably destabilizes them and enables them to enter into basins of attraction that are appropriate of the percept. In this way the cortical response to a stimulus is "newly constructed"...rather than retrieved from a dead store.[51]

---

[d]Prigogine[50] addresses the dynamics inherent to Bion's "selected fact" non-metaphysically, with the idea of "nucleation," a self-organizing process that occurs within sufficiently complex systems of "*metastable*" states. There, transient particles (such as liquid droplets in supersaturated vapor) either do or do not arise as a joint outcome of particular pairings of instantaneous system and environmental conditions rather than as a result of metaphysical elements.

4 Conclusion

Regression, then, cannot possibly lead to the recollection of individual history since nonlinear functions approached "backwards" can only, by accident, lead to the same locus as when approached forwards. "You can't go home again," but you can create transient historical homes consistent with other conscious contents in the moment! Since no two individuals, by virtue of their non-identical initial conditions and history (sensitivity to initial conditions), can experience the same reality, we cannot get into each other's heads—empathy is at best a wish, at worse, a delusion—countertransference is ubiquitous. Differential equations and rules derived from them cannot describe universal developmental trajectories—the devil is in the nonlinearities. Interpretation, in terms of such trajectories, cannot reliably link contemporary content to the lived history of one who produces it nor can it be tied to universal developmental trajectories. The selected fact no longer reflects a gift of the "godhead," but *marks the process of nonlinear biological change as the unique organism interacts, in the moment, with its own particular environment giving rise, thereby, to the bifurcation of chaotic attractors into limit cycles.*

These notions, among others, require psychoanalysis to take particular forms that are consistent with human biology, a biology that, at its chaotic core, is identical for patient and analyst. Whatever pathology is and whatever psychoanalysis changes, it does not change this fact! Structures that depend upon history given up by regression, independence of analyst and patient, constancy of memory, universal developmental trajectories, etc., can no longer remain candidates, while structures such as D. W. Winnicott's "squiggle game"[52] coupled with Wilfred Bion's admonition that, as analysts, we approach each analytic encounter "without memory, desire or understanding," as equals-in-the-moment meeting and interacting through play in transitional space, are. In play, *a necessary implication of nonlinear feedback within dynamic chaos*, they give rise to an instantaneously expanding manifold that creates unique individual and joint historical trajectories with unpredictable consequences for all individuals immersed in the transitional field, rather than to predictable ones defined within hierarchical relationships inherent in classical, linear, analytic paradigms.

5 Acknowledgements

I am grateful to Gail S. Young and William Sulis for constructive suggestions in earlier versions of this paper.

6 References

1. S. Pepper *World Hypotheses: Prolegomena to Systematic Philosophy and a Complete Survey of Metaphysics, 91-2* (University of California, Berleley, 1946).
2. E. Mayr *The Growth of Biological Thought: Diversity, Evolution, and Inheritance,* 88 (Harvard University Press, Cambridge, MA, 1982).
3. P. J. Bowler *Evolution: The History of an Idea,* 5-6 (University of California, Berkeley, 1983).
4. S. Freud. in *The Standard Edition of the Complete Psychological Works of Sigmund Freud,* ed. and trans. J. Strachey, **21**, 36 (W.W. Norton, New York, 1966).
5. J. Goldstein in *Proc. 1990 Int'l System Dynamics Conf.*, **1**, 440-454 (TheMIT Press, Cambridge, MA,1990).
6. J. Goldstein in *Chaos Theory in Psychology,* eds. R. Robertson and A. Combs (Lawrence Erlbaum, Hillsdale, forthcoming)
7. S. Freud in *The Standard Edition of the Complete Psychology Works of Sigmund Freud,* ed. and trans. J. Strachey, **1**, 281-397 (W.W. Norton, New York, 1966).
8. S. Freud in *The Standard Edition of the Complete Psychological Works of Sigmund Freud,* ed. and trans. J. Strachey, **13**, 1-162 (W.W. Norton, New York, 1966).
9. S. Freud in *The Standard Edition of the Complete Psychological Works of Sigmund Freud,* ed. and trans. J. Strachey, **19**, 3-66 (W. W. Norton, New York, 1966).
10. F. J. Sulloway, *Freud, Biologist of the Mind: Beyond the Psychoanalytic Legend,* 125-6 (Basic Books, New York, 1979).
11. S. Freud in *The Standard Edition of the Complete Psychological Works of Sigmund Freud,* ed. and trans. J. Strachey, **1**, 324 (W. W. Norton, New York, 1966).
12. A. E. Jackson, *Perspectives on Nonlinear Dynamics,* **1**, 324 (Cambridge Univ. Press, Cambridge, 1991).
13. W. Singer, C. Gray, A. Engel, P. König, A. Artola and S. Bröcher *Cold Spring Harbor Symposia on Quantitative Biology,* **55**, 946 (1990).
14. W. Singer in *The Principles of Design and Operation of the Brain,* ed. J. C. Eccles and O. Creutzfeldt, 128 (Pontificia Academia Scientarum, Vatican City, 1990).
15. W. J. Freeman and C. A. Skarda, *Brain Res. Rev.,* **10**, 161 (1985).
16. M. Bowie, *Lacan,* 192 (Harvard Univ. Press, Cambridge, MA, 1991).
17. J. Lacan, *Scilicet,* **4**, 5-51 (1973).

18. S. Kauffman, *The Origins of Order: Self-organization and Selection in Evolution,* 177-81 (Oxford Univ. Pess, New York, 1993).
19. J. B. P. Lamarck, *Zoological Philosophy: An Exposition with Regard to the Natural History of Animals,* trans. H Eliot, 355-62 (Univ. of Chicago Press, Chicago, 1984 <1809>).
20. F. J. Sulloway, *Ibid.,* 274-5.
21. E. C. Zeeman, *Catastrophe Theory,* 6 (Addison-Wesley, Reading, MA, 1977).
22. W. Singer in *The Neurosciences—Fourth Study Program,* eds. F. O Schmitt and F. G. Worden (MIT Press, Cambridge, MA, 1979).
23. W. Singer, *Exp. Brain Res.,* **47**, 209-222 (1982).
24. W. Singer, *Eur. Arch. Psychiatry Neur. Sci.,* **236** (1), 4-9 (1986).
25. W. Singer, *Prog. Brain Res.,* **95**, 462 (1993).
26. W. Singer, *Acta Morphologica Hungarica,* **31** (1-3), 251 (1983).
27. W. Singer, *Arch. Biol. Med. Exp. (Santiago),* **16,** 323 (1983).
28. W. Singer in *The Principles of Design and Operation of the Brain,* eds. J. C. Eccles and O. Creutzfeldt, 131-2, 161 (Pontificia Academia Scientarum, Vatican City, 1990).
29. W. J. Freeman, *Biol. Cyber.,* **33** (August), 237-247 (1979).
30. W. J. Freeman, *Biol. Psychiatry,* **18** (10), 1107-1125 (1983).
31. W. J. Freeman in *Rethinking Neural Networks: Quantum Fields and Biological Data,* ed. K. H. Pribram, 507-14 (Lawrence Erlbaum and INNS Press, Hillsdale, NJ, 1993).
32. C. A. Skarda and W. J. Freeman, *Beh. Br. Sci.,* **10** (2), 161-195 (1987).
33. A. Modell, *Other Times, Other Realities: Toward a Theory of Psychoanalytic Treatment* (Harvard Univ. Press, Cambridge, MA, 1990).
34. W. J. Freeman and C. A. Skarda, *Brain Res. Rev.,* **10**, 161 (1985).
35. S. H. Kellert, *In the Wake of Chaos: Unpredictable Order in Dynamical Systems,* 1-14 (Univ. of Chicago Press, Chicago, 1993).
36. G. L. Baker and J. P. Gollub, *Chaotic Dynamics: An Introduction,* 3 (Cambridge Univeristy Press, Cambridge, 1990).
37. W. R. Bion, *Learning from Experience* (William Heinemann, London, 1962).
38. A. Stein, *Splitting, Induction and the Politics of Borderline Object Relations* (unpublished, 1989).
39. M. Klein in *Envy and Gratitude and Other Works 1946-1963,* 71-80 (Hogarth Press, London, 1980).
40. W. R. Bion, *Elements of Psychoanalysis,* 42 (William Heinemann, London, 1963).

41. W. R. Bion, *Learning from Experience,* 42 (William Heinemann, London, 1962).
42. W. R. Bion, *Ibid.,* 26.
43. M. Klein in *Envy and Gratitude and Other Works 1946-1963,* 1-24 (Hogarth Press, London, 1980).
44. U. Eco, *A Theory of Semiotics,* 242 (Indiana Univ. Press, Bloomington, 1979).
45. D. Meltzer in *Do I Dare Disturb the Universe: A Memorial to Wilfred R. Bion,* ed. J. S. Grotstein, 530-5 (Caesura Press, Beverly Hills, 1981).
46. W. R. Bion, *Elements of Psychoanalysis,* 72-73 (William Heinemann, London, 1963).
47. M. Eigen, *Int. J. Psych.-Anal.,* **62**, 413-433 (1981).
48. W. R. Bion, *Attention and Interpretation* (Tavistock, London, 1970).
49. W. R. Bion, *Learning from Experience,* 43-4 (William Heinemann, London, 1962).
50. I. Prigogine, *From Being to Becoming: Time and Complexity in the Physical Sciences,* 145-7 (W. H. Freeman, New York, 1980).
51. W. J. Freeman in *Rethinking Neural Networks: Quantum Fields and Biological Data,* ed. K H Pribram, 511-2 (Lawrence Erlbaum and INNS, Hillsdale, NJ, 1993).
52. D. W. Winnicott, *Therapeutic Consultations in Child Psychiatry,* 3, 7, 12-24 (Basic Books, New York, 1971).

QUALITY OF LIFE – A DYNAMIC PERSPECTIVE

P. LEMAY, J-P. DAUWALDER, V.POMINI, M. BERSIER
University of Lausanne, Department of Psychology
BFSH 2, Dorigny, CH-1015 Lausanne - Switzerland

In an empirical pilot-study we explored the psycho-social dynamics of quality of life based on an eco-behavioral paradigm and complex dynamical systems theory. Ten subjects received a programmed pocket computer and reported cognitive, emotional and behavioral variables in relation to their natural environment and answered corresponding questions every four hours over a 3-month period.
Results have shown the importance of a bio-psycho-social conception of the individual at the structural as well as the dynamical level. Psycho-social dynamics were first explored through appropriate phase spaces and Karnaugh Maps. Structural nonlinearities between variables have been shown, pointing to severe limitations of linear methods such as correlations and factor analyses. An index measuring the complexity of transitions indicates which variables and configurations are the most predictable and which are not. Results indicate a strong tendancy of subjects to evolve around an attractor state (psycho-socially defined), a strong indication of self-organizational processes. Multivariate basins of attraction are observable.

1 INTRODUCTION

Despite the central role of "quality of life" in our daily experiencing, the concept itself is not well defined. Current operationalizations emphasize symptomatic complaints, related diagnostic criteria and negative outcomes, such as prevalence and/or incidence of illness. "Risk factors" are identified by probabilistic relationships to distant outcomes, and mediating variables such as "knowledge", "coping-strategies" or resources from "social support" are hypothetically linked in between.

The term "quality of life", used in titles of publications, showed an eight fold increase over the last decade, from 31 papers in 1980 to 246 in 1990 [1]. Most available instruments for assessment, however, focus exclusively on negative features. Traditional approaches to the assessment of quality of life are either close to psychometric procedures, limited by lack of test-retest reliability or criterion-related validity in a constantly moving context, or close to decision theory, limited by reduction of outcome criteria to one single dimension [1]. More recent developments, such as the "multi-attribute health index (MAHI)" approach [2], require focused applications and, thereby, suffer from lack of generalizability. More qualitatively oriented investigations revealed three different directions of lay conceptions about health related quality of life: quality of being healthy, quality of having health and quality of actively maintaining health.

The most serious critics, however, concern the focus on a single point-in-time assessment, which characterize most quality of life investigations.

Thus the question arises whether current research on quality of life is not going into the wrong direction. How can we assess the positive features of health ? How to get a global and yet differentiated view on quality of life ? How to overcome the single point-in-time and correlational design of studies in this field ?

In our opinion we need a completely different approach. At the individual level the question of self-organizational processes appear to be of utmost importance. At the contextual level the emerging eco-behavioral paradigm aims at a more complete understanding. We have to analyze reciprocally determined ecological systems and their complex dynamics. These might be mainly nonlinear, and positive quality of life might be the result of a "strange attractor", rather than the result of a fixed point attractor. Quality of life, as a phenomenon of complex system's dynamics, might then be understood in a global and yet qualitatively differentiated way.

The present paper summarizes a first attempt to capture subjective quality of live in its natural context. Specifically, we tried to develop new methods for its assessment, relying on self-observational techniques, in order to describe and analyze quality of life as a subjective and evolutive process. In contrast to the usual top-down approach in the study of complex dynamic systems, we privileged a bottom-up approach, where empirical data constitute the starting point for ulterior model construction and simulation. Given the poor quality of psychological data, as compared to precisely measurable physiological or physical data, the search for adequate methodologies, especially those being able to reflect complex systems dynamics, was the main challenge for the pilot study to be presented here.

Three global hypotheses were to be explored by this pilot study. First, we supposed the dynamics of subjective quality of life to be nonlinear. Second, we supposed the corresponding psychosocial configurations to be organized in a non-random way, as to their structure and sequence. And finally, we supposed these configurations and patterns of evolution to be different for each subject.

2 PILOT STUDY

These broad lines of interest led to the following operationalization of our pilot study. The frequent gathering of bio-psycho-social data in a natural context over a longer period of time implies the use of self-observational techniques, and more precisely, since attempts of such scope are scarce, the elaboration of an adequate instrument for observation.

The **assessment** (by way of a self-ratings) was designed in **three parts**, in order to answer different questions. The **first part** was composed of eight questions related to the global assessment of the last four-hour period the subject had lived. These included the following topics which proved to be of interest:

- one's **internal state**, including mood, emotionality (intensity of emotions), stress and well-being [3,4,5,6,7,8,9];
- one's **environment**, such as material support, social tolerance, and familiarity of encountered situations [10,11];
- one's **interactions** with his environment, characterized here by familiarity of behavior.

The **second part** focused on the complete detailing of an event that had triggered positive or negative change of one's mood, or an event that was significatively intense, during the same four-hour period. It was inspired by a behavioral-cognitive framework issued from experimental and clinical psychology, where behavioral analysis models have proven useful [12,13,14,15]. Emphasis was put on questions of tolerance (three questions related to one's evaluation of initial and final level of tolerance regarding the event, and one's behavior during it), whereas other aspects included goals (one's self-efficacy and outcome-efficacy [14]), characteristic emotions and their level, behavior, attribution [16,17,18,19] and self-appraisal (in terms of reproaches).

The **third** and last part had to give clues as to which types of coping were preferentially employed and also the kinds of environment-protecting behaviors subjects adopted. It was answered once every day.

Most of the items were designed on a six-point scale, chosen in order to be answered rapidly, to be reliable and robust enough yet to obtain sufficient variability and discrimination. Furthermore such a scale facilitated the use of discrete and boolean mathematical methods. A small number of open questions required keyed answers.

The assessment took place four times a day, over a 3-month period (from end of July to end of October 1993). It was programmed on a pocket computer (palmtop Hewlett-Packard 95LX), to facilitate answering and consequent analyses.

Ten voluntary subjects recruited by advertisement at the University of Lausanne participated in the study. Three of them were males, 7 were females; four were married. The age varied between 22 and 38 years. In order to maintain motivation and to insure data control, subjects were paid SFr 50.- every two weeks.

3 METHODS

New approaches require new methods. The current trend of theoretical developments in psychological and biological sciences has relied on the (more or less metaphoric) investigation of concepts such as chaos, nonlinearity, self-organization, bifurcation and the likes [20,21,22]. Some analytical methods have emerged to translate intuitions into objective demonstrations: phase spaces, bifurcation diagrams or Grassberger-Procaccia algorithms are by now familiar tools for the complex systems scientist.

If such mathematical tools are appropriate for continuous data, such as psychophysiological measures, they bring however certain limitations for less precise and reliable measures obtained in research focusing on self-observational evaluations in a natural setting.

We addressed this problem by developing new methods based on **Boolean algebra**. Kauffman [23,24] and Thomas [25,26] use this approach to simulate evolution of genetic and biochemical systems. Their goals are to understand the stability, phase transitions and possible evolutions of complex systems. Their work is mainly top-down oriented: they construct a system of logical equations supposed to represent the studied phenomenon, derive the corresponding table of transitions and plot a graph of its evolution. Inspired by their work, we tried to build up ways of grasping the evolution of the natural complex system formed by a human being and its interactions with his environment. Our approach privileges the bottom-up direction, where empirical data constitute the starting point and lead to ulterior model construction.

Our method allows to:

- take simultaneously into consideration many variables, so to describe existing complex psychosocial configurations;
- represent graphically the *structure* of psychosocial configurations encountered during observations in a compact, understandable manner;
- represent graphically the *evolution* of such psychosocial configurations;
- derive *logical equations* in order to later on simulate deduced processes.

Boolean method is based on binary calculus and logical connectors. Values of variables are reduced to a 0/1 dichotomy. In our case, the six-point scale was transformed so that values from 1 to 3 took the binary value 0, and values from 4 to 6, the binary value 1. Interpretation of booleanized variables corresponds to each polarization of the six-point scales. For example, the well-being question is then strictly interpreted as well-being or uneasiness, the emotionality question is considered as low or high level of emotion, instead of a gradation of the two characteristics. While it is clear that part of the shades are lost, one gains useful insight in this integration of multiple sets of variables, offering advantages largely compensating the loss of local precision.

Psychosocial configurations are defined as a multivariate combination of psychological and social variables having emergent properties. Such configurations are composed out of 2 to 8 variables, while keeping representations manageable and understandable. We have worked so far with 3 and 4 variables. Any N-variable configuration has 2^N possible (theoretical) combinations (i.e. 4 variables have $2^4 = 16$ possible combinations). The configuration is simply the appending of the individual binary values into a single number[a]; the order of the variable in the configuration has no importance, as long as the order is preserved throughout the analyses.

Adapted representation of observed data is a necessary further step toward thorough understanding of the underlying dynamics of psychosocial phenomena. For this purpose we use a technique of computer science that usually helps to derive logical functions from truth tables: the **Karnaugh map**[27]. In our context, each cell of the Karnaugh map corresponds to a given psychosocial configuration. Figure 1 shows a Karnaugh map for four variables. It is structured so that the first two variables are displayed on rows and the two other on columns, each combination in the map made up by their combined values. It should be noted that the sequence of configurations does not follow the usual progressive order; the particular order is arranged so that any two adjacent configurations differ from each other by the change of only one variable. For example, one observe that neighboring configurations 1101 (#13) and 1111 (#15) differ only by the third variable. Moreover the disposition of the map makes the sides also adjacent (top with bottom, left with right); the states 0001 (#1) and 1001 (#9) are almost identical excepted for the first variable, having the value 0 for state #1 and 1 for state #9.

These important characteristics make the Karnaugh map a powerful tool having the properties of a phase space. It serves as the basis for the representation of the evolution. Like in any phase space, one simply has to connect the successive observed states by a line (from the state at time *t* to state at time *t+1*). As will be shown later the evolution of a complex system is easily plotted and tracked, as its bifurcations and phase transitions appear in an integrated

[a] A configuration may be represented by a decimal number for a matter of concision. For example, using 3 variables the binary combination 101 is equivalent to 5 in decimal, since $5 = 1 * 2^2 + 0 * 2^1 + 1 * 2^0$. Because of the arbitrariness of the order of variables, the decimal number is not an indicator of the degree of anything, e.g. configuration #14 (1110) is not better than #7 (0111).

	00	01	11	10
00	0	1	3	2
01	4	5	7	6
11	12	13	15	14
10	8	9	11	10

Figure 1: Karnaugh map of four variables.

and economical fashion. Last but not least, this tool has shown definite marks of self-organizational processes for the subjects participating in our study.

4 RESULTS

According to our hypotheses, emphasis was put on the analysis of the dynamics of psycho-social configurations reflecting quality of life. Specifically we were interested to see, how the empirically observed configurations were structurally organized and how they evolved over time. The presentation of the results will follow the two main themes: structure and dynamics for full-scale (discrete) data and their Boolean equivalent.

4.1 Structural relations of psychosocial configurations

Traditional statistical analyses were performed on the data in order to reveal their structure: correlations, multiple regression, factor analyses. They all permit to spot a variety of relationships between variables[b]. The first important characteristics found is that although there was some low to moderate correlations between the three internal variables of stress, well-being and emotionality ($r_{STRE-WB} = 0.589$, $r_{EMOT-STRE} = -0.165$ and $r_{EMOT-WB} = -0.239$), we found a rather interesting link between the three variables, as observed by plotting the frequency of occurrence of the combined variables. The relation is essentially linear, showing that when the stress intensifies the emotionality intensifies as well and the well-being decreases and reverses to uneasiness.

[b]For a matter of concision these abbreviations are used (in parentheses are signification of the variables along with the two poles of the six-point scale): MOOD (change of one's mood, negative-positive), EMOT (emotionality, scale: 0-100), STRE (stress, no-high), SIT (familiarity of situation, unfamiliar-familiar), MSUP (material support, obstacle-support), TOL (social tolerance, intolerance-tolerance), BHV (familiarity of behavior, unfamiliar-familiar) and WB (well-being, well-being - uneasiness).

But there are two exceptions to this general rule: the first shows that when stress is absent (STRE=1 on the six-point scale), the emotionality and the well-being are mainly high (instead of having a low emotionality if the relation were strictly linear) and the second, when stress is at its maximum, emotionality is also at its highest, but there were as many periods of well-being as of uneasiness, as shown by an almost equal frequency of these configurations. The first exception seems to correspond to moments of intense pleasure, while the second confirms what stress studies have shown, that stress may be a source of pleasure, not only of discomfort.

Another surprising result comes from the analysis of familiarity of situations and familiarity of behavior (see figure 2). The linear correlation is $r_{SIT-BHV} = 0.152$, which would straightforwardly mean there exists no significant relationship between them. However, detailed data inspection reveals that while the relation is mainly linear (behavior is increasingly familiar when the situation is evaluated as such, and vice-versa), there exists also a striking nonlinear exception: when the situation is completely unfamiliar (SIT=1 on the six-point scale), behavior is almost exclusively highly familiar (BHV=6). This might be interpreted as if, in the absence of situational landmarks, subjects look for other landmarks and find them in behavior. This is also obvious when the distribution is further analyzed: occurrences of unfamiliar behaviors in familiar situations are rather scarce (yielding a one-sided distribution where there are almost no occurrences for which SIT $>$ BHV). Therefore one finds a type of coping with unfamiliar situations that is not cognitive but behavioral, an active coping strategy being triggered by perception of the situation.

Similar nonlinear phenomena also appeared in the analyses of the variables of the second part of the assessment (evaluation of particular episodes). Most of the occurrences of the tolerance variables have similar evolutions (initial tolerance with behavioral tolerance, initial intolerance with behavioral intolerance). There are almost no cases of "crossed" patterns, that is, initial intolerance with behavioral tolerance, and initial tolerance with behavioral intolerance. One also observes that tolerance at the beginning of an episode is a good predictor of the tolerance at the end (almost no crossed patterns having initial tolerance and final intolerance, or initial intolerance and final tolerance). These results might reflect what in game theory is called the tit-for-tat strategy, or more popularly "an eye for an eye, a tooth for a tooth". If this were true, our subjects would have a serious lack of coping regarding the tolerance aspect (i.e. not being able to change a sequence beginning with intolerance by behaving tolerantly). More specific observations at a microscopic level should therefore be undertaken to explore further these intertwined patterns of social cognition and behavior.

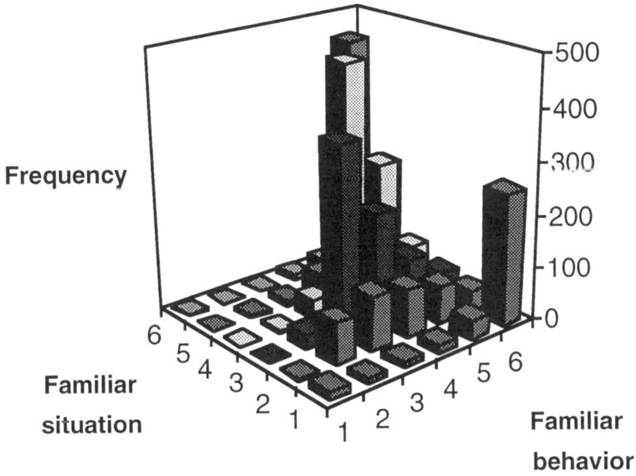

Figure 2: Frequencies of familiarity of situation & behavior co-occurrences.

As a *preliminary conclusion*, detailed graphical inspection appear to be of utmost importance. Particularly here plotting frequency of co-occurrences helped point out mechanisms of adaptation. Data also revealed a series of underlying structural nonlinear relationships. These nonlinearities bring to light severe limits of traditional statistics: first, traditional statistical procedures might not be able to identify them and second, nonlinearities blur the results obtained (correlations, regression analyses, factorial analyses, to name but a few, rely exclusively on linear relationships). This limitation is still more important, if one tries to identify multivariate connections.

4.2 Psychosocial phase spaces

Phase space representations play an increased role in the study of dynamical natural systems. We used them to explore the multivariate dynamics of psychosocial configurations.

A series of factor analyses performed upon the observational data repeatedly exhibited three fundamental dimensions: well-being, emotionality and tolerance. Phase space representations plotted on these dimensions were done for all ten subjects. The resulting graphics were similar for eight of them. Each of these typical subjects evaluated most periods as low to average on

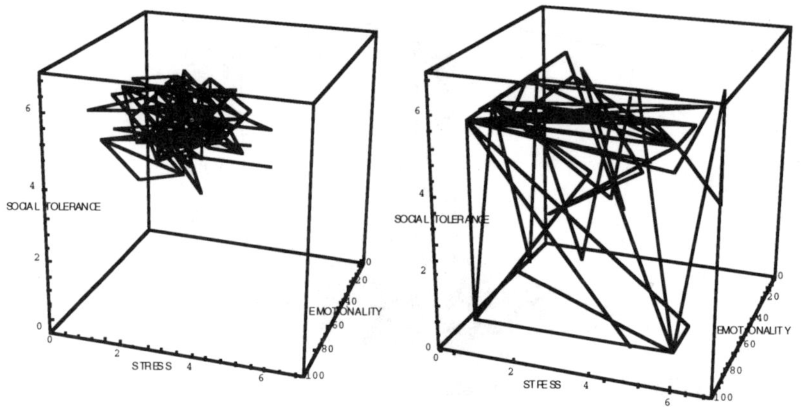

Figure 3: Phase space of a typical subject (a, left) and an atypical subject (b, right).

emotionality, low on stress and average social tolerance. The corresponding phase spaces are structured around a more or less compact region of the space, which implies a coherent mode of coping and living during this three-month period. Figure 3 (left) represents one of the typical subjects.

The two other subjects structured their periods quite differently. Consider figure 3b, which shows the phase space of an atypical subject. It is much more scattered across the space; there is no central core around which other combinaisons are arranged. There is in fact one, but it is "centered" at high emotionality (EMOT= 80–100), no stress (STRE=1) and high social tolerance (TOL=6). The phase transitions are more radical and abrupt than for the typical subjects, evolving to states involving strong changes of values of variables. When looking in detail at this subject's situation, we found that she had just divorced a few weeks before the beginning of the experiment and most radical transitions corresponded to moments of meeting her ex-husband.

As a further preliminary conclusion, phase space representations revealed typical patterns of subjective quality of life. In our empirical data, the quality of life phenomenon is probably best described by (metaphoric) strange attractors of different scopes, reflecting lower stress, more well-being and experienced tolerance for typical "normal" subjects, than for vulnerable atypical ones.

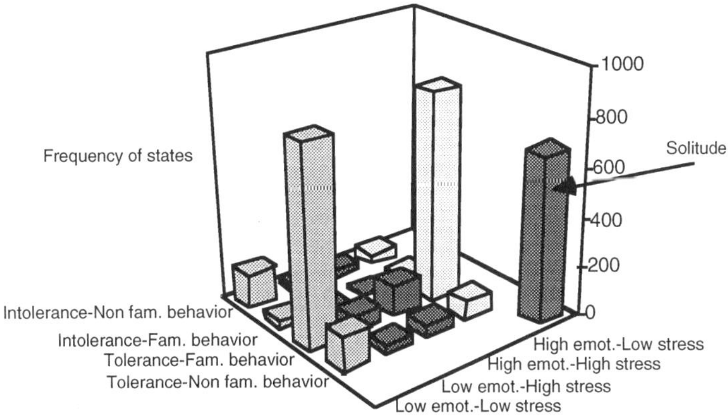

Figure 4: Frequency of Boolean psychosocial configurations.

4.3 Boolean analyses

Structures of psychosocial configurations

The specific power of Boolean algebra develops when multivariate combinaisons are considered. The four-variable case is presented here. The variables of social tolerance (TOL), familiarity of behavior (BHV), emotionality (EMOT) and level of stress (STRE) were chosen in order to understand the interactions between psychological (internal state and behavior) and social variables.

Figure 4 shows the frequencies of these multivariate combinaisons (variables transformed on the 0/1 scale; all ten subjects included). Three psychosocial configurations emerge to form the majority of all configurations. One includes social tolerance, familiar behavior, low emotionality and low stress (highest column on the left) , another includes social tolerance, familiar behavior, high emotionality and low stress (highest column on the center-right), and the last configuration represents periods of solitude. These results are indicative of relatively easy-going and smoothly passing periods; the summertime (when the pilot study took place) may be a relevant artefact explaining it. But self-organizational processes are surely another strong explanation, which we explore in the following sections.

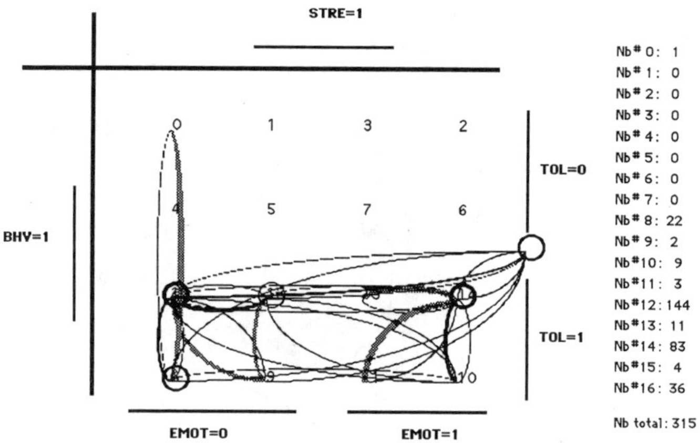

Figure 5: K-map of a typical subject for tolerance, behavior, emotionality and stress.

Boolean phase space

These representations of frequency of co-occurrences only partially reflect the characteristic individual dynamics over time. We therefore further explored phase transitions as displayed on Karnaugh maps. Lines are added to connect one configuration to the next one appearing in the course of observations. So the principles are the same as with phase space representations but displayed on a different mapping.

The two following figures show the evolution of two subjects during the three-month observation. Let us consider the Karnaugh map of a typical subject (figure 5, same subject as on figure 3a). He went through 242 periods of social tolerance and behaved in a familiar manner (TOL=1 and BHV=1, third row; cf frequencies on the right), so that 77% of all periods were characterized by these two particular values. One also observes 145 periods of high emotionality and low stress (EMOT=1 and STRE=0, fourth column), for 46% of all configurations. More precisely this subject visibly lived periods of social tolerance, behaved in familiar ways, felt low stress and had a few more periods of high than low emotionality (respectively state #14, corresponding to line 11 and rows 10, and #12, line 11 and row 00). One also sees that there is a region where no states but one were transited through, a region including states #1 to #7; this subject went through no periods but one where social intolerance was present (only state #0, once). These statistics could be included inside the Karnaugh map but they are confusing at first.

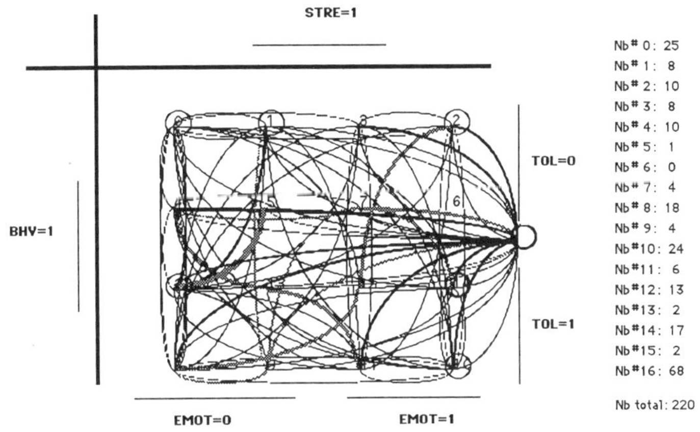

Figure 6: K-map of an atypical subject for tolerance, behavior, emotionality and stress.

Regarding the *transitions* between psychosocial configurations the given picture shows a structured dynamics. Two conventions are used concerning the direction of transitions from one configuration to the next: first the flow of transitions is clockwise, and second, circles around a configuration means a transition to itself. Codes have been used to distinguish the importance of states and their transitions (e.g. color and width of lines), but will not be detailled here. One sees the number of transitions from a particular state to other ones to be relatively few. We later introduce an index of the complexity of the transitions which accounts for the organization of the patterns of transitions.

For the second subject (cf figure 6) the organisation of the periods differs from the previous one. The transitions seem really not to be as structured as the typical subject. The configurations seem to be experienced an almost equal amount of time, and the sequence of states looks close to random. From the calculated basic statistics, we see that this student mostly behaved in an unfamiliar manner, during as many periods of social tolerance as intolerance. Visible are the many transitions to the configuration labelled #16 (center-right of the map), which corresponds to a state of solitude, experienced in 31% of all periods. This foreign female probably used these solitary moments to cope with moments resulting from difficulties of adaptation. Despite these relatively few states, experienced transitions were unstable; after each state followed a high number of states. We will later see that her index of complexity is much

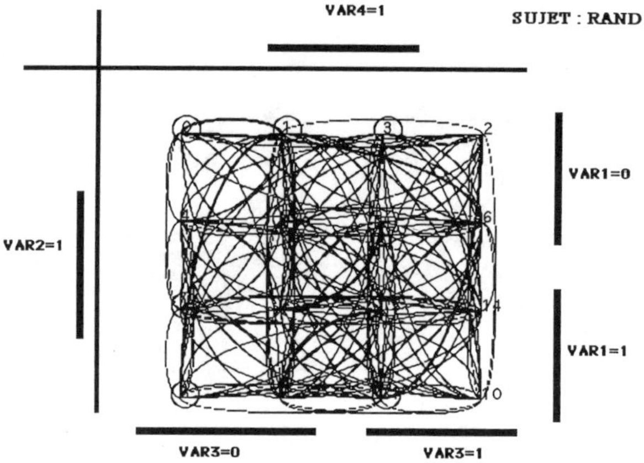

Figure 7: Random simulation of four variables.

higher than the first subject.

In order to examine more closely our assumption that self-organizational processes underlie these individual dynamics, random simulations were performed and plotted on a Karnaugh map. Random values were generated on a six-point scale, then reduced to binary values 0 or 1 and combined four at a time to yield a four-variable Karnaugh map. Figure 7 clearly indicates there are no organized patterns of transitions. Frequencies among states are relatively equal and there are transitions from and to almost every state, which is not the case of our subjects, even for the less organized.

4.4 Index of complexity of transitions

One may quantify the degree of complexity of the transitions with an index originating from information theory (cf [28]); the use of Boolean algebra here is the perfect bridge with the theory. The formula of this index of complexity of transitions is:

$$I_{ct} = -\sum_{ij}^{N} \pi_i * p_{ij} * log_2 p_{ij}$$

where π_i is the relative weight of configuration i (number of configurations i divided by total number of configurations observed) and p_{ij} is the relative frequency of transitions from configuration i to configuration j. The index ranges from 0 to $log_2 m$, where m is the number of possible different configurations;

here $m = 16$ or $m = 17$, depending if there is a supplementary configuration meaning "no answer" (or the likes), yielding respectively a maximum of 4.000 or 4.087. The index mainly indicates the determinism of the process: with an index of 0, the process is completely deterministic, there is one and only one state j that follows state i. With a maximal index, the number of states following each state i is maximal.

Other interpretations of the index are possible. In connection with information theory the smaller the index the less information and surprise is brought by the next state (and vice-versa, a greater index means that the next state brings more information). A deterministic system ($I_{ct} = 0$) is less informative regarding its transitions (however complex the system is) than a random one ($I_{ct} = log_2 m$) where each transitions come as a surprise. It may also represent the stability of the system, where a low index indicates a stable system, and a high index, an unstable one.

The calculated index of the first subject is 1.514, and for the second, it is 2.749. It indicates that transitions of the first subject are more structured than those of the second. The Karnaugh Maps were effectively different at first glance; even though the two subjects went through a limited number of states, the next states of the first subject were relatively fewer than those of the second, which is what the index really measures. The index is therefore representative of the organization of the transitions.

The index for the eight other subject ranges from 0.897 to 2.380. It means that the patterns of transitions are neither deterministic nor completely random, but that there is individual self-organization in the evolution of the psychosocial systems. In regard of the value of the index we may infer two possibilities concerning the three month period the subjects lived, not exclusive of course: some subjects have better coping strategies than others, and/or they live less stressful and demanding events (eg. the first subject vs the second). Since one deals with psychosocial systems, self-organization also concerns the organization of the environment, which may be fluctuating differently or may be more or less perturbed from person to person. More detailed information is of course needed to distinguish between the two possibilities.

Individual dynamics using most frequent transitions

Having understood that each person is a complex adaptive system in interaction with an evolving environment, one faces difficulties when trying to model evolution of such systems. Transitions could fundamentally not be deterministic because of bifurcations, namely encountered transitions from one state to more than one subsequent state. What if not all transitions from each state

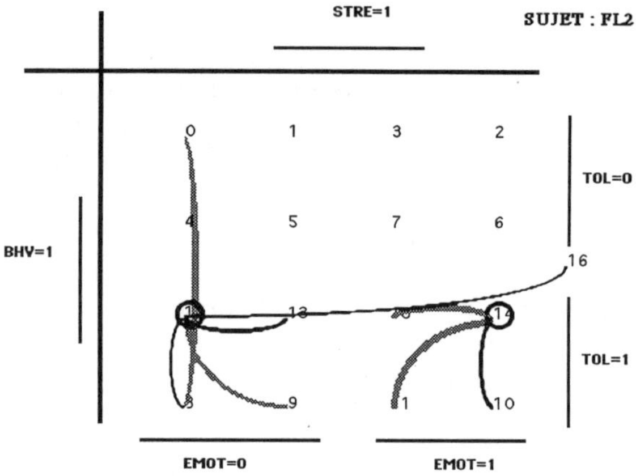

Figure 8: Most frequent transitions for a typical subject.

were shown? We tried to plot only the most frequent transition from each state and we found interesting insights.

Looking at the most frequent transitions (MFT) for each subject, one remarks the following transitions and infers some general mechanisms of evolution and adaptation: subjects tend to reduce their level of stress (or remain into state of low stress), go to/remain into a state of well-being, return to/keep familiar schemas of behavior, maintain their level of emotional intensity and seek situations of social tolerance. These mechanisms can be inferred because the most frequent transitions of most subjects go into states having these characteristics. Figure 8 represents the most frequent transitions for a typical subject (same as figure 5) for the same four variables: social tolerance, familiarity of behavior, emotionality and stress. There are three attractors: states #12, #14 and #16 (state of solitude), the set of states going to them form their basin of attraction. Reduction of stress is seen by transitions from the center columns of the map (STRE=1) to those of the edges (STRE=0) as shown by transitions from state #13 to state #12 and from state #15 to state #14, or transitions to states having the same low level of stress (from state #8 to #12 and from state #10 to #14). Stable emotional intensity and other characteristic phenomena are also deduced by analyzing such patterns represented on the Karnaugh map.

If the representation of most frequent transitions allows to infer mecanisms of self-organization and adaptation, then it should differenciate the seemingly

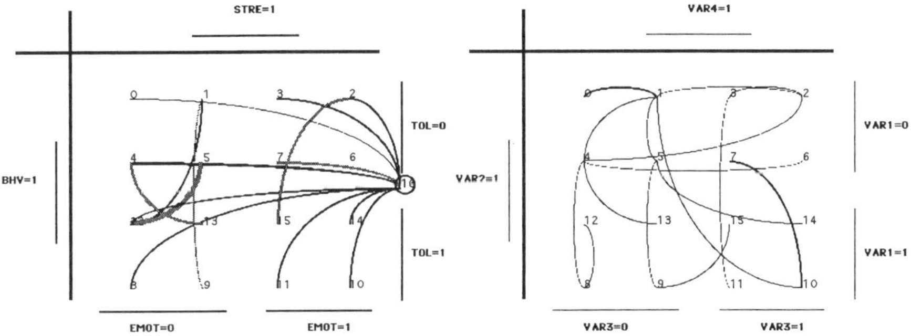

Figure 9: MFT for the atypical subject (left) and random values (right).

random Karnaugh map of the second subject presented with the randomly-generated values (compare figure 6 and 7). One may see with figure 9 that it is exactly the case. On the left side is the subject's most frequent transitions K-map and on the right is the one obtained from random values. The difference is striking: the subject almost always returns to a state of solitude (#16), whereas no definite pattern of transitions is found on the second K-map. The method of plotting only most frequent transitions seems therefore a rather informative one for investigating self-organizational processes.

One of the problem one faces when trying to simulate processes from empirical data is the presence of these bifurcations, the fact that in course of time many different states followed at least one particular state. How to resolve the problem? One proposed idea [29] is to consider the sequence of states as a succession of deterministic models. Another method is the consideration of most frequent transitions, which seems to be promising because they are supposed to be representative of the individual mechanisms of adaptation. One may simulate the process with logical equations and then add some perturbations to the system and study what are the resulting evolution of the system. This procedure is still under investigation.

5 DISCUSSION

The methods presented (histograms, phase spaces, Karnaugh maps) have advantages and disadvantages that need to be critically reviewed. Such efforts clarify the strengths and limitations of each method, provide guidelines and indicate steps towards the appropriate use of these analyses. The following remarks mainly concern analyses performed on ecological data, but they are sufficiently general to cover other types of data.

Histograms and **phase spaces** are both representational tools revealing the organization (structural and temporal) of many variables. They both serves as a first global contact with data. **Histograms** of multivariate co-occurrences show if the corresponding curve is normal or not, if nonlinearities are present or not. Such characteristics need to be known, for subsequent traditional statistical analyses might be distorted. Variables used with this graphics may include one variable at time t related to itself at time $t+1$, consequently yielding an account of its dynamics. Its utilisation is thus not limited to structural relationships. Its disadvantages concern the lack of direct quantitative evaluations (e.g. absence of any parameter of reference) needed to make more explicit statements about phenomena under study.

The **phase space** allows for the visual exploration of both structural and temporal organization of variables. If data are gathered around a certain region or not, one observes possible and impossible (or at least very improbable) configurations. If data are compact or diffuse across the phase space one may infer the variability of the systems' behavior; in a psychosocial context it gives clues as effectiveness of coping and/or fluctuations of environment. Attractors (fixed points, cycles or strange attractors) may then be identified. Because of the many inherent bifurcations of naturally evolving psychosocial systems, the real direction of evolution can not be directly determined from such a graph (in a deterministic system evolution is obvious since a system never passes twice through the same state unless it locks into a cycle). Refinements of this kind of graphics are necessary; for example adding arrows could show the direction of evolution and colors or symbols could also be added to mark the frequency of given configurations, considering one cannot know if a given point represents an important configuration or only a transient one. Another limitation concerns the maximum of three variables that can be plotted on the phase space.

The **Boolean methods** through the adapted use of Karnaugh maps provide structural as well as dynamical information about psychosocial processes. It was shown that boolean histograms (on a Karnaugh map) first render the structure of data then giving a clue concerning possible attractors. Transitions are then introduced onto the map; most frequent transitions may be displayed, therefore capturing what we believe are general mechanisms of adaptation and coping. Possibilities of coding both frequencies of each state and their transitions exist but have not been shown here for a matter of clarity. The advantages have been constantly underlined in this paper. Its limitations comes from the simplification of information and correlative loss of fine discrimination.

Different techniques are thus available to the complex systems scientist interested in understanding and modelling the evolution of a reciprocally determined psychosocial system. The presented methods have proven their usefulness and offer interesting **perspectives**. The next steps being undertaken by our research team concern more quantitative methods. We intend to apply mathematical procedures that determine the order of transitions, for knowing to what extent past states influence the present state (influenced by states at time $t-1$? $t-2$? etc). Animations of transitions on Karnaugh map associated with subjects' description of events have shown that they generated specific patterns of transitions at specific moments. Quantitative measures exist that precisely determine such "windows" of specific patterns by indicating when significative differences of transition probabilities occur (omnibus test[30]). Another step is the simulation of systems of Boolean equations that can be derived from the most frequent transitions. A last step regards the extension of psychosocial configurations to more than 4 variables (an Karnaugh map of 8 variables can be obtained using a "fractal" perspective, i.e. by integrating a Karnaugh map of four variables inside each (16) cell of a four-variable Karnaugh map).

6 CONCLUSION

Various relevant points at multiple levels (empirical, methodological and mathematical) were exposed in this paper. We showed that Quality of Life is a complex phenomenon, which can be evaluated positively, ecologically and dynamically. Our study gives a clear indication that well-being (for example), an implicit variable associated with quality of life, presents multiple relationships with other psychological and social variables. In fact one must take into account both these variables and their particular relationships if he is to understand the phenomenon. Nonlinearities were shown to be present at many places. Presence of individual attractors and structured patterns of transitions confirm self-organizational (non-random) processes in an ecological study.

Such ecological studies relying on self-observational techniques must deal with somewhat (sometimes) unreliable data. Self-evaluation of psychological and social variables are necessarily subjective, non continuous and difficult to analyze (basic conditions for statistical procedures are not always present). The need for methods robust enough to overcome these limitations is all too evident. Boolean methods seem promising in that respect.

The mathematical developments mentioned throughout this paper clearly represent efforts to create new methods for analyzing psychosocial processes. The Boolean methods really open innovative perspectives. Coupled with appropriate numerical analyses and the further use of simulation, it allows not only to *tickle* psychosocial complexity but really to *tackle* it.

Acknowledgments

Part of this research project was sponsored by Philip Morris SA (Switzerland).

References

1. S. Wright, in *Psychology and the Promotion of Health*, ed. J.-P. Dauwalder (Hogrefe & Huber Publishers, Seattle, 1994).
2. M. H. Boyle and G. W. Torrance, *Medical Care*, 22, 1045–1057, 1984.
3. M. S. Clark and A. M. Isen, in *Cognitive Social Psychology*, ed. A. Hastorf and A. Isen (Elsevier North-Holland, New York, 1982), p. 73–108.
4. M. Carlson, V. Charlin, and N. Miller, *Journal of Personality and Social Psychology*, 55, 211–229, 1988.
5. K. Fiedler, *Psychologische Rundschau*, 40, 127–140, 1989.
6. J. A. Gray, *The Psychology of Fear and Stress* (Cambridge University Press, Cambridge, 1987).
7. A. M. Isen, in *Advances in Experimental Social Psychology*, ed. L. Berkowitz (Academic Press, New York, 1987), Vol. 20, p. 220–253.
8. R. S. Lazarus and S. Folkman, *Stress, Appraisal and Coping* (Springer, New York, 1984).
9. S. Taylor, *Health Psychology* (Random House, New York, 1986).
10. S. Kennedy, J. K. Kiecolt-Glaser, and R. Glaser, in *Social Support. An Interactional View*, eds. B.R. Sarason, I.G. Sarason and G.R. Pierce (John Wiley & Sons, New York, 1990), p. 253–266.
11. S.J. Lepore, *Journal of Personality and Social Psychology*, 63, 857–867, 1992.
12. I.-M. Blackburn, K. Davidson, *Cognitive Therapy for Depression and Anxiety. A Practioner's Guide* (Blackwell Scientific Publications, Oxford, 1990).
13. F.H. Kanfer, & W.R. Nay, in *Contemporary Behavior Therapy : Conceptual and Empirical Foundations of Clinical Practice*, ed.G.T.F. Wilson, C. M. (Guilford, New York, 1982), p. 367–402.
14. A. Bandura, *Social Learning Theory* (Prentice Hall Inc, Englewoods Cliffs, NJ, 1977).
15. F.H. Kanfer, H. Reinecker and D. Schmelzer, *Selbstmanagement-Therapie. Ein Lehrbuch für die klinische Praxis* (Springer-Verlag, Berlin,

1991).
16. H. Betancourt and I. Blair, *Personality and Social Psychology Bulletin*, 18, 343–350, 1992.
17. J.-C. Deschamps and A. Clémence, *L'explication quotidienne* (DelVal, Cousset, 1987).
18. C.J. Haferkamp, *Current Psychology Research and Reviews*, 10, 227–240, 1992.
19. J. Jaspars, and M. Hewstone, in *Psychologie sociale*, ed. S. Moscovici (PUF, Paris, 1984), p. 309–329.
20. J.S. Nicolis, *Dynamics of Hierarchical Systems. An Evolutionary Approach* (Springer, Berlin, 1986).
21. H. Haken, *Synergetics. An Introduction. Nonequilibrium Phase Transitions and Self-Organization in Physics, Chemistry and Biology* (Springer, Berlin, 1983).
22. W. Tschacher, G. Schiepek, and E. J. Brunner, *Self-Organization and Clinical Psychology. Empirical Approaches to Synergetics in Psychology* (Springer, Berlin, 1992).
23. S.A. Kauffman, Scientific American, 264, 1991.
24. S.A. Kauffman, *The Origins of Order. Self-Organization and Selection in Evolution* (Oxford University Press, Oxford, 1993).
25. R. Thomas, *Kinetic Logic. A Boolean Approach to the Analysis of Complex Regulatory Systems* (Springer, Berlin, 1979).
26. R. Thomas and R. D'Ary, *Biological Feedback* (CRC Press, Boca Raton, 1990).
27. P. Leussler and A. van Ham, in *Kinetic Logic. A Boolean Approach to the Analysis of Complex Regulatory Systems*, ed. R. Thomas (Springer, Berlin, 1979).
28. C.E. Shannon and W. Weaver, *The Mathematical Theory of Communication* (University of Illinois Press, Urbana, 1949).
29. P. Lemay, *The Modelling of the Evolution of Psychiatric Patients' States using two Non-statistical Methods* (University of Lausanne, Lausanne, 1993).
30. J. M. Gottman and A. Kumar Roy, *Sequential Analysis. A guide for behavioral researchers* (Cambridge University Press, Cambridge, 1990).

LOGICAL ATTRACTORS: A BOOLEAN APPROACH TO THE DYNAMICS OF PSYCHOSIS

Z. KUPPER, H. HOFFMANN
University Psychiatric Services Berne, Mid and West Sectors, Laupenstrasse 49, CH-3010 Berne, Switzerland

A Boolean modeling approach to attractors in the dynamics of psychosis is presented: Kinetic Logic, originating from R. Thomas, describes systems on an intermediate level between a purely verbal, qualitative description and a description using nonlinear differential equations. With this method we may model impact, feedback and temporal evolution, as well as analyze the resulting attractors. In our previous research the method has been applied to general and more specific questions in the dynamics of psychotic disorders. In this paper a model is introduced that describes different dynamical patterns of chronic psychosis in the context of vocational rehabilitation. It also shows to be useful in formulating and exploring possible treatment strategies. Finally, some of the limitations and benefits of Kinetic Logic as a modeling tool for psychology and psychiatry are discussed.

1 Kinetic Logic

1.1 Introduction

In this paper we introduce a logical, Boolean approach to attractors and apply this view to the dynamics of psychosis. More generally, we suggest that in order to understand the systems' dynamics, we should 'think logically'. Why should we think logically? Are we normally not thinking logically? In the view we present here, there is nothing wrong with our theoretical or clinical thinking. We might even have quite good ideas about how things work in our field, as we know the phenomena often from first hand and long standing experience. Moreover we have developed or adopted theories that have some explanatory value. On the other hand, our knowledge seems to be partial and not fully appropriate. There is need for a dynamical, more comprehensive view of the phenomena, their interdependence and evolution in time. For schizophrenia and mental disorders in general such a change in view has been proposed[1 2 3] and possible applications of chaos theory have been discussed[4 5]. In empirical research we can use sophisticated methods of modeling and time series analysis and eventually apply algorithms looking for chaos in our data (see Scheier and Tschacher in this volume). Models of this type are complex and need considerable mathematical skills for understanding. Furthermore, the quality of our quantitative data as well as the theoretical knowledge might often not be sufficient for this type of modeling[6].

In the approach we are presenting here, we suggest starting by *thinking logically* first. We propose to integrate in a dynamical perspective, what we know from theory and experience about the systems of interest. Thus, one develops verbal descriptions that are accessible to most of us, but are far from being formalized models. Kinetic Logic (Thomas [7]) is a modeling tool which can then help us, to bring those qualitative verbal descriptions to a more formal mathematical level, still preserving and using the knowledge that is contained in the verbal descriptions. Having done this, it is possible to model impact, feedback, and the temporal evolution of the variables. As the systems evolve in time, we are able to detect resulting attractors and to explore the effect that changes in the model might have on the dynamics. Models built with Kinetic Logic can be evaluated using the same criteria, e.g. fit with 'real' data or theoretical consistency [8], as used for other dynamical models.

In this paper we first give a short introduction to Kinetic Logic. This method has been developed for modeling in biology by Thomas [7]. We cover Kinetic Logic both in its basic 'naive' and advanced 'generalized' form. The main part of this paper focuses on the application of 'Generalized Kinetic Logic' [9]. We propose a model of the dynamics of chronic psychosis and apply this model to two types of interventions in the vocational rehabilitation of the chronically mentally ill.

1.2 The 'naive' form of Kinetic Logic

Kinetic Logic states that the elements in a system interact positively or negatively with a strength of impact depending on the levels of the elements. This is common to most methods in qualitative and quantitative modeling. The crucial assumptions of Kinetic Logic are: (1) The system elements have little effect on each other until they reach a certain threshold. (2) At high levels the effect tends to reach a plateau. Supposing this is true, we can say that an element is 'absent' when it is under the threshold level and 'present', when above the threshold level. In this way, Kinetic Logic approximates a sigmoid curve of interaction with a step function as shown in Figure 1.

Thus, a variable can have two *logical values*, 0 for absent, or 1 for present. *Logical values* are denoted by small letters: x, y, z. On the other hand, *logical functions* reflect the *evolution* of variables and are symbolized by capital letters (X, Y, Z). Logical functions should not be confused with the derivative. Instead they describe towards which states the variables in the systems tend to, given the actual state of the system. In a typical genetic application, for example, $x = 0$ means 'gene product is absent', $x = 1$ means 'gene product is present' while $X = 0$ means 'gene is off' and $X = 1$ means 'gene is on' [10].

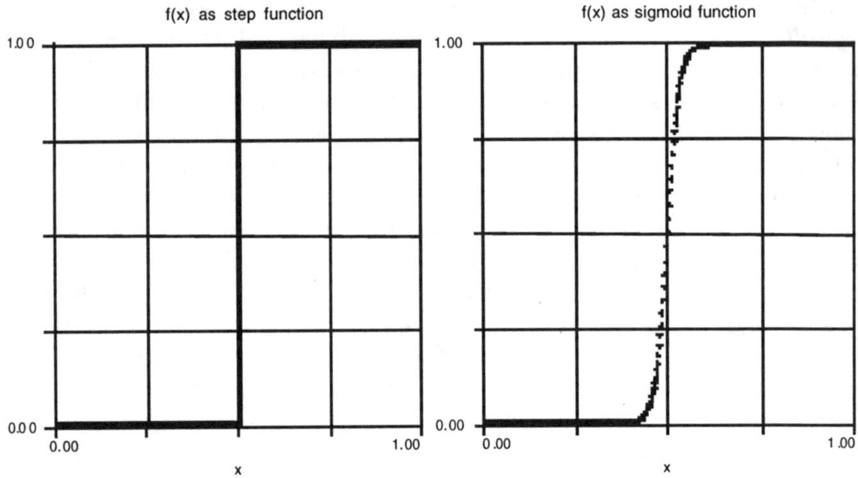

Figure 1: Approximating sigmoid functions by step functions

In psychology, $X = 1$ could be for instance a tendency or readiness to show a certain behavior, while $x = 1$ could represent overt behavior. In this case, X is the logical value of that readiness, its presence or absence, while x stands for the presence or absence of the behavior itself. As a basic example of Kinetic Logic reasoning, consider the two loop structures and the resulting attractors in Table 1. The first example shows a *negative feedback loop*, where $X = 1$ if $y = 0$ and $Y = 1$ if $x = 1$, expressing that X is 'on' when y is not present, while Y is 'on' when x is present.

The *graph* of positive and negative interactions can be rewritten in the form of logical relations. The *logical relations* describe the possible evolution (X, Y) of the system as function of its actual state (x, y). $Y = x$ states, for instance, that y has the tendency to stay or become present, if x is present. Variables that have the same value in the *state* (x, y) as well as in the *logical function* (X, Y) will not change in the next step, while variables which have logical functions different from the actual state, are about to change. In the state table, 'about to change' is expressed by a '+' or '-' sign above that state, depending on the direction of change [11]. Distinguish the '-' in this notation from the dash above a variable in the logical functions where it means 'not'. From the state table one can obtain the temporal evolution of the system, called the *graph of the sequence of states*. Consider that Kinetic Logic - unlike Kauffman's approach[12] - assumes an asynchronous evolution which means that only one variable is allowed to change at a time. In this first example (see Table

Table 1: Description of two basic attractor types by Kinetic Logic

Example 1: *Negative feedback loop*		Example 2: *Positive feedback loop*
$x \underset{-}{\overset{+}{\rightleftarrows}} y$	*graph of interactions:*	$x \underset{-}{\overset{-}{\rightleftarrows}} y$
$X = \bar{y}$ $Y = x$	*logical relations:*	$X = \bar{y}$ $Y = \bar{x}$

state table:

Example 1:

x	y	X	Y
$\overset{+}{0}$	0	1	0
0	$\bar{1}$	0	0
$\bar{1}$	1	0	1
1	$\overset{+}{0}$	1	1

Example 2:

x	y	X	Y
$\overset{+}{0}$	$\overset{+}{0}$	1	1
(0	1)	0	1
$\bar{1}$	$\bar{1}$	0	0
(1	0)	1	0

graph of the sequence of states:

$$\begin{array}{ccc} \overset{+}{0}0 & \rightarrow & 1\overset{+}{0} \\ \uparrow & & \downarrow \\ 0\bar{1} & \leftarrow & \bar{1}1 \end{array} \qquad\qquad \begin{array}{ccc} (10) & \leftarrow & \overset{+}{0}\overset{+}{0} \\ \uparrow & & \downarrow \\ \bar{1}\bar{1} & \rightarrow & (01) \end{array}$$

<u>*periodic attractor*</u> | <u>*multistationarity*</u>

1 left side, bottom), the graph of the sequence shows the pattern of a *periodic attractor* which is typical for a negative feedback loop.

The second example in Table 1 is a positive feedback loop, where the logical function is 'on' for both variables, if the other variable is not present. The states marked by a circle in the state table are stable, as both variables have the same value as the corresponding evolution functions. As to be expected, for a positive feedback loop we find *multistationarity*, i.e. two stable states.

When thinking about applying this method to psychology, such a Boolean reduction of the levels of variables may seem simplistic at first glance. How could such a model be able to reflect the richness of psychological phenomena? Theoretical and clinical reasoning, however, often seem to be similar to threshold models. Many theories, when defining the interactions between variables, treat the variables as having two different states or levels. Family atmosphere in schizophrenia, for instance is discussed in terms of high expressed emotion (critiques or overinvolvement) versus low expressed emotion[13]. When describing the state of a schizophrenic patient, a clinician may observe that the patient

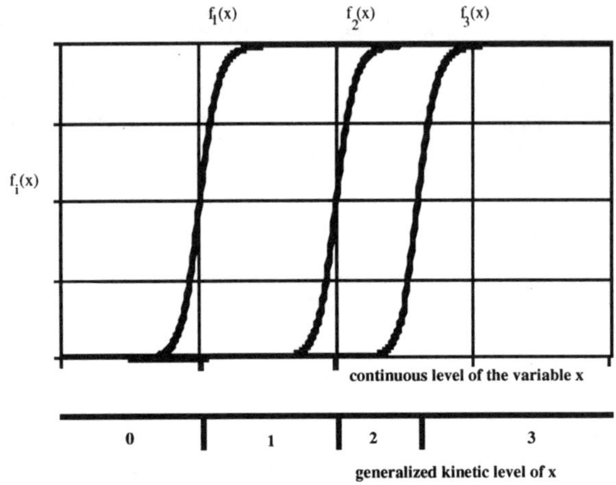

Figure 2: Step functions in 'Generalized Kinetic Logic'

is 'no longer psychotic' or 'psychotic again'. These thoughts both imply a two state model, just as 'Naive Kinetic Logic' does.

1.3 'Generalized Kinetic Logic', an advanced form of Kinetic Logic

In 'Generalized Kinetic Logic' [9], a variable can have more than two levels. The number of 'qualitative' levels of a variable is derived from the number of elements n that a variable influences in the graph of interactions. $n+1$ gives then a intuitively convincing estimate of the number of qualitative levels of a variable.

Consider a variable that interacts at three locations in the graph of interactions. For this variable, level '0' implies that the level of the variable is so low that it has no impact at all. Level '1' denotes a level beyond the first threshold, and an impact on one variable, whereas level '2' implies an impact upon two variables. At level '3', the variable has reached its qualitative maximum, where it influences three variables in the system (see Figure 2).

2 Application of Kinetic Logic to the Dynamics of Psychosis

2.1 General and more specific models - levels of application

In psychology and psychiatry the use of Kinetic Logic has begun only recently [14 15 16 17]. Our main interest has been in the application of the method to

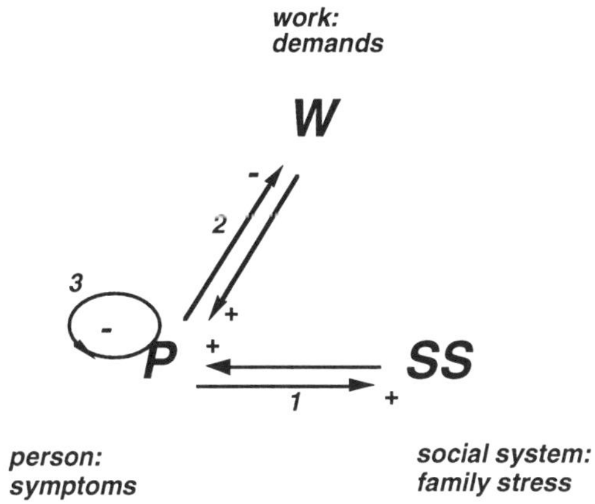

Figure 3: A model of chronic psychosis (for explanations see text)

the dynamics of psychotic disorders, especially of schizophrenia. Schizophrenic psychosis can be modeled on different levels with different theoretical backgrounds and on different level of formalization [18 19 20 21 22 23 24]. One can take a more general or a more individual perspective and start from a biological, a psychological, or a social frame of reference. In this paper we focus on the psychosocial dynamics of chronic psychosis. In previous research the so called 'naive form' of the method was applied to the general dynamics of schizophrenia [15] or to a personalized model of chronicity and rehabilitation in psychosis, exploring the successful rehabilitation of a given patient [14].

2.2 A model for chronic psychosis

In the following an application of 'Generalized Kinetic Logic' is presented, describing the dynamics of chronic psychosis in the context of vocational rehabilitation. In this model (see Figure 3), variable P stands for psychopathological symptoms and behavioral disturbances of a patient.

The numbers 1, 2, 3 in Figure 3 indicate, that the symptoms have different threshold levels regarding the different possible interactions and the levels of symptoms. In chronically mentally ill persons we find heterogeneous patterns of symptomatology where so called 'positive symptoms' (e.g. hallucinations, delusions) and 'negative symptoms' (e.g. emotional withdrawal, motor retardation) as well as coping strategies are present at the same time. Both primary symptoms and coping efforts are often sources of distress for patients and others. This symptomatic behavior might exert stress on the Social System (e.g.

the family, or other relevant people), and foster considerable concern or criticism. If this takes the form of an enduring attitude of families it is termed as 'high-expressed-emotion'. It is known that this can worsen the problems of the patients, specifically promoting relapse [13 25]. Given these findings, we assume positive feedback between the patient's problems and the stressful reactions of his social environment (see Figure 3).

Work demands are another form of psycho-social stress. Wing, Bennet and Denham [26] studied the effect of intense vocational training efforts on chronic long-stay patients. Since that study it is known that excessive work demands have the potential to bring about symptoms that had been absent for years. In older studies symptoms had rather been undervalued in their effect on work performance (see the review of Anthony and Jansen [27]). Recent studies point to a higher importance of symptomatic behavior for future vocational functioning [28 29]. Accordingly we propose to model interaction of symptoms and the demands of work as a negative feedback loop. An increase in symptoms results in a reduction of work demands (often in the form of a dismissal). In return this causes a decrease in the symptoms, considered in medium-term, probably over a period of weeks or months.

The model refers primarily to chronic mentally ill patients, living in the community, showing less symptoms than long-term hospitalized patients. However, to get a comprehensive idea of the possible developments, we also included extreme peaks of the symptoms in our model (level '3' of symptoms). We assume that those peaks are reduced relatively fast in our population by different factors (such as withdrawal, increase of medication, short-term hospitalization). For reasons of clarity and frugality of the model, we formalize this process as a negative feedback loop on the symptoms. At the lowest level (level 1, which stands for 'mild symptoms') symptoms increase the social stress (e.g. negative emotional responses from relatives), at level 2 (= 'moderate symptoms') they have an additional reducing impact upon the demands at work (e.g. in form of a dismissal), where at the highest level (level 3 = 'severe symptoms') corrective negative feedback becomes active on symptoms.

The duration of the states and the time needed for a given transition can be left unspecified in a Kinetic Logic model. It is also possible that these times are different in the same model depending on the relevant transition. Nevertheless, all possible transitions can be studied in this way. However, one always has to bear in mind that interactions might depend considerably on the time scale. This is also known from empirical time series research, where time lagged cross-correlations are studied between variables. Therefore, in developing a Kinetic Logic model its time frame should be specified. As mentioned above, the model presented here applies to the development on a time scale of weeks or months.

If the model should focus on the dynamics in time scales of hours or days, other interactions would eventually be relevant. Furthermore the interactions could change with the age of patients[30]. Our model applies to young psychotic patients.

2.3 Kinctic Logic analysis of the model of chronic psychosis

The assumptions of our model for chronic psychosis can be formalized in a first step in the simple form of 'naive' logical equations:

$$P = \overline{p} + ss + w$$
$$W = \overline{p}$$
$$SS = p$$

This means: P has a reducing effect on itself and is increased by SS and W. In these 'naive logic' equations, the threshold levels are missing. The postulated effects, their directions, and the threshold levels for effects can be summarized in a interaction matrix as follows:

$$\begin{array}{c} \\ P \\ W \\ SS \end{array} \begin{array}{c} \begin{array}{ccc} p & w & ss \end{array} \\ \begin{pmatrix} -3 & 1 & 1 \\ -2 & 0 & 0 \\ 1 & 0 & 0 \end{pmatrix} \end{array}$$

The first row represents the impact on the symptoms (P), the second row on the work demand (W) and the third one stands for the impact on stress in the social system (SS); e.g. the first column vector states: 'P has a reducing effect on itself on level 3, 'P has a reducing effect on W, beginning on level 2 and 'P has an increasing effect on SS, beginning on level 1.

The description of generalized kinetic systems is completed by defining the so called 'K- values'. A K-value describes the level towards which a given variable tends, if a certain combination of interactions are effective on that variable. K-values can therefore be higher than 1 only for variables which can reach levels higher than 1. In Table 2 two possible sets of K-values are presented. The K-values in the left part of Table 2 imply a high impact on symptoms by family stress and work demands. We assume that this is given in the group of severely impaired patients. The right part of Table 2 shows an example of lower impact on the symptoms as it might be given in less impaired chronic outpatients. These patients occasionally apply for vocational rehabilitation. The numbers below the small letters (p, w, ss) represent all

Table 2: Two different sets of K-values (Note that not all combinations of impact need to be meaningful for all variables. Such K-values are left out and denoted by dots.)

high impact on symptoms
(severely handicapped patients)

p	w	ss	K_P	K_W	K_{SS}
0	0	0	.	.	.
0	0	1	1	.	.
0	1	0	2	.	.
0	1	1	2	.	.
1	0	0	0	1	1
1	0	1	2	.	.
1	1	0	3	.	.
1	1	1	3	.	.

lower impact on symptoms
(less handicapped patients)

p	w	ss	K_P	K_W	K_{SS}
0	0	0	.	.	.
0	0	1	1	.	.
0	1	0	1	.	.
0	1	1	1	.	.
1	0	0	0	1	1
1	0	1	1	.	.
1	1	0	1	.	.
1	1	1	2	.	.

possible combinations of influences by the variables P, W, SS, where 1 stands for a positive influence.

The numbers under Kp, Kw and Kss are K-values. As an example for the meaning of K-values consider the vector (1 1 1 — 3 . .) in Table 2, left part, lowest row. This vector states that if all three variables (p, w, ss) have a positive influence on the symptoms, the K-value of the symptoms Kp is 3, i.e. the symptoms will tend to level 3 (= 'severe symptoms'). In other words: if the symptoms are below the level, where they have negative feedback on itself *and* both work demands/having a job and family stress are given, the symptoms will tend to become severe.

Having defined all necessary elements of the model, we transform the interaction matrix into a state table, including all possible states and the corresponding evolution functions, or 'images' as Thomas[11] calls them. As shown in the simple examples in Table 1, we can derive the resulting dynamics from such a table. State tables are suppressed in this and all following examples. For models with few variables calculations can be done manually. However help from computers is preferable. [a]

Figure 4 shows the dynamics in our model of chronicity using the K-values for 'high impact on symptoms' (see Table 2 left part). It is a three-dimensional table with symptoms on y-axis, work and family stress on combined x- and z-axis. Hence the table includes all possible states of the system, i.e. the 'state space'. The figure can also be imagined as a landscape on which a ball is

[a] An algorithm performing the basic calculations using SAS/IML (1989) matrix langague is available from the first author. For use the SAS/IML software is prerequisite.

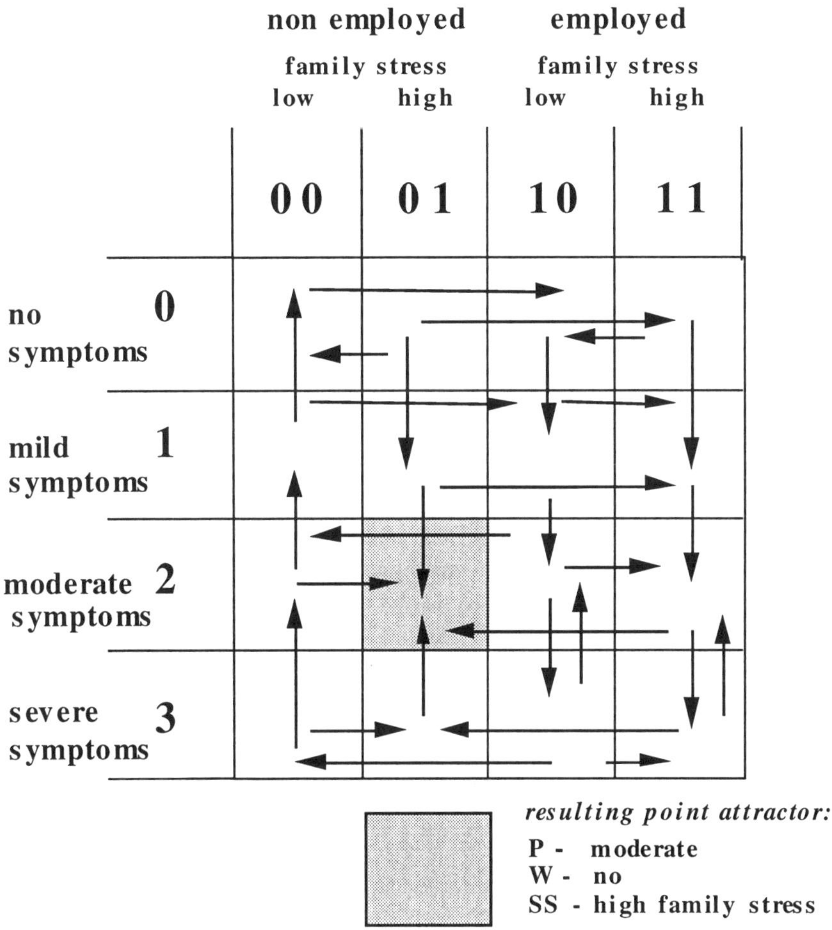

Figure 4: Chronicity as a point attractor

moving. The arrows represent the direction of the ball's movement, i.e. the transition between different combinations of symptoms (P), work situation (W) and family stress (SS) which have to be expected according the assumptions. As demonstrated, the presented transitions are no arbitrary settings but they are the *clear consequence* of the variables' interactions and of a certain setting of their power (K-values). In this form of the model bifurcations can occur, i.e. certain states can develop in more than one pathway. If it is known which transitions are faster, these separations of pathways can be dissolved by the setting of so-called t-values. Otherwise we imply that all of them are equally probable. Most realistic is to see these transition times t as stochastic variables with given means and variations (see Thomas and D'Ari[10]). Thus, attractors emerging out of the analysis are in fact the dominating dynamical structures in the system.

The most important result of our application is, as shown in Figure 4, that under the assumptions of the presented model, only one *stable state* is possible: the combination of moderate symptoms, together with no work and high family stress. In this case, chronicity has the dynamical pattern of a *point attractor*, as the system has the tendency to return always to the one undesired state. Even if external factors, as e.g. therapeutic interventions with only short-term effect, move the system to another, more desired state, this tendency persists. This corresponds in the landscape analogy to a ball which always returns to the hollow after being pushed. The improvement of the working capacity in a vocational rehabilitation program followed by an employment but without additional interventions on P, W or SS, would be doomed to failure.

In the second example, the assumed effects of the system (the matrix of interactions) are kept constant, but the power of the undesired effects on P (the K-values) are reduced, as shown in Table 2 right part ('lower impact on symptoms'). Given these K-values, the dynamics presented in Figure 5, will result.

The person circles within four states: the state 'no work, mild symptoms, high family stress' is followed by the state 'work, mild symptoms, high family stress', i.e. with mild symptoms, the person looks for a job and gets it. As a next step, however, the symptoms increase, then the job will be lost, which finally results in a reduction of the symptoms. Thus the patient is back at the first state of the circle. In this model chronicity is not a single state but equals a periodic attractor. If external factors (such as therapeutic interventions with only short-term effects) are included, the system can also develop other states for a short period of time (e.g. work, no symptoms, low family stress), but there is a strong tendency to return to the attractor of chronicity.

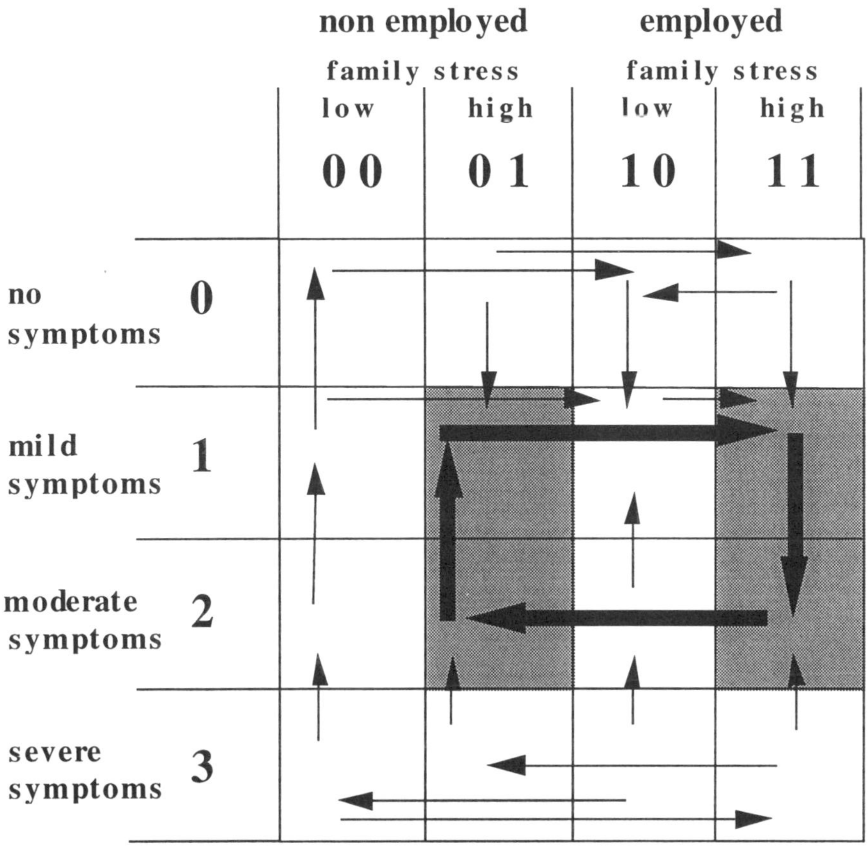

Figure 5: Chronicity as a cyclic attractor

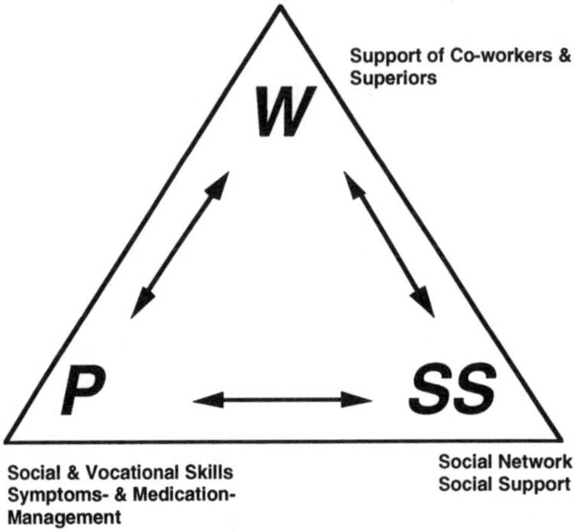

Figure 6: System-related vocational rehabilitation

2.4 Implications for vocational rehabilitation: a systemic perspective

This approach can stimulate a new view of chronicity and new intervention strategies in rehabilitation. Chronicity has to be seen not as a state but as a pattern resulting from a system's dynamics. The same is true for rehabilitation and its goals. It is not the primary and only goal in rehabilitation, to reach a desirable state (e.g. to get a job), but to create conditions which always let the system return to a desired state (e.g. not to loose the job in a crisis). Dauwalder and Hoffmann[14] have described this in a single case with Kinetic Logic. Finally, many intervention strategies result from the interactions postulated in Figure 3. Crucial in this view is that the patient's symptoms (P) are one, but not the only variable that could be addressed by interventions. As the system's 'chronic' behavior results from the interaction of all elements, attempts for change should be tried either on the social stress (SS) or the work situation (W). If for example the increase of symptoms caused a dismissal in former times, the question arises whether the patient can develop a different approach to his symptoms. On the other hand one should consider if the vocational context can react more helpfully on symptomatic behavior. According to this approach, in our rehabilitation program we include interventions on the level of the patient as well as at work and in the social system (see Figure 6).

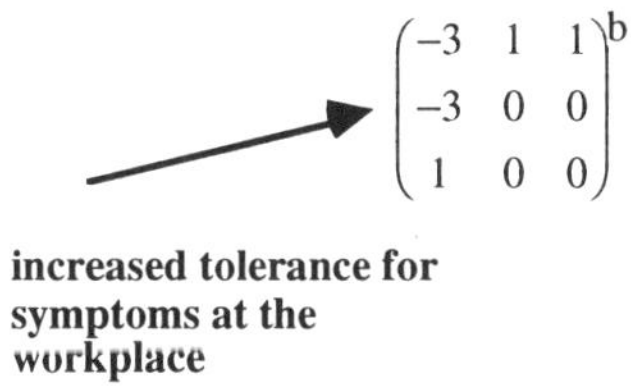

Figure 7: Workplace intervention

2.5 Exploring treatment strategies

After having built a model, one generally has to evaluate it, using specific criteria or 'tests'. Levine et al.[8] outlined tests for the validity of models as being either (1) 'structural tests', (2) 'behavioral tests' or (3) 'policy tests'. One could label the corresponding criteria (1) theoretical consistency, (2) empirical consistency and (3) consistency with changes from interventions. In the following two examples we explore the reactions of our model to two interventions. This type of test can be called test of *behavior sensitivity*: 'Do plausible changes in parameters values only lead to changes observed in the real system?' (p. 217) [8]. We first introduce an intervention on the work site and go then to family intervention.

First type of intervention: sheltered job

As a starting point we take our model of chronicity (Figure 3) with the K-values denoting a lower impact on the symptoms (Table 2), which results in chronicity as cyclic attractor (Figure 5). As mentioned above, this seems realistic for many patients applying for vocational rehabilitation. An increased tolerance for symptoms, or problematic behavior at the workplace in generally is an essential part of a sheltered job intervention. Formalizing this in the language of the model, the thresholds have been changed (see Figure 7). [b]

Workplace situation now only changes if the patient shows severe symptoms (level 3). After having defined all necessary parameters of the system, the resulting dynamical changes can be analyzed in the pathways of Figure 8.

We see that the system shifts from the old cyclic attractor of chronicity to a new point attractor, representing in fact a stable result of the rehabilitation

[b]Notice that this matrix marks a slight deviation from classical kinetic logic as the levels are 'frozen' at a distinct real value (1=mild symptoms, 2=moderate symptoms, 3=severe symptoms) and levels are not collapsed in only two categories, as classical Kinetic Logic would do.

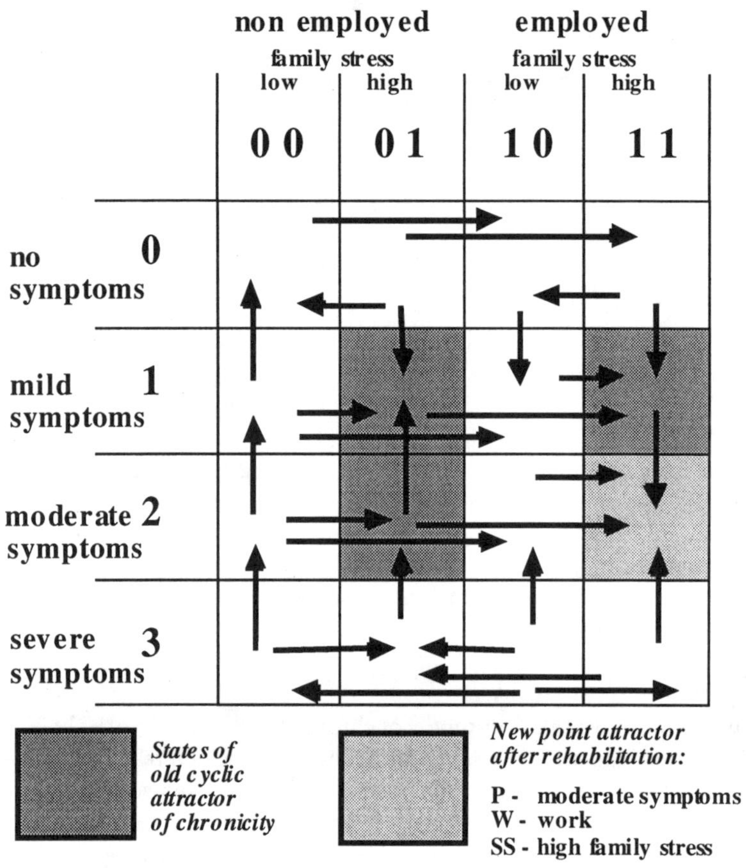

Figure 8: Dynamical impact of the first type of intervention (sheltered job)

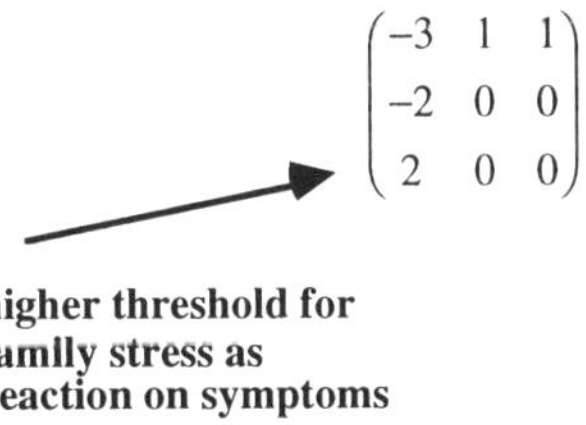

Figure 9: Family intervention

process. This state, however, is not a ideal one, as the patient still suffers from moderate symptoms and family stress is still high.

Second type of intervention: family intervention
In a second example we test the impact of a family intervention on our system of chronicity (Figure 3). Again we start from the system analyzed in Figure 5. Here, the intervention consists in a higher threshold for family stress as reaction on symptoms (see Figure 9). This effect might be common in different forms of family therapy, such as systemic[31] or behavioral[32 33] approaches, but can be introduced and understood from different theoretical backgrounds. On the formal level of the model, family stress is now only increased, if the patient shows at least moderate symptoms.

The dynamical result is again that the cyclic attractor of chronicity changes to a point attractor. In this case, however the resulting stable state after rehabilitation is more desirable as in the last example (see Figure 10). The patient remains employed, symptoms are low and the family stress has been reduced to a low level too. Thus, one could assume a quite successful rehabilitation.

3 Discussion

Starting from theory and having systematized and formalized the interactions in Kinetic Logic, the resulting models are fundamentally different from purely verbal or graphic models. Formalized models as Kinetic Logic models allow clear statements about possible developments of the system.

Having tested two interventions on our model of chronicity, we can conclude that it reacts in a consistent and understandable way on two standard procedures in rehabilitation. From the last dynamical pattern we gained the following new hypothesis: there is only one 'way out' of the cyclic attractor of chronicity. Only if the patient does not immediately accept new demands

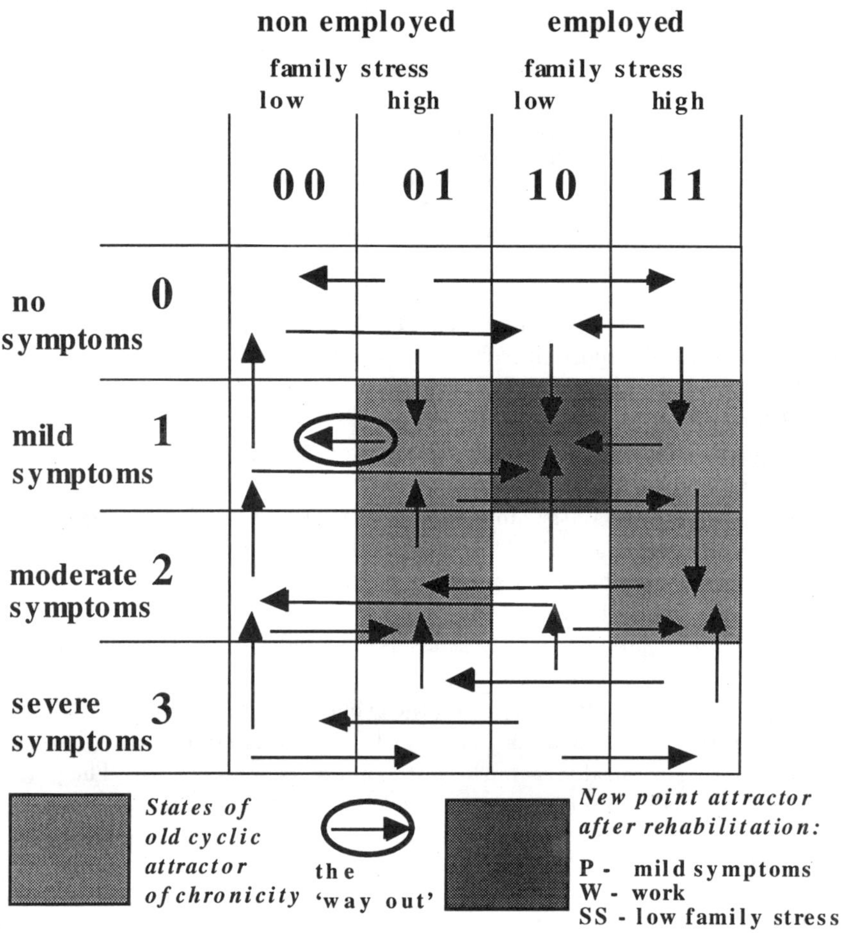

Figure 10: Dynamical impact of family intervention

after recovering, but waits until the family interaction pattern has changed, a new pathway arises. We could now test empirically if this prediction holds for processes of recovery from chronic psychosis. It has to be expected that this dynamical pattern might only be found in a subgroub of patients. Furthermore the dynamics might not be strictly deterministic. Accordingly statistical procedures are needed. Sequential analysis [34] might be helpful for testing such a hypothesis using discrete levels of quantitative variables.

The appeal of Kinetic Logic is first of all that complex dynamical processes, stated by theory or observable in the clinical field, can be transferred into relatively simple models, showing distinct dynamical patterns and 'logical' attractors. Secondly, single observations or theoretical statements can be integrated in a consistent model. Third, the predictions which result from the model can be compared with observations made so far. If the model is comparable with the theoretical assumptions and former results, modified predictions can be compared with empirical time series. Following these lines our model of chronic psychosis will be explored further and eventually be modified.

To get a better idea of the benefits and limitations of Kinetic Logic as a modeling tool in psychology and psychiatry, considerable work has to be done. We have to compare the possibilities of Kinetic Logic with the features of other methods for modeling. Having done this, we might conclude that ordinary modeling in differential equation is superior to Kinetic Logic, because it seems to be more flexible and open for data input from empirical research. On the other hand, Kinetic Logic is more accessible for verbal, qualitative input and gives an *overall picture* of the possible dynamics and attractors which we can never get from a differential system without trying different starting values and parameters. Compared with methods from qualitative modeling (see Levine [35]) such as *pulse processes*, Kinetic Logic seems to be far more adequate for modeling nonlinear systems, as its striking simplicity stems from its conception of nonlinearity. As Thomas points out, the system has all ingredients to produce Rössler chaos [36]. Hence Kinetic Logic has the capabilities to express chaos.

The use of Kinetic Logic for modeling in a wide variety of psychological systems could yield insights about the relevance of the method. Some steps in this direction have already been taken. Dauwalder [37] has applied Kinetic Logic to the psychological management of ecological risks while Dauwalder and Kupper [16] extended Kinetic Logic to health psychology. Finally, we might conclude that this approach is simplistic and limited and therefore turn away from 'logical attractors'. Or we might find that Kinetic Logic is attractive *because* it is simple and based on profound logic.

References

1. J. S. Strauss *et al.*, American Journal of Psychiatry **142**, 289-296 (1985).
2. J. S. Strauss, British Journal of Psychiatry **155**, 22-28 (1989).
3. L. Ciompi, British Journal of Psychiatry **155**, 15-21 (1989).
4. W.J. Freeman, Biological Psychiatry **31**, 1079-1081 (1992).
5. G. G. Globus and J. P. Arpaia, Biological Psychiatry **35**, 352-364 (1994).
6. P. E. Rapp, Biologist **40**, 89-94 (1993).
7. R. Thomas, *Kinetic Logic. A boolean approach to the analysis of complex regulatory systems* (Springer, Berlin, 1979).
8. R. L. Levine *et al.* in *Analysis of Dynamic Psychological Systems*, ed. R. L. Levine and H. E. Fitzgerald (Plenum Press, New York, 1992).
9. E. H. Snoussi *et al.* in *Biological Feedback*, ed. R. Thomas and R. D'Adri (CRC Press, Boca Raton, 1990).
10. R. Thomas and R. D'Adri, *Biological Feedback* (CRC Press, Boca Raton, 1990).
11. R. Thomas, Journal of Theoretical Biology **153**, 1-23 (1991).
12. S. A. Kaufmann, *The origins of order: Self-organization and selection in evolution* (Oxford Universitiy Press, New York, 1993).
13. C. E. Vaughn and J. P. Leff, British Journal of Psychiatry **129**, 125-137 (1976).
14. J. P. Dauwalder and H. Hoffmann, Psychopathology **25**, 139-146 (1992).
15. L. Ciompi *et al.*, System Familie **5**, 133-147 (1992).
16. J. P. Dauwalder and Z. Kupper, Psychoscope **14**, 4-7 (1993).
17. Z. Kupper *et al.* in *Sozialpsychiatire in der Praxis*, ed. B. Eikelmann and T. Reker (LIT, Münster, 1994).
18. A. E. Scheflen, *Levels of schizophrenia* (Brunner and Mazel, New York, 1981).
19. P. A. Schmolling, Behavioral Science **28**, 253-267 (1983).
20. K. H. Nuechterlein and M. E. Dawson, Schizophrenia Bulletin **10**, 300-312 (1984).
21. H. D. Brenner, British Journal of Pychiatry **155**, 74-83 (1989).
22. G. B. Schmid, Psychopathology **24**, 185-198 (1991).
23. E. Aebi *et al.*, Zeitschrift für Klinische Psychologie, Psychopathologie und Psychotherapie **41**, 18-30 (1993).
24. Z. Kupper and H. Hoffmann, Swiss Journal of Psychology **54**, 211-224 (1995).
25. K. H. Nuechterlein, *et al.*, British Journal of Psychiatry **161**, 88-96 (1992).
26. J. K. Wing *et al.*, *The industrial rehabilitation of long-stay schizophrenic*

patients: a study of 45 patients at an industrial rehabilitation unit (Her Majesty's Stationery Office, London, 1964).
27. W. A. Anthony and M. A. Jansen, American Psychologist **39**, 537-544 (1984).
28. W. A. Anthony *et al.*, Psychiatric Services **46**, 353-359 (1995).
29. P. Lysaker and M. Bell, Acta Psychiatrica Scandinavica , 205-208 (1995).
30. H. Hoffmann *et al.*, Psychiatric Services **46**, 404-406 (1995).
31. M. Selvini Palazzoli, Journal of Marital and Family Therapy **12**, 339-349 (1986).
32. K. Hahlweg *et al.*, British Journal of Psychiatry **155**, 112-116 (1989).
33. I. R. H. Falloon, in *Handbook of schizophrenia, Vol.4: Psychosocial treatment of schizophrenia*, ed. M. I. Herz *et al.* (Elsevier, Amsterdam, 1990).
34. J. M. Gottman and A. Roy, *Sequential analysis: A guide for behavioral researchers* (Cambridge Univ. Press, Cambridge, 1990).
35. R. L. Levine in *Analysis of Dynamic Psychological Systems*, ed. R. L. Levine and H. E. Fitzgerald (Plenum Press, New York, 1992).
36. O. E. Rössler, Physics Letters **57A**, 397-398 (1976).
37. J. P. Dauwalder in *Psychology and promotion of health*, ed. J. P. Dauwalder (Hogrefe and Huber, Seattle, 1994).

Acknowledgments

This research was supported by the Swiss National Science Foundation, grant 3200-028795.

Part IV

APPLICATIONS TO CREATIVITY

CHAOS THEORY, SCIENTIFIC REVOLUTIONS AND CREATIVITY IN LEARNING

ROBERT E. KAHN
Department of Language and Literacy, School of Education, University of Missouri-Kansas City, 5100 Rockhill Road, Kansas City, MO 64110, USA

This paper seeks to clarify the vocabulary of chaos theory, analyze its popularization and acknowledge certain problems. On this foundation, an analysis is presented of creativity in the learning process, as well as how chaos theory can lead to a better understanding of the complexity of learning.

The Vocabulary of Chaos Theory

It has been suggested that modern physics has "three ultimate frontiers: . . . the very small, the very large and the very complex"[1]. On that basis, the three key concepts are quantum mechanics for the very small, relativity for the very large, and chaos theory for the very complex. Indeed, these three areas have sometimes been called the three "scientific revolutions" of the twentieth century[2]. However, debate rages about whether chaos theory is "the beginning of the third scientific revolution in this century"[3], simply a small part of this "'third revolution' . . . across a broad front"[1], or only part of "the computer revolution"[4]. Of course, both Shlesinger and Gleick have failed to mention the revolution in geology--the development of plate tectonic theory and the seafloor spreading hypothesis[5,6]. Furthermore, I would suggest that it is quite possible to see chaos theory as merely a tool, such as calculus, useful in many scientific disciplines. The question has also been provocatively raised: "Is physics conquering chaos, or chaos undermining physics?"[7,8]

I am not a physicist, and having raised these questions, I do not propose to answer them. However, I do wish to define three phrases--*the science of chaos*, *chaotic phenomena* and *chaos theory* --and then argue that the central issue is how well contemporary chaos theory defines chaotic phenomena, both quantitatively and qualitatively, not whether the science of chaos can be clearly defined or constitutes a revolution. In evaluating chaos theory, Larry Laudan's framework of "theory pursuit" and "theory acceptance" is, I believe, far more relevant than Thomas Kuhn's misleading distinction between "normal" and "revolutionary" science[9,10]. Within Laudan's framework of a pursuit of rationality (rather than revolution), chaos theory enables us to conceptualize creativity in learning in new and exciting ways.

An agreed vocabulary is essential for communication. As working definitions, consider "the science of chaos," as a quest for a distinct "interdisciplinary science . . . the science of complexity, the science of process"[11]. "Chaotic phenomena", whether in mathematics, nature, life, or the brain are, in the strict technical sense, apparent randomness[3,12], although consideration of those phenomena are generally in the context of how apparently random phenomena contain hidden order, as well as how ordered phenomena become random. For example, Harrison begins with the arrangement of the letters "A" and "B":

> BAABABBBABAABBABAA⪵BBBBBBBBBAAAAAAAAA
> Here we have two kinds of hypothetical objects in a disorderly arrangement and in an ordered one, with transitions between the two in either direction: left to right, order out of chaos; right to left, chaos out of order. Both kinds of changes occur in natural systems. What is it, in the initial state of a system, which determines its destiny: being ordered to stay ordered; being chaotic to stay chaotic; to go from order to chaos; to go from chaos to order[13]?

To tackle these questions, "chaos theory" has developed as a scientific theory whose subject matter is dynamic systems, regardless of the entities within the system. Some prefer the term "complexity theory" to express the relationship between chaotic and ordered phenomena, but there is also a tendency to treat complexity as an "emerging science at the edge of order and chaos", leading to further confusions[14]. However, it is important to note that both "chaos theory" and "complexity theory" are concerned with precisely the same scientific problem:

> Levels of emergent order . . .originate at the localest grass-roots rules and cascade up, yielding self-organization at increasing scales of complexity. . . . One can intuitively see how, with a clever choice of laws, local rules can govern global behavior. The hard part is understanding how global behavior can govern local rules[15].

The Popularization of Chaos Theory: 1963-1993

As Hayles has commented, the vague term, "chaos" has "now become so thoroughly deprofessionalized that its use is regarded as a signal that one is in the presence of a dilettante rather than an expert"[16]. However, she retains the term "precisely because of the ambiguous meanings that inhere within it," and no doubt many others will do the same. The ambiguities surrounding the term "chaos" have arisen in large part because of the popularization of that word by Gleick in his *Chaos: The Making of a Science*[2] and by Li and Yorke in their earlier article, "Period Three Implies Chaos," in *The American Mathematical Monthly*[17].

In popularizing chaos as a new science, Gleick accepted Kuhn's division between "normal science" and "revolutions"[2, 10, 18, 19]. Gleick's study on the origins of chaos as a science certainly emphasized the vision of chaos as a science in its own right. However, Kuhn's revisions of *Structure of Scientific Revolutions*, as Musgrave and others have pointed out, made the initial approach far less radical, far more ambiguous[20], and Kuhn's work is now seen by Laudan and many others as "deeply flawed, not only in its specifics but in its central framework assumptions"[21, 9]. In brief then, Gleick's reliance on Kuhn's philosophy of science is not sound, especially the distinction between "normal" and "revolutionary" science.

Justifying the ambiguous use of the word "chaos" in the title of their article, Li and Yorke develop their equations with the viewpoint that irregularities and chaotic oscillations of complicated phenomena may sometimes be understood in terms of the simple model, even if that model is not sufficiently sophisticated to allow accurate numerical predictions[17]. They are, of course, following the approach of Lorenz, but offering new terminology, as Lorenz had not used the term "chaos" in his earlier work on "deterministic nonperiodic flow", and even in the later personal review of his work in 1984, preferred the term "irregularity" to describe "a fundamental property of the atmosphere"[22, 23, 24, 25].

As Schwartz has commented, it is quite relevant to ask "how an analysis of dynamic instabilities in physical systems received the compelling name of chaos theory"[26]. Yorke's reply was:

> I was trained [in] the abstract, terminology bound traditions of pure mathematics and I very much wanted to get away from that. So I chose something a bit more glamorous[26].

Certainly, Li and Yorke's use of the term "chaos" did a great deal to popularize work on dynamic instabilities, especially among applied mathematicians, and physical, biological and social scientists, as well as the general public, including those who reject scientific determinism. Even Lorenz now acknowledges the relevance of the term "chaos" to "refer to processes that may be quite orderly, but appear on casual inspection to be disordered"[27, 28].

Reflecting on the manner in which the vague term "chaos" has been expanded, Schwartz claims:

> ...whatever may have been going on for Jim Yorke in his choice of chaos as a name for dynamical instability, the phrase has struck a deep social chord. Chaos theory? Now *that* sounds interesting. Perhaps--so we think--we can learn something from the scientists about our present condition. The danger is in trying to extract meaning for social relationships from an analysis of our experience of nature is that we tend to project our experience of society onto our experience of nature in the first place. . . . The value of chaos theory as a framework for understanding human social relationships is minimal. The origin of chaos in human relationships as we commonly understand it usually rises from the failure of social structures to meet basic human needs[26].

Schwartz is, in my view, unaware of the relevance of mathematical chaos theory to other disciplines, especially economics and organizational theory. The subtitle of his book, *The Creative Moment: How Science Made Itself Alien to Modern Culture* indicates his aim, but, in fact, chaos theory is, better seen as "a recombining of the analysis and rigor of the physical sciences with the vision of the social scientists and humanists"--the words which George Cowan used to characterize the purpose of the Santa Fe Institute for the Study of Complexity[14].

Those of us interested in utilizing chaos theory have a responsibility to consider Schwartz's closing argument:

> Chaos theory, that spectacularly named development of the century-old theory of dynamical instabilities, represents a genuine intellectual advance in its escape from the confines of Newtonian mechanical determinism. But it is still not clear whether chaos theory is simply a fad with a flashy name or a genuinely new conceptual framework. When one recalls the hype that accompanied the entrance of catastrophe theory into the media in the 1970s, one realizes how much can be attached to a name. Catastrophe theory has vanished without a trace. Chaos theory, as a theory, has an uncertain future[26].

Yet in practice, since Lorenz's work in 1963, chaos theory has impacted not only weather forecasting, but a large number of other disciplines[3, 29]. Recently, Harrison has suggested that chaos theory is an important bridge in linking physical scientists and experimental biologists in their search for how living organisms acquire shape[13, 30], while Gregersen and Sailer have applied chaos theory to social science research[31]. Furthermore, Kellert has examined four stages of determinism and claimed:

> Chaos theory not only argues against the predictability of certain systems, but that when combined with quantum-mechanical considerations it leads us to grave doubts about the doctrine of determinism itself[32].

In terms of the history of ideas, chaos theory has completed its first generation of popularization (1963-1993), but is still developing rapidly, and yet there is a real need to look backwards and develop a much fuller understanding of how and why chaos theory has developed to its present position[32].

Problems in Evaluating Chaos Theory

The mathematician, Stanislaw Ulam, has pointed out the impossibility of deciding what is "important" within the discipline of mathematics when any one mathematician sees so few of the 200,000 new theorems in academic journals each year: "One cannot have survival of the fittest if there is no interaction"[33, 34]. The difficulty is not simply the increasingly narrow specialities within mathematics and many other physical, biological and social sciences, but that thinking itself "is possibly a mirror of the prodigality of nature which produces a million species of different insects"[33, 34]. These issues are compounded for chaos theory which is certainly an interdisciplinary pursuit where all the rules of interdisciplinary research apply[35, 36, 37].

To determine for each area of science whether chaos theory has produced better solutions than non-chaotic models of analysis is a daunting task, but precisely what needs to be done if one is to evaluate its success. Clearly, some of the solutions produced by chaos theory are now being judged better than non-chaotic solutions in such areas as meteorology, cardiology, neuroscience, ecology and a host of other areas[3, 38, 39, 40, 41]. It would be worthwhile to produce a detailed list of the specific scientific problems where chaos theory has proved superior to more traditional approaches, but beyond the boundaries of this paper. However, what can be done is to delineate an approach to science within which such a perspective fits.

For Larry Laudan, the initial question of whether any theory should be accepted is based on whether "it provides satisfactory solutions to important problems", rather than whether the theory is in some sense "true", "corroborated", or "well-confirmed"[9]. However, it should be noted that within Laudan's wider model of cognitive progress, he follows Duhem in insisting that "individual theories *in isolation* generally solve no problems" since it is "complexes of theories which are involved in problem solution"[42]. In fact, to use Laudan's taxonomy of scientific problems more than one theory needs to be considered. The resulting emphasis upon contextualization creates what might be termed a seamless web in a gradual movement from a scientific problem to a theory which purports to solve that problem, to a research tradition based on a number of theories, to a world-view which underpins the research tradition.

It is entirely keeping within Laudan's approach to consider the movement from the problem of how to explain chaotic phenomena, to theories about chaotic phenomena in a number of branches of science and social science, to a research tradition based on chaos theory, to chaos as a scientific world-view. With his emphasis upon rationality as a basis for theory acceptance (i.e. why scientists "choose to accept one among a group of competing theories and research traditions"), Laudan opts for "the theory (or research tradition) with the highest problem-solving adequacy"[9]. However, Laudan's approach requires consideration of more than one theory in order to measure a rate of cognitive progress, and despite his awareness of the importance of a diochronic and developmental perspective, his own model really lacks such a perspective. Even if individual scientists lack knowledge of *all* the theories in a vast field, they will still need to know the relevant competing theories that attempt to solve a particular scientific problem in order to make a judgment about which theory should be accepted. The need to consider competing theories is especially important when it is not clear to which discipline a particular scientific problem belongs or how particular disciplines interact in solving a particular problem[9].

From an historical perspective, Laudan points out that:

> Whether we look to Copernicanism, the early stages of the mechanical philosophy, the atomic theory in the first half of the nineteenth century, early psychoanalytic theory, the preliminary efforts at the quantum mechanical approach to molecular structure, we see the same pattern: scientists often begin to pursue and to explore a new research tradition long before its problem-solving success (or its inductive support, or its degree of falsifiability, or its novel predictions) qualifies it to be accepted over its older, more successful rivals[9, 43].

Often, scientists will have "good reasons for working on theories that they would not accept", especially if a particular research tradition "has a higher rate of progress than its rivals"[9]. That is precisely the position that chaos theory as a research tradition has now reached, and why many scientists may soon become known as "chaoticians"--those who analyze chaotic phenomena--within their own branch of science or social science. In brief then, despite a number of problems with Larry Laudan's philosophy of science which have not been presented here, his basic understanding of "theory acceptance" and "theory pursuit" does appear to be relevant to the development of chaos theory as a research tradition[43, 21].

A final problem with evaluating chaos theory is the ubiquitous nature of mathematics in chaos theory, whatever the particular branch of chaotic phenomena that is being considered. Furthermore, the less confident the scientist or philosopher of science is with mathematics, the greater the problem. Ulam relates a story by the Nobel Prize physicist, Chen Ning Yang about the intellectual relationship between mathematicians and physicists:

> One evening a group of men came to a town. They needed to have their laundry done so they walked around the city streets trying to find a laundry. They found a place with the sign in the window, 'Laundry Taken in Here.' One of them asked: 'May we leave our laundry with you?' The proprietor said:"No. We don't do laundry here.' 'How come?' the visitor asked. 'There is such a sign in your window.' "Here we make signs," was the reply. This is somewhat the case with mathematicians. They are the makers of signs which they hope will fit all contingencies[33].

The fact that mathematicians have done a very good job of making all kinds of signs about chaos theory does not mean they will do "the dirty work" of actually washing the laundry. Individual scientists and philosophers of science will need to wash their own laundry, to construct their own theories of how to understand chaotic phenomena. Sometimes those theories will need to use mathematics, and sometimes the scientist will be entitled to explore different avenues and point out that mathematics is "only one of many possible ways of looking at the world" [34]. After all, mathematics can be seen as "only a language, a rich language with a structure and internal relationships worth exploring, but nevertheless only a language" [26].

Given all these problems in evaluating chaos theory, the philosopher of science has an important role to play in clarifying the terms of the debate, suggesting possible ways forward and, as Bruno Latour suggests, "protecting against one factor's hegemony" by questioning in the Socratic manner[44,45]. However, there is a danger that "philosophy *of* science has now become philosophy *for* science", that philosophers have simply become what John Locke termed "under-laborers" for the scientific "master-builders" and what Fuller calls "dangerously close to ... apologists for the masters whose houses they so dutifully clean"[45].

It seems important then not to be sucked into the debate about the nature of the overall ramifications of chaos theory, where one loses sight of the fact that chaos theory, like other scientific theories, claims to solve problems. That is why I now wish to turn to the issue of how chaos theory might link to creativity in learning. In this search for creativity, one is seeking a relationship between order and disorder, between structure and freedom--striving for that dynamic point of balance where creativity occurs, where in the midst of constant movement and disagreement, self-organization occurs in such a way that a new idea is born.

Creativity in Learning

Both in art and in science, Bronowski argues that a person becomes creative by finding "a new unity", "a likeness between things which were not thought alike before"[46]. Indeed, "the creative mind is a mind that looks for unexpected likeness" finding "a more profound unity", "a pattern" behind chaotic phenomena.[46] In the context of scientific discovery, it can also be argued, that creativity is linked to a "search in a space of hypotheses and a space of experiments"[47]. This "Dual Search" involves the formation of new hypotheses and new experiments which are then linked by "a comparison of the predictions derived from a hypothesis with the results obtained from the experiment"[47, 48, 49]. Thus even after the creative mind has produced a new hypothesis--a flash of insight-- experiments follow which confirm or deny that initial insight.

Root-Bernstein has one of the characters in his dialogues suggest that the very nature of science involves a process of discovery in which:

> We actually *transform* ideas from one way of thinking to another. We take numbers and turn them into pictures, or an apparatus and turn it into a mental abstraction. 'Transformational thinking,' I call it. So the more different mental skills--tools of thought--you can utilize, and the more ways you can transform ideas, the better your chances of solving problems[50]. [emphasis in original]

Thus both Bronowski and Root-Bernstein would seem to be arguing that there are creative processes applicable to both art and science in which a transformation--a new idea--is formed in the mind of a single individual. I believe these processes can best be called "chaotic thinking", which involves a movement toward attractors and their basins, especially cascading attractors when two or more attractors couple and blend together into a new self-organizing structure. However, to anchor this kind of thinking empirically it is important for both scientists and artists to engage in Dual Search, as Klahr and Dunbar have suggested--to formulate and test new hypotheses with new experiments[48].

It can be argued (and most scientists would) that such tests should take place "in simplified and tightly controlled laboratory situations," rather than "in uncontrolled naturalistic situations"[51]. However, the problem remains of applying the laboratory research to specific real-life problems. Therefore, it is sensible to adopt Jenkins' suggestion that: "The laboratory profits from the stimulation of the world and real life problems as much as the real problems profit from the laboratory research"[52]. This balanced approach has the further advantage that creativity in learning is no longer seen as solely a cognitive process, but also as a social practice in which the focus shifts "from the individual as learner to learning as participation in the social world": Learning then involves not simply "the acquisition of propositional knowledge" but more significantly, the construction of a new personal identity [53, 54].

The change in perspective is significant because creativity in learning can now be considered in the light of "improvisation, actual cases of interaction, and emergent processes which cannot be reduced to generalized structures"[54]. Indeed, as Hanks comments upon the work of Lave and Wenger:

> The challenge, it would seem, is to rethink action in such a way that structure and process, mental representation and skillful execution, interpenetrate one another profoundly[54, 53].

In other words, learning arises from participation, "based on situated negotiation and renegotiation of meaning in the world", linked to the interaction of understanding and experience[53]. Just as "peripheral participation" becomes a legitimate path to learning for the newcomer to a specific community, so chaos theory may offer some important clues to creativity in learning even if initially what chaoticians have to offer appears somewhat peripheral to other approaches to learning. In the investigation of how creativity in learning occurs, it is helpful to consider earlier work in psychological theory on the nature of attention. Although Csikszentmihalyi has popularized the term "flow"[55, 56, 57], the phenomenon being investigated is broadly what William James defined in 1890:

> Everyone knows what attention is. It is the taking possession by the mind, in clear and vivid form, of one out of what seem several simultaneously possible objects or trains of thought. Focalization, concen-tration, of consciousness are of its essence. It implies

> withdrawal from some things in order to deal effectively with others, and is a condition which has a real opposite in the confused, dazed, scatterbrained state which in French is called *distraction*, and *Zerstreutheit* in German. . . . The abolition of this [dazed] condition is what we call the awakening of attention. One principal object comes then into the focus of consciousness, others are temporarily sup-pressed. The awakening may come about either by reason of a stimulus from without, or in consequence of some unknown inner alteration[58, 59, 55].

This process by which "one principal object comes then into the focus of consciousness" has been outlined by James's contemporary, the French mathematician, Henri Poincaré, who stressed that the "appearance of sudden illumination" is "a manifest sign of long, unconscious prior work"[60, 61]. In brief, Poincaré argued that this sudden emergence of a key, clarifying idea (in the context of "mathematical creation" or "mathematical discovery") only occurred when a period of sustained conscious work both preceded and followed the unconscious reflection. Within the unconscious ("in the subliminal self"), there is an on-going search for useful combinations of ideas. Poincaré calls this principle, "liberty...the simple absence of discipline...the disorder born of chance...this disorder [which] itself permits unexpected combinations"[60, 61].

Furthermore, in seeking "to find a better definition of chance" Poincaré recognized the importance of "the initial conditions":

> To find a better definition of chance we must examine some of the facts which we agree to regard as fortuitous, and to which the calculus of probabilities seems to apply; we then shall investigate what are their common characteristics. . . . A very slight cause, which escapes us, determines a considerable effect which we can not help seeing, and then we say this effect is due to chance. If we could know exactly the laws of nature and the situation of the universe at the initial instant, we should be able to predict exactly the situation of this same universe at a subsequent instant. But even when the natural laws should have no further secret for us, we could know the initial situation only *approximately*. If that permits us to foresee the subsequent situation *with the same degree of approximation*, this is all we require, we say the phenomenon has been predicted, that it is ruled by laws. But this is not always the case; it may happen that slight differences in the initial conditions

> produce very great differences in the final phenomena; a slight error in the former would make an enormous error in the latter. Prediction becomes impossible. . . .[60] [emphasis in original French]

In the context of learning, "the initial instant" can be considered as the first presentation of a new idea or problem, when the attitude of the student (whether in self-learning or instruction) becomes crucial, especially in the context of what is happening at an unconscious level.

In order to clarify initial attitude to learning, the linguist Peter Strevens applied Mihaly Csikszentmihalyi's insights to language learning[62]. On the basis of considerable teaching experience, Strevens agreed with Csikszentmihalyi that the relationship between challenge and skill would determine whether a particular individual was in one of four states--in "flow", bored, anxious or apathetic. However, "flow" is more of a psychological state than an attitude, and the term "engagement" is perhaps a better description of the particular attitude to learning which is being sought[63]. Furthermore, "competence" seems to express the learning objective better than "skill", while the level of "challenge" will be a function of competence and task difficulty.

With these adaptations to Csikszentmihalyi and Strevens' approach, the initial response to a learning problem will be characterized by one of four basic attitudes: (1) engagement--when students of high competence are confronted with difficult tasks; (2) boredom--when students of high competence face easy tasks; (3) anxiety (or frustration)--when students of low competence face difficult tasks; and (4) apathy--when students of low competence face easy tasks. (See Figure 1.) Furthermore, since learning is a continuing process over time, an individual student might well experience more than one of these attitudes, even in the space of a single class. It should also be noted that this conceptual framework has been developed independently of the similar work at the University of Milan of Massimini and Carli, as cited by Csikszentmihalyi and Csikszentmihalyi, but their work making use of z-scores to measure "flow" does have an attractive empirical foundation[64, 55].

In a purely linear context, if students with high anxiety about learning mathematics have low competence in tackling a particular problem, their initial attitude to learning is going to be dominated by anxiety, and it will be necessary for the teacher to confront that anxiety and draw the students toward engagement in order for learning to take place. In a nonlinear context, each of these four attitudes--apathy, anxiety, boredom and engagement--can be viewed as attractors in a dynamic system, with our interest centered on the bifurcations (or major changes in attitudes to

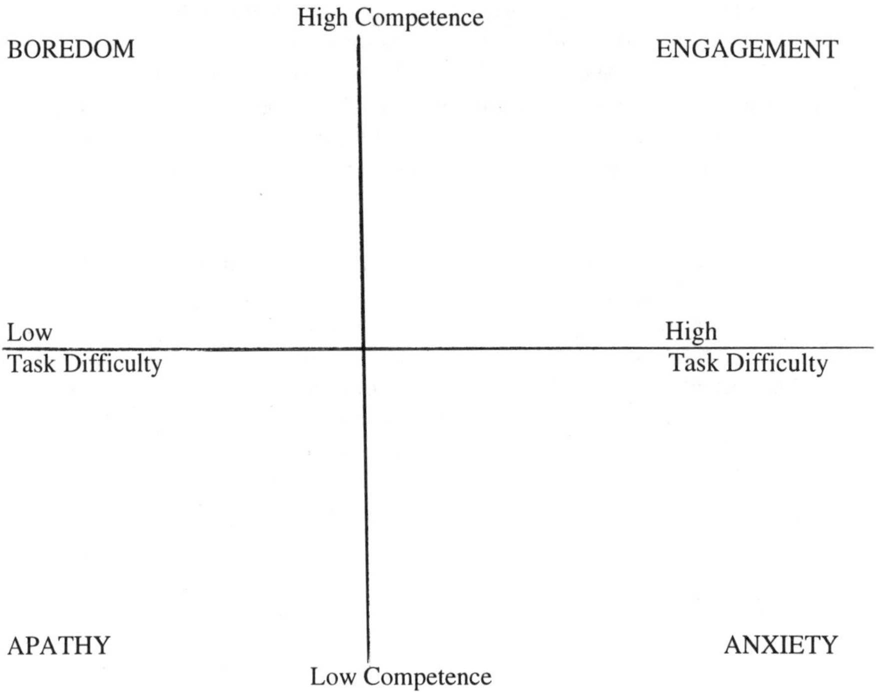

Figure 1. The Relationship of Competence to Task Difficulty in the Learning Process

learning) which create a new learning environment. To a considerable extent, the relationship between competence and task difficulty is a search for that moment when understanding can take place, as prior knowledge and new ideas mesh together in "the teachable moment" or the "a-ha! experience"--that critical point where understanding occurs.

Further reflection suggests that within a dynamic framework, learning can be represented as a dialectic between competence and task difficulty in which interaction with another person or a text is crucial. Even students of low competence can become engaged effectively in learning if the task difficulty matches or is only slightly above their present level of competence. It may well be that the attitudes each individual brings to this learning process have a complexity that can be dissected and understood, at least partially. For example, anxiety may be a product of earlier frustration and stress, just as apathy may be linked to a sense of defeat and depression. Whatever the techniques of analysis, engagement within a learning framework with an instructor should be a step toward the intrinsic motivation essential for self-regulated learning[65, 66, 67, 68].

In seeking to understand the inherent complexity of attitudes the work of J. Richard Eiser is of particular interest[69, 70]. Eiser has pointed out that consistency

> can only be achieved through a *selective* definition of one's context or frame of reference. Such consistency is relative, not absolute; it depends on what we choose to try to relate to what (that is, how we define the context). However, it is far from clear how we can make such a choice (or any choice, perhaps). Faced with a multi-dimensional world, how do we choose what to relate to what[69]?

This is precisely the problem with which Poincaré wrestled in *Science and Method*, in considering "the choice of facts":

> ...what we ought to aim at is less the ascertainment of resemblances and differences than the recognition of likenesses hidden under apparent divergences. Particular rules seem at first discordant, but looking more closely we see in general that they resemble each other; different as to matter, they are alike as to form, as to the order of their parts. When we look at them with this bias, we shall see them enlarge and tend to embrace everything. And this it is which makes the value of certain facts which come to complete an assemblage and to show that it is the faithful image of other known assemblages[60].

For Poincaré, the solution lay in "an intuition of mathematical order", in seeking out the small number of useful combinations; in brief, "invention is discernment, choice"[60, 61]. For Eiser, a change in attitudes is linked to a change in behavior, building on the research of Ajzen and Fishbein[71, 72, 73]. However, the problem remains of precisely how choices are made in specific situations, as well as "how the system is supposed to develop itself and sustain itself over time"[69]. Where Poincaré saw "liberty" within "the subliminal self", and Lave and Wenger stressed the social value of "legitimate peripheral participation", Eiser suggests that "freedom" lies "in the complexity of attitudes" and that: "We are not just spectators, but active participants in the natural world, and our attitudes arise from that participation" [53, 69].

In considering precisely how attitudes develop over time, Poincaré's insight about the importance of the unconscious in the learning process is an important benchmark. An interesting four-stage framework has been suggested by Beeler, after many years of working with university students[74, 75]. Although both Beeler and Poincaré used the terms "consciousness"/"unconsciousness", I shall use the simpler dichotomy, "awareness"/"lack of awareness", in order to avoid a complex series of philosophical and psychological issues. Beeler views learning as a progressive process with four stages: (1) lack of awareness of incompetence --when students with low ability don't know how little they know; (2) awareness of incompetence--when students realize how little they know; (3) lack of awareness of competence--when

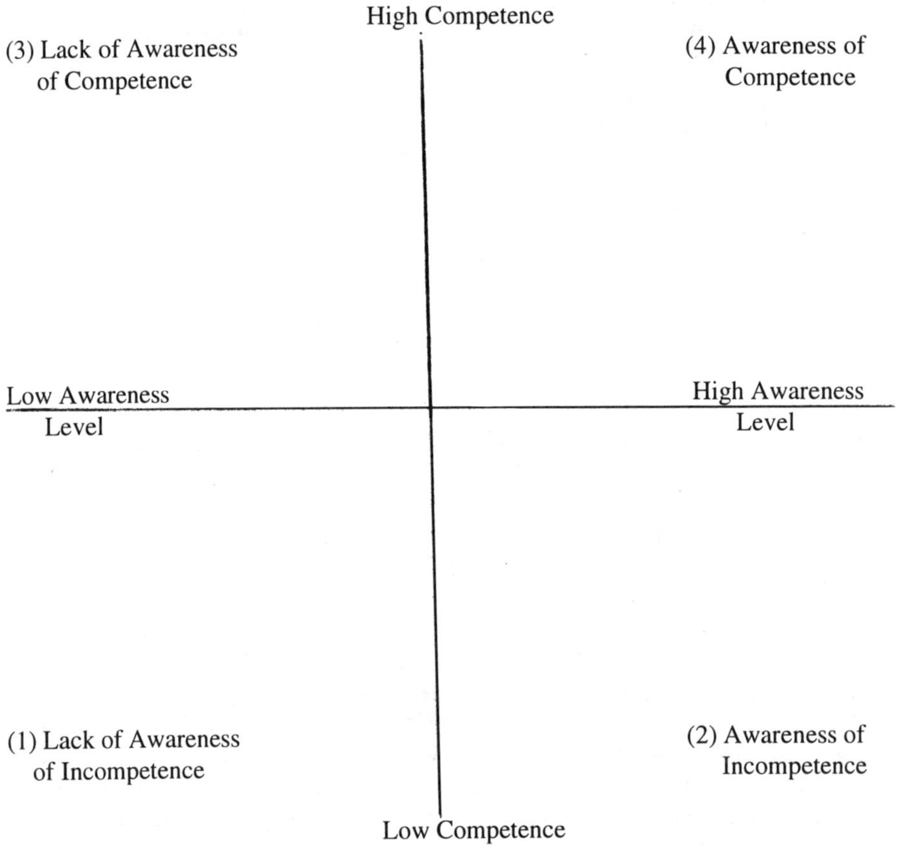

Figure 2. The Relationship of Competence to Awareness in the Learning Process (after Beeler, 1991)

students don't know how much they know; and (4) awareness of competence--when students realize how much they know. (See Figure 2.)

In the context of graduate students, Beeler suggested that his paradigm is often linked to specific time periods, with Phase 1 covering the first half of the first semester, Phase 2 the second half of the first semester, Phase 3 the first half of the second semester, and Phase 4, hopefully, the final half of the second and concluding semester of the academic year. However, this approach is primarily concerned with emotional and academic adjustment, rather than the formation of intrinsically motivated, self-regulated learners facing tasks of varying difficulty. In the context of creativity, it is appropriate to view competence as "the ability of the individual to make meaning", whatever the degree of personal awareness that might be involved [76]. Furthermore, in moving from a "measured" to a "transformative" curriculum, there is a self-organizing process at work within each individual, as "order emerges internally, through interaction" with other individuals and texts[77]. In other words, it would be useful to adapt Beeler's essentially linear perspective to a more dynamic system in order to understand creativity in learning.

Chaos Theory in Learning

The fact that both the natural world and the brain are "inherently nonlinear" suggests that chaos theory might have an important role in learning[78, 79]. However, the possible contribution of chaos theory to learning is still much disputed. Ennis has offered important reflections and sources for "reconceptualizing learning as a dynamic system," but she is concerned primarily with values as attractors and constraints in the educational system[80]. Doll sees chaos theory as central to future curriculum development in the context of post-modern structuralism, but my concern is with the more basic issue of how competence, task difficulty and degree of awareness are related [81].

Cziko has argued that precisely because human behavior is so complex and unpredictable, researchers should be content to seek descriptions and interpretations, rather than predictions of behavior[82]. Certainly, in some of the applications of chaos theory to adult education and composition studies, the interest has been primarily in using chaos theory as a metaphor or aid to reflection at the descriptive level suggested by Cziko[83, 84]. However, there is increasing awareness that with the use of sophisticated non-linear statistics, chaos theory can be applied effectively to reading research[85], developmental disabilities[86] , the nature of human action itself[87] , and widely within the field of social psychology [88].

Precise tasks and competences will need to be carefully defined if chaos theory is to make a useful contribution to learning. Carroll's remarkable exploration of cognitive abilities in more than 450 factor-analytic studies assumes the use of purely linear models with respect to narrow, broad and general cognitive abilities[89]. However, Carroll points out that:

> . . . successful performance on a task may require at least a certain minimum level on each of two abilities [or competences]. The possibility of discovering such a way in which abilities combine is normally excluded by the linear methodology of factor analysis. This methodology can be modified in special studies, however, or replaced by an entirely different methodology. These special studies and methodologies would in no way conflict with the basic three-stratum theory of abilities; they would only address different ways in which the abilities defined in this theory operate or show their effects[89].

In essence, Carroll claims that even if the linear assumptions which underlie his analysis are abandoned, his three-stratum theory of cognitive abilities might be reinterpreted in the context of both social practice (as suggested earlier in this monograph) and chaos theory. Given the scale of Carroll's investigations, this is a daunting task, but the basic data is available to tackle the work. However, is the factor-analytic research methodology appropriate for the study of nonlinear systems?

In contrast to Carroll's work and the traditional factor-analytic research methodology that seeks to identify independent, isolated variables, a dynamic systems approach (whether or not there is chaotic phenomena) stresses "the feedback among the relevant factors and the tendency of the system to become self-organized on the basis of patterns of such feedback"[90]. This is a profound change in methodology based upon a recognition that the factor-analytic research methodology is inadequate for the study of dynamic systems. In this context, it is relevant to note that chaos theory itself is just as likely to develop into an important, new aspect of statistics as a new science of chaos [91, 92].

It is appropriate then to take some first steps in the application of chaos theory as a statistical tool to improve our understanding of the nature of learning and various attitudes to learning. Two approaches appear promising: (1) the logistic equation and further applications of the work of Edward Lorenz[22, 23, 24, 25]; and (2) a more basic systems theory approach seeking to understand the processes involved in the emergence of a dynamic system. Since Lorenz's work is very much concerned with understanding the characteristics of a dynamic system, the line between these two approaches is at times blurred, especially as both are concerned with an analysis of process rather than structure. Whatever the particular statistical technique used to understand the chaotic phenomena, the dynamic system needs to be studied as "a system with a rule of evolution"[93].

As a first step, consider the learning process as analogous to motion in the physical sciences: from one or more clearly delineated points, operations take place which draw the person or the physical system to a new position through a process which can be precisely described. While it is widely accepted that "chaotic motion is not a rare phenomenon"[94], chaotic learning is not widely accepted. Yet the same necessary conditions for the existence of chaotic motion also apply to chaotic learning. As Baker and Gollub note:

> Several necessary conditions for chaotic motion are that (a) the system has at least three independent dynamical variables, and (b) the equations of motion contain a nonlinear term, that couples several of the variables....While these conditions do not guarantee chaos, they do make its existence possible[94].

In brief, if the key variables and "rule of evolution" within the learning process can be clearly defined, then it should be possible to make use of chaos theory as a statistical aid in understanding the learning process.

In reflecting on the analogy of motion to learning, the work of Edward Lorenz is of particular assistance. In showing how weather patterns could be simulated in models based on twelve and then on only three equations, Lorenz "captured the essence of chaotic behavior in a wide range of natural systems"[95]. Furthermore, Lorenz later demonstrated the essence of chaos in one simple equation:

> $x_{new}=Rx_{old}(1-x_{old})$, possesses unexpectedly complicated behavior analogous to climate and weather. The equation works as follows: Choose a value of R and an initial value of x_{old}. Use these to calculate x_{new}. Then update x_{old} by letting it equal x_{new}. The procedure is then repeated as many times as you wish.
>
> The equation contains the crucial physical ingredients of forcing, R; dissipation, (1-x); and nonlinearity (x^2) that make it a candidate for irregular or chaotic behavior. R could represent a temperature gradient in a dishpan experiment or a heating rate for a convecting fluid while x represents the energy or momentum of the fluid. The dissipation term, (1-x), keeps the solution bounded.[95]

In the learning process what would be analogous to "forcing"?

Remembering that R could represent a heating rate which is forcing action in the fluid, in the context of learning it would be challenge--the relationship of task difficulty to competence--that is pushing an individual to increase prior knowledge. Completing the analogy, if x is the state of knowledge, then x_{old} represents the individual's content knowledge, which is being increased, so that the movement from x_{old} to x_{new} represents the learning process. Furthermore, the starting point of this dynamic learning system is x_{old} which represents the prior knowledge an individual brings to each learning challenge.

For example, in seeking to understand the reading process, many qualitative inventories have been developed that measure a student's prior knowledge and reading competence with texts of different degrees of difficulty[96]. Since these texts are graded by year for both task difficulty and reading competence, a vast amount of data is available that could be analyzed in the context of reading as a self-organized dynamic system for each student. In the context of Lorenz's equation "1" would represent perfect knowledge; alternatively, a numerical system could be utilized in which a certain grade level (such as 10) could be set as a point at which the student would be judged to be a proficient reader.

While building upon the present linear analyses of reading, this Dynamic Model of Reading would have a number of advantages: (1) the importance of "teaching up" would be emphasized, so that teachers would be encouraged to give students texts that were one to two years above their independent reading levels; (2) the importance of relating task difficulty to competence would also be stressed, so that teachers would not give students texts that were much too difficult for their existing level of competence; (3) reading would be seen as a dynamic system in which prior knowledge, task difficulty and competence interacted within each student; (4) the feedback among these three variables could be studied in terms of how each student organized the act of reading as a personal self-organized system; and (5) the necessity of longitudinal studies would be emphasized in order to see how each student was progressing over time. Furthermore, particular stages in the reading process such as the initial act of "learning to read", as well as the later step of "reading to learn" could be analyzed in some detail, in order to clarify the feedback among the variables at different levels of competence.

There is increasing awareness that the language arts should be seen as a unity with reading, writing, listening, and speaking understood to be related variables in a dynamic system. Here again, the relationship of the variables could be studied using Lorenz's single equation, especially in terms of the feedback of the various measures of literacy upon each other. In a sense, the Dynamic Model of Reading proposed above is one possible implementation of Eiser's call for "a dynamic view of mind" based upon William James's awareness that "thought is in constant change"[69, 70]. The recognition that reading and thinking can be taught together has already led to the development of an Independent Reading and Thinking Inventory[97] which would provide a useful tool on which to build a Dynamic Model of Reading.

The underlying idea in relating chaos theory to creativity is that literacy itself is "in motion"--that the movement from ignorance to knowledge can be defined in different way in different disciplines, but the basic pattern of movement has an interdisciplinary foundation. The self-awareness of the individual is also crucial to the development of literacy, particularly because self-awareness of competence builds self-esteem. As the key variables in literacy development are defined, it should then be possible to build relevant one-dimensional maps which will identify the key attractors, constraints and boundaries[98, 99, 100]. One key hypothesis to be tested is whether awareness itself (or perhaps self-awareness of competence defined as self-esteem) is the crucial "forcing" agent in the development of literacy. In other words, when awareness is added as a further dimension to the development of literacy, the interaction between task difficulty, prior knowledge, competence, and self-awareness will provide an exciting formulation to understand and enhance literacy in each student. However, self-esteem should not be seen as a purely cognitive variable because of the relevance of emotional and evaluative factors[101].

To document these ideas for individual students is the next step, recording the sensitivity to initial conditions which characterizes all dynamic systems. For example, it should be possible to demonstrate the impact of changes in instructional techniques on learner motivation and achievement. Each student and each instructor will have a highly personal preference for different types of teaching and learning styles. It is important to recognize that such preferences will differ not only for each student, but often for each subject studied by a student, at changing levels of prior knowledge, task difficulty, levels of competence, and degrees of awareness of competence. Furthermore, it should be noted that this approach is not a plea for further development of the misleading Myers-Briggs Inventory with its various personality categories[102], but rather an assertion that the act of learning is itself part of a dynamic system, both within each person and in the interaction between student and teacher. To conceive of a person as having only one major personality type or learning style is dangerously simplistic when learning itself is such a dynamic system.

Of course, much work remains to be done in testing these ideas empirically, but respecting the individual student's metacognitive awareness of the learning process should enable us to gather data about how each student is experiencing learning. It should be possible to determine how learning based upon instruction can draw an individual into self-regulated learning, since engagement itself becomes an attractor that transforms the individual from simply trying to understand the instructor into developing new ideas. The key step may be to link learning theory to connectionism--"the idea that many simple structures exhibit complex collective behavior because of connections between the structures" in a neural context[103, 104, 105, 106, 69].

Although these ideas do have a complexity which is not initially inviting, if learning itself is a dynamic system then complexity cannot be avoided. As H. L. Mencken once said, "For every complex problem, there is a simple solution--and it's always wrong [107]. Creativity itself is a complex phenomenon. How much of this complexity is of the variety amenable to chaos theory is still an open question, but given the chaotic nature of the brain itself, the inherent complexity of the learning process is not in itself surprising[40, 79, 80]. Perhaps the term "chaotic thinking" will prove to be a correct designation of how we learn to think creatively, and it may prove possible to improve our understanding of creativity and literacy through the use of nonlinear statistical models.

Acknowledgments

I should like to acknowledge the substantial assistance and encouragement of Professor Henry Frankel, Department of Philosophy, University of Missouri-Kansas City, in the preparation and organization of this paper. Comments received from William Doll, Jr., Jim Horn, Terry Marks-Tarlow, Anthony Manzo and Denise Wilburn were also most helpful.

References

1. P. Davies in *The New Physics*, ed. P. Davies (Cambridge University Press, Cambridge, 1989), 4, 1.
2. M. Schlesinger quoted in J. Gleick, *Chaos: Making a New Science* (Penguin, New York, 1987), 6, 35-39.
3. J. Ford in *The New Physics*, ed. P. Davies (Cambridge University Press, Cambridge, 1989), 353, 350.
4. J. Franks, *The Math. Intelligencer*, **11**(1), 65 (1989).
5. L. Laudan in *Scientific Discovery: Case Studies*, ed. T. Nickles (Reidel, Dordrecht and Boston, 1980), 331.

6. H. Frankel in *Scientific Discovery: Case Studies*, ed. T. Nickles (Reidel, Dordrecht and Boston, 1980), 345.
7. R. V. Jensen, *American Scientist*, **75**, 168 (1987).
8. R. Poole, *Science*, **245**, 26 (1989).
9. L. Laudan, *Progress and Its Problems: Toward a Theory of Scientific Growth* (University of California Press, Berkeley, 1977), 136, 13-14, 7, 115 110
10. T. S. Kuhn, *The Structure of Scientific Revolutions*, 2nd. ed. (University of Chicago Press, Chicago, 1970).
11. W. E. Doll, Jr, *Educational Leadership*, **47**(1), 66 (1989).
12. J. P. Crutchfield *et al, Sci. American*, **255**(6), 46 (1986)
13. L. G. Harrison, *Kinetic Theory of Living Pattern* (Cambridge University Press, Cambridge, 1993), 24.
14. W. M. Waldrop, *Complexity: The Emerging Science at the Edge of Order and Chaos* (Simon and Schuster, New York, 1992), 357.
15. K. Kelly, paraphrasing P. Maese in *Doing Science*, ed. J. Brockman (Prentice Hall, New York, 1991), 31.
16. N. K. Hayles in *Chaos and Order: Complex Dynamics in Literature and Science*, ed. N. K. Hayles (University of Chicago Press, Chicago and London, 1991), 2.
17. T. Li and J. Yorke, *American Math. Monthly*, **82**, 985 (1975).
18. M. W. McRae in *The Literature of Science: Perspectives on Popular Scientific Writing*, ed. M. W. McRae (University of Georgia, Athens, Georgia, 1993), 1.
19. J. W. Slade, *Science*, **264**, 980 (1994); review of ref. 18.
20. A. Musgrave, *British J. for the Phil. of Science*, **22**, 287 (1971). Reprinted in *Paradigms and Revolutions: Appraisals and Applications of Thomas Kuhn's Philosophy of Science*, ed. G. Gutting (University of Notre Dame Press, Notre Dame, 1980), 39.
21. L. Laudan, *Science and Values: The Aims of Science and Their Role in Scientific Debate* (University of California Press, Berkeley), xiii.
22. E. N. Lorenz, *J. of the Atmos. Sciences,* **20**, 130 (1963).
23. E. N. Lorenz, *J. of the Atmos. Sciences,* **20**, 448 (1963).
24. E. N. Lorenz, *Tellus*, **16**, 1 (1964).
25. E. N. Lorenz, *Tellus*, **36A**, 98 (1984).
26. J. Schwartz, *The Creative Moment: How Science Made Itself Alien to Modern Culture* (HarperCollins, New York, 1992), 29, 205, 31-32.
27. E. N. Lorenz, *The Essence of Chaos* (University of Washington Press, Seattle; UCL Press, London, 1993), 5.
28. E. N. Lorenz, *Theoretical and Applied Chaos in Nursing*, **1**, 5 (1994).
29. A. Wolf, *Nature*, **305**, 182 (1983).
30. E. Cox, *Science*, **264**, 1010 (1994); review of ref. 13.
31. H. Gregersen and L. Sailer, *Human Relations*, **46**, 777 (1993).
32. S. H. Kellert, *In the Wake of Chaos: Unpredictable Order in Dynamical Systems* (University of Chicago Press, Chicago and London, 1993), 50, 158.
33. S. Ulam, *Adventures of a Mathematician* (Scribners, New York, 1976), 288, 289, 292.
34. P. J. Davis and R. Hersh, *The Mathematical Experience* (Birkhauser, Boston, 1981), 21, 22.
35. J. T. Klein, *Interdisciplinarity: History, Theory and Practice* (Wayne State University Press, Detroit, 1989).
36. D. E. Chubin *et al, Interdisciplinary Analysis and Research: Theory and Practice of Problem-Focused Research and Development* (Lomond Publications, Mt. Airy, MA, 1986).
37. R. E. Kahn, presented at the 16th Annual Conference of the Association for Integrative Studies, Pittsburgh, PA, September 29-October 2, 1994.
38. R. F. May and G. F. Oaster, *The American Naturalist*, **110**, 573 (1976).

39. A. V. Holden, *Nature*, **305**, 183 (1983).
40. E. Basar, *Chaos in Brain Function* (Springer-Verlag, Berlin and New York, 1990).
41. A. Nowak and M. Lewenstein in *Dynamical Systems in Social Psychology* (Academic Press, San Diego, 1994), 17.
42. P. Duhem, *The Aim and Structure of Physical Theory* (Princeton University Press, Princeton, 1954).
43. R. Laudan *et al*, *Scrutinizing Science: Empirical Studies of Scientific Change*, ed. A. Donovan, L. Laudan and R. Laudan (Kluwer, Dordrecht, Boston and London, 1988), 3. 44. W. Callebaut, *Taking the Naturalistic Turn: Or, How Real Philosophy of Science Is Done* (University of Chicago Press, Chicago and London, 1993), 218.
45. S. Fuller, *Science*, **264**, 982 (1994); review of ref. 44.
46. J. Bronowski in *Scientific Genius and Creativity* (W. H. Freeman, New York, 1982), 7. Reprinted from *Scientific American*, **xxx**, xx (1958).
47. D. Klahr, A. L. Fay and K. Dunbar. *Cog. Psych.*, **25**, 111 (1993).
48. D. Klahr and K. Dunbar, *Cog. Science*, **12**, 1 (1988).
49. K. Dunbar, *Cog. Science*, **17**, 397 (1993).
50. R. S. Root-Bernstein, *Discovering* (Harvard University Press, Cambridge, 1989), 245.
51. W. K. Estes in *Stevens' Handbook of Experimental Psychology*, ed. R. C. Atkinson *et al*, 2nd ed. (John Wiley, New York, 1988), II, 405.
52. J. J. Jenkins in *Abilities, Motivation, and Methodology: The Minnesota Symposium on Learning and Individual Differences*, ed. R. Kanfer, P. L. Ackerman and R. Cudeck (Erlbaum, Hillsdale, 1989).
53. J. Lave and E. Wenger, *Situated Learning: Legitimate Peripheral Participation* (Cambridge University Press, Cambridge and New York, 1991), 43, 53, 115, 51-52.
54. W. F. Hanks in *Situated Learning: Legitimate Peripheral Participation* by J. Lave and E. Wenger (Cambridge University Press, Cambridge and New York, 1991), 14, 16.
55. M. Csikszentmihalyi in *Optimal Experience: Psychological Studies of Flow in Consciousness*, ed. M. Csikszentmihalyi and I. S. Csikszentmihalyi (Cambridge University Press, Cambridge and New York, 1988), 17-19, 261.
56. M. Csikszentmihalyi, *Flow: The Psychology of Optimal Experience* (Harper and Row, New York, 1991).
57. M. Csikszentmihalyi, *The Evolving Self: A Psychology for the Third Millennium* (HarperCollins, New York, 1993).
58. W. James, *Principles of Psychology* (Holt, New York, 1890), 403-405.
59. D. R. Davies and R. Parasuraman, *The Psychology of Violence* (Academic Press, New York, 1982), 1.
60. H. Poincaré, *The Foundations of Science; Science and Hypothesis; The Value of Science; Science and Method*, trans. G. B. Halsted (The Science Press, New York, 1913), 389, 394, 397-398, 362, 366, 385-386.
61. H. Poincaré in *The Creative Process: A Symposium*, ed. B. Ghiselin (University of California Press, Berkeley, 1952), 38, 42, 35.
62. P. Strevens in *An Introduction to Applied Linguistics*, ed. W. Grabe and R. Kaplan (Addison-Wesley, Reading, MA, 1992), 13.
63. Personal communication from H. Frankel (June 1994).
64. F. Massimini and M. Carli in *L'esperienza quotidiana*, ed. P. Inghilleri (Franco Angeli, Milan, 1986), 65. In Italian.
65. T. Marks, *Creativity Research Journal*, **2**, 204 (1989).
66. D. H. Schunk & B. J. Zimmerman, eds., *Self-regulated Learning and Academic Achievement: Theory, Research and Practice* (Springer-Verlag, Berlin & New York, 1989).
67. D. H. Schunk & B. J. Zimmerman, eds., *Self-regulation of Learning and Peformance* (Erlbaum, Hillsdale, NJ, 1994).

68. R. E. Kahn, presented at the Midwest Regional TESOL Conference, St. Louis, MO, November 3-5, 1994.
69. J. R. Eiser, *Attitudes, Chaos and the Connectionist Mind* (Blackwell, Oxford and Cambridge, MA, 1994), 16-17, 23, 243, 213.
70. J. R. Eiser in *Dynamical Systems in Social Psychology*, ed. R. R. Vallacher and A. Nowak (Academic Press, San Diego, 1994), 197, 216.
71. I. Ajzen and M. Fishbein, *Psych. Bulletin*, **84**, 888 (1977).
72. I. Ajzen and M. Fishbein, *Understanding Attitudes and Predicting Social Behavior* (Prentice-Hall, Englewood Cliffs, NJ, 1980).
73. I. Ajzen, *Organ. Behavior and Human Decision Processes*, **50**, 179 (1991).
74. K. D. Beeler, *NASPA Journal*, **28**, 163 (1991). Published by the National Association of Student Personnel Administrators.
75. K. D. Beeler, *College Transitions: A First-Year Guide for Parents and Students* (Kendall/Hunt Publishing, Dubuque, IA, 1993).
76. Personal communication from W. E. Doll, Jr. (July 1994).
77. W. E. Doll, Jr., *Theory into Practice*, **25**(1), 10 (1986).
78. T. Mullin, ed., *The Nature of Chaos* (Oxford University Press, Oxford, 1993), xvii.
79. C. A. Skarda and W. J. Freeman, *Behavioral and Brain Science*, **10**, 161 (1987).
80. C. D. Ennis, *J. of Curriculum and Supervision*, **7**(2), 115 (1992).
81. W. E. Doll, Jr., *A Post-Modern Perspective on Curriculum*, (Teachers College Press, New York and London, 1993).
82. G. A. Cziko, *Ed. Researcher*, **18**(3), 17 (1989) and *Ed. Researcher*, **21**(9), 10.
83. R. Domaingue, *New Horizons in Adult Education*, **2**(1), 55 (1988). Eric Document ED 330780.
84. D. Hesse, *Strange Attractors: Chaos Theory and Composition Studies*. Eric Document 342010.
85. R. Robinson and D. B. Yaden, *Reading Research and Instruction*, **32**(4), 15 (1993).
86. D. Guess and W. Sailor, *The J. of Special Ed.*, **27**(1), 16 (1993).
87. A. Neuringer and C. Voss, *Psych. Science*, **4**(2), 113 (1993).
88. R. R. Vallacher and A. Nowak, eds., *Dynamical Systems in Social Psychology* (Academic Press, San Diego, 1994).
89. J. B. Carroll, *Human Cognitive Abilities: A Survey of Factor-Analytic Studies* (Cambridge University Press, Cambridge and New York, 1993), 632.
90. R. R. Vallacher and A. Nowak in *Dynamical Systems in Social Psychology*, ed. R. R. Vallacher and A. Nowak (Academic Press, San Diego, 1994), 8, 13, 4.
91. R. Abraham, cited by O. E. Rossler in *Applied Chaos*, ed. J. H. Kim and J. Stringer (John Wiley, New York, 1992), 457.
92. A. Combs, *Nonlinear Dynamics Workshop*, presented at 4th Annual Conference of the Society for Chaos Theory in Psychology and the Life Sciences, Baltimore, MY, June 28-29, 1994.93. R. Abraham, as cited by A. Combs, ref. 92.94. G. L. Baker and J. P. Gollub, *Chaotic Dynamics: An Introduction* (Cambridge University Press, Cambridge and New York, 1990), 2, 3.
95. S. D. Gedzelman, *Weatherwise*, **47**(4), 25, 26 (1994).
96. See L. Leslie and J. Caldwell, *Qualitative Reading Inventory* (HarperCollins, New York, 1990).
97. A. V. Manzo and U. C. Manzo, *Independent Reading and Thinking Inventory* (Harcourt Brace Jovanovich, Fort Worth, 1995).
98. H. O. Peitgen, H. Jurgens and D. Saupe, *Chaos and Fractals: New Frontiers of Science* (Springer-Verlag, Berlin and New York, 1992).
99. D. Holton and R. M. May in *The Nature of Chaos*, ed. T. Mullin (Oxford University Press, Oxford, 1993), 95.
100. C. H. Wu and N. F. Rul'kov, *IEEE [Institute of Electrical and Electronics Engineers] Transactions on Circuits and Systems--I. Fundamental Theory and Applications*, **40**, 10, 707 (Oct 1993).

101. R. D. Hoge and J. S. Renzulli, *Rev. of Ed. Research*, **63**, 449 (1993).
102. D. J. Pittenger, *Rev. of Ed. Research*, **63**, 467 (1993).
103. D. K. Campbell in *From Cardinals to Chaos: Reflections on the Life and Legacy of Stanislaw Ulam*, ed. N. G. Cooper (Cambridge University Press, Cambridge and New York, 1989), 218.
104. J. A. St. Julien, presented at the Annual Meeting of the American Educational Research Association, New Orleans, April 4-8, 1994.
105. J. T. Townsend in *From Learning Theory to Connectionist Theory: Essays in Honor of William K. Estes*, eds. A. F. Healey, S. M. Kosslyn and R. M. Shiffrin (Erlbaum, Hillsdale, NJ, 1992), 65.
106. J. Dinsmore, ed., *The Symbolic and Connectionist Paradigms: Closing the Gap* (Erlbaum, Hillsdale, NJ, 1992).
107. R. D. Anderson, *J. of Research in Science Teaching*, **29**, 861 (1992).

THE CREATIVE CHAOS: SPECULATIONS ON THE CONNECTION BETWEEN NON-LINEAR DYNAMICS AND THE CREATIVE PROCESS

TOBI ZAUSNER
The New School for Social Research, New York

Chaos theory may provide models for creativity and for the personality of the artist. A collection of speculative hypotheses examines the connection between art and such fundamentals of non-linear dynamics as iteration, dissipative processes, open systems, entropy, sensitivity to stimuli, autocatalysis, subsystems, bifurcations, randomness, unpredictability, irreversibility, increasing levels of organization, far-from-equilibrium conditions, strange attractors, period doubling, intermittency and self-similar fractal organization. Non-linear dynamics may also explain why certain individuals suffer mental disorders while others remain intact during a lifetime of sustained creative output.

1 Introduction

Chaos theory may provide models for creativity and for the personality of the artist. Non-linear dynamics may also explain why certain individuals suffer mental disorders while others remain intact during a lifetime of sustained creative output.

In the following collection of speculative hypotheses about the connection between art and chaos theory, I use the word "painting" generically for other visual arts such as drawing and sculpture because of my background as a painter. It is possible that the analogies between non-linear dynamics and the visual arts may also apply to other activities such as writing and composing music.

In this paper the word "chaos" is used in the way the word "complexity" is used by the Santa Fe Institute scientists, to designate a state of activity that calls forth new combinations of order. At the Santa Fe Institute, the term "chaos" designates a highly active state which does not produce new or lasting organization. I call this state "disorder."

2 Dissipative Processes, Open Systems, and Entropy

Non-linear systems are characterized by dissipative processes, open systems, and entropy [1,2]. In nonequilibrium states entropy can give rise to order.[3]

Creativity also appears to be a dissipative process, one organized by the open system of the creative person. Artists take in energy and information from their environment and discharge entropy as local examples of order which we call art. Ordering energy is creating information and as such is negentropic [4], making creativity a negentropic process.

Like dissipative processes, creativity is fueled by an exchange of energy with the outside world and exists because it is open to this exchange. If an artist were not open to outside influences, disordered stagnation would occur.

As each dissipative structure is a separate entity, so every period of creativity is unique and every work of art should be unique as well. When the works of an artist resemble each other too closely, there is a lack of creativity, and what might have been the chaos of creation is only the linearity of repetition.

3 Iteration

Non-linear processes are built through iteration.[5] Creating a work of art can also be seen as an iterative process. The painting in its current state reacts with the painting in its previous state which is in the artists memory. The painting is always being compared and referred to itself and this comparison lays the trajectory for the evolution of a work of art.

In iterative processes small changes may become amplified.[6] A decision to affect one part of the painting, no matter how small, changes the painting. Becoming incorporated into the whole, it is the basis for future iteration and affects the direction of the painting's progress.

4 Irreversibility

Chaotic processes are characterized by irreversibility [2]. Creativity is also an irreversible process. As a macro process, creativity is time bound. A painting cannot be un-painted. Even though parts of it may be obliterated or taken out, that is still part of the process of bringing it to its final state. Overpainting, scraping or turpentining a work off the canvas may expunge its image but cannot erase the past of its brushstrokes and how it looked. Obliteration can be image obscuring, but is not time reversing. Irreversibility appears to be fundamental to the organization of a work of art.

5 Randomness and Unpredictability

Chaotic process have both randomness and unpredictability.[6] Because there is a degree of randomness in every work of art there is no way of completely predicting its outcome. Without the random unexpected element, art can become rote and lifeless. If a work of art were only the result of randomness it would be a product of disorder. Determination, purpose and intent form the framework or attractors within which the random occurs. Randomness increases the irreversibility of a work of art.

6 Sensitivity to Stimuli

Chaotic systems react to stimuli that would be inconsequential in processes approaching equilibrium. In this sensitivity, a system produces a spontaneous adaptive organization to its environment.[3]

Extreme reaction to stimuli is also found in creative individuals, who are known for their sensitivity to the external world. This reaction can be a positive or a negative adaptation. As a positive adaptation, an artist may be struck by an idea or by the beauty of a scene and be inspired to produce a work of art. Another form of positive adaptation may be increased creative output as a response to encouragement or to critical acclaim.

A negative response to an artist or the works of art can produce a negative adaptation. Both Vincent Van Gogh[7] and Edvard Munch[8] were tormented by the negative response of their neighbors and responded by going deeper into mental illness. Positive critical response aided the career of Titian who went on to become the leading painter in sixteenth century Venice.[9]

7 Autocatalysis

Chaotic processes are autocatalytic.[10] The creative process also appears to be autocatalytic. Artists, through their moods and emotions react with themselves either positively or negatively to enhance or inhibit their productivity. A balance of positive and negative feedback is necessary for the production of a work of art.

Responding positively to their own work, as in a positive feedback loop, artists can increase their excitement and spur their creativity. But with unbridled positive feedback, an artist may paint without adequate assessment of the work produced. Responding negatively, an artist can examine the work and detect flaws. However, if the negative feedback loop is not interrupted, it is possible to become overly critical. Artists can inhibit themselves and their creativity, stopping the work.

8 Sub-Systems and Bifurcations

Chaotic systems contain subsystems which are in constant fluctuation. A single fluctuation can become strong enough to break the existing order and bifurcate the system into a new order.[3]

As systems are composed of subsystems which are constantly in motion, so the artist has layers of consciousness and emotions, which like the resting chaos of brain cells, [11] are ready for excitation. A painting in progress also contains sub-systems. As it is being worked on, its parts are in a constant state of adjustment demonstrating the fluctuations of non-linear dynamics. Very often, in fact always, in my personal experience, a part or subsystem of the canvas can fluctuate in such a way that it bifurcates from the existing composition and takes the whole work of art in a new direction.

One might ask, doesn't the artist have plans for a painting? Usually, and especially in figurative art, the artist does have a plan in the form of a drawing, smaller composition, or just an idea. Such a plan can be seen as the attractors within which the non-linear dynamics of a work of art begin to take form. However in translating the plan to the canvas many things happen. For example, there is a scale change. Things that look good small may not work large. Some artists compensate for this by making a full size drawing of the composition, but even then there are always changes to be made in the final work of art. Sometimes an adjustment is better than the original idea, and the artist has a whole new route to take. Creativity is in fact stopped if the artist is not open to the constant bifurcations that happen in the painting process.

9 Increasing Levels of Organization

Chaotic systems are characterized by increasing levels of organization.[3] In creating a work of art, disparate elements are brought together into a synthesis creating a higher organization than that of the parts by themselves. On the physical level of the painting, the artist combines paints, canvas, and solvents to create a finished work of greater complexity than its individual components. The artist's ideas, intentions, and emotions generate higher organizations of meaning in the work of art.

10 Far-From-Equilibrium Conditions and Strange Attractors

Chaos is a far-from-equilibrium condition [5] whose turbulence is contained within the limits of the strange attractors.[2] Creativity is also a far-from-equilibrium condition. The most chaotic part of the creative process is that of inspired activity. Inspiration takes the artist into in a state of hyper-excitability. In this condition, innovation becomes available because the system can move away from repetitive old patterns and establish something new and unique. Inspiration, with its flood of ideas, can at times, be characterized by turbulence.

The matrix of the personality acts like attractors holding the artist within the parameters of sanity during periods of chaotic inspiration. Artists need to have a strong foundation. The non-equilibrium that produces sensitivity and instability can also create disorder in certain individuals. Some artists have breakthroughs while other artists have breakdowns.

11 The Period Doubling Cascade to Chaos

When a system is pushed beyond stability it may exhibit period doublings which appear as bifurcations cascading into chaos.[5] It is possible that period doublings into chaos may provide a model for the mental instability found in some creative individuals. During inspiration the different ideas or options available to an artist can be modeled as attractor basins in a phase state. As inspiration is fueled autocatalytically by the artist's response to the excitement of creativity, the non- linearity of the situation increases, which may be seen as a multiplication of period doublings.

In a mentally healthy artist, the strange attractors of the personality may act to limit the number of period doublings to ensure a creative chaos that will produce works of art. In a mentally disordered person, the already weak attractors of the personality may not be able to limit the number of period doublings. The nervous energy of inspiration may flood the system with too much non-linearity until the chaos achieved is not creative but disordered.

An example of excessive linearity creating disorder is found in Langton's quantification of Wolfram's work with cellular automata. These computer simulations of life processes were originally found in the work of Conway.[12] Wolfram [13] described four classes of cellular automata: Class I, where every cell died after one or two moves; Class II, with static oscillation of groups of unconnected cells; Class III with groups of over-energized cells that never produced anything; and Class IV with cells that combined, produced new structures, and recombined in a complex manner. When Langton [14] assigned parameters to Wolfram's four states, he found that Class IV was actually a state before, not after Class III, which contained the most non-linearity. He discovered that increasing the non-linearity of a system can push it from complexity into disorder. Again, there is a difference of semantics. What Langton has called complexity, I call a creative chaos, and what he termed chaos, I call disorder.

Huberman [6] also found that too much non-linearity caused a system to behave erratically. He analyzed the constant extraneous eye movements of schizophrenics and their relatives who were unable to track the motion of a pendulum.

Another possible model for mental instability in a creative individual may found be Abraham's [15] buckling column diagram. This phase space with two attractor basins can be seen as two aesthetic options set before an artist. A healthy person is capable of making a choice. One attractor basin is chosen while the other one vanishes.

An unhealthy artist with obsessive tendencies may not be able to choose between the attractor basins. A crippling obsession without resolution might flood the system with too much non-linearity. Obsessive rumination may over-examine each possibility dissecting it into almost infinite nuances. If this happens it is possible that the two attractors may period double into myriad small attractors, becoming a model of too much non- linearity in a system and of mental disorder in an individual.

12 Intermittency and Self-Similar Fractal Organization

Chaotic processes are intermittent and exhibit a structure that is fractal and self-similar.[16] In the mosaic of a chaotic system like the Mandelbrot set, there are islands of order which resemble each other but are not identical. These have been called "Mandelbrots," and are connected to one another by thin filaments.[10]

Creativity also appears to have a fractal nature. Like the Mandelbrots, periods of creativity appear to be self-similar but not identical. Works of art produced by an single artist also tend to be similar but are not identical. As the Mandelbrots appear separate but are connected to each other through a fine filament, so the periods of creativity and the works of art are connected in the threads of the painter's unconscious processes.

From a social/art historical standpoint, the Mandelbrots are analogous to the paintings of a particular school or time which also resemble each other but are not exactly alike. Here the filaments would be the social or art theoretical issues that identify the style. The Mandelbrots can also be seen as the intermittent periods of time when a specific painting is being worked on. Creativity is not a constant process. Not only because some ideas develop over time, but also because the flow stops. Artists sometimes take years to finish a painting and to compensate for this often work on more than one piece at a time. As periods of chaos and order alternate in systems,[5] so do periods of creativity and reflectivity alternate in the life of an artist.

13 Summary

Certain aspects of non-linear dynamics appear to be relevant to the process of creativity and to the personality of the artist. Examining analogies between non-linear dynamics and creativity may provide some understanding of the artist and of the process of making art, and in doing so may strengthen the connection between science and the arts.

References

1. S.H. Kellert, *In the Wake of Chaos*, (University of Chicago Press, Chicago, 1993).
2. D. Ruelle, *Chance and Chaos*, (Princeton University Press, Princeton, 1991).
3. I. Prigogine and I. Stengers, *Order Out of Chaos*, (Bantam Books, New York, 1984).
4. C. Shannon and W. Weaver, W. *The Mathematical Theory of Communication*, (University of Illinois, Urbana, 1971).
5. I. Stewart, *Does God Play Dice, The Mathematics of Chaos.* (Blackwell Publishers, Cambridge, 1989).
6. J. Gleick, *Chaos*, (Bantam Books, New York, 1987).
7. V. Van Gogh, *Dear Theo*, ed. I. Stone, (Penguin Books, New York, 1969).
8. N. Grimes, *Art News*, **Summer**, 29, 1994.
9. D. Rosand, *Titian*, (Abrams, New York, 1978).
10. J. Briggs and F.D. Peat, *Turbulent Mirror*, (Harper and Row, New York, 1989).
11. C.A. Skarda and W.J. Freeman, *Behavioral and Brain Sciences*, **10**(2), 161-195, (1987).
12. M. Waldrop, *Complexity*, (Simon and Schuster, New York, 1992).
13. S. Wolfram, *Scientific American*, **Sep**, 188-203 (1984).
14. C. Langton, in *Artificial Life II*, eds. C. G. Langton, C. Taylor, J. D. Farmer, and S. Rasmussen, (Addison Wesley, Reading, MA, 1992).
15. F.D. Abraham, R.H. Abraham, and C. Shaw, (1989) *A Visual Introduction to Dynamical Systems Theory for Psychology*, (Aerial Press, Santa Cruz, 1989).
16. B.B. Mandelbrot, *The Fractal Geometry of Nature*, (W.H. Freeman, San Francisco, 1983).

THE ICONOGRAPHY OF CHAOS IN A RENAISSANCE PAINTING

TOBI ZAUSNER
The New School for Social Research, New York

Titian's Presentation of the Virgin to the Temple (1534-38) is a large oil painting on canvas (11'x 25'4"), hanging in the Gallerie dell'Accademia in Venice. Underneath its overt Christian imagery, the painting appears to have a hidden humanist symbolism encoding self-transformation through initiation. Humanist disciplines such as Neoplatonism, the ancient mystery religions, Graeco-Roman paganism, and the Jewish Kaballah are suggested in Titian's painting. Certain figures and groups of figures appear to symbolize the underworld and the period of chaos necessary for transformation. Additional evidence for the chaos in self-transformation is found in the painting's geometry and in its astronomical references.

1 The Painting and its Iconography

At last year's conference, several people asked me "What is iconography?," so this year I thought I would speak about it, and specifically about the iconography of chaos in a Renaissance painting. The chaos I will refer to is the emotional chaos that comes with personal growth. Self-transformation isn't easy. If it were, we would all be immediately transformed. Unfortunately this is usually not the case because our old patterns hold us back.

In the Renaissance and in antiquity, old patterns were viewed as an old self. Change was seen as a death of the old self so that a new self could be born at a higher level of functioning. Archaic civilizations saw the emotional stress accompanying self- transformation as a trip to the underworld, a place of chaos, death, and rebirth.

Current psychology believes that transformation comes from accessing the unconscious,[1] but as Harding [2] points out, "the ancients knew no inner or psychological realm, to them the inner world was conceived of as the underworld."(p. 114) The underworld and the unconscious are both characterized by chaos, and it is in chaos that radical change becomes possible.

As people of western industrialized societies use psychotherapy in an effort to transform themselves, so peoples of the past used initiation.[3] The ritualized disequilibrium of initiation symbolized a trip through the underworld as the person re-enacted the death and resurrection of a deity. It is in references to initiation, to self-transformation, and to the underworld that the iconography of chaos becomes evident in the *Presentation of the Virgin to the Temple* by Titian.

Fig. 1 *Presentation of the Virgin to the Temple* by Titian (1534-1538) 11' x 25'4" (345 x 775 cm) Venice, Gallerie dell'Accademia.

This large oil on canvas is considered to be a masterpiece of the artist's midcareer. Titian worked on it from 1534 to 1538, when he was approximately 46 to 50 years old. The painting, measures 345 x 775 centimeters, or approximately 11 feet in height by 25 feet four inches in length. It was commissioned by a men's fraternal organization, the Scuola Grande di Santa Maria della Carita, for the Albergo or Guest Room of their headquarters in Venice. When the brotherhood decided that they needed to accommodate increased traffic, a door was cut into the room in 1572, obscuring part of the lower left portion of the painting.[4] In the early nineteenth century the building was converted into the Accademia Museum and the Albergo of the confraternity became Room XXIV of the Accademia. The *Presentation of the Virgin to the Temple* still hangs on the same wall it was commissioned for and this paper will examine its iconography.

2 Levels of Iconography

According to Webster's Dictionary,[5] (1983) "iconography," which comes from the Greek "eikon," meaning image and "graphia," meaning to write or draw, is the study of images in art. Panofsky[6] believed that iconography has three levels of interpretation each increasing in depth. The first level is formal analysis and concerns form and emotion, such as the recognition of a person and the expression on a person's face. At this level, the *Presentation of the Virgin to the Temple* shows an admiring crowd watching a little girl climbing stairs in a setting that resembles sixteenth century Venice. For the first level all you need is to see the painting, but the following two levels of iconography require a key.

The second level of iconography is concern with subject matter and social conventions. Here, the little girl is recognized as Mary, and the painting is shown to illustrate a story of her childhood found in *The Golden Legend*[7] and in *The Apocryphal New Testament*.[8] Mary, who has been dedicated to God by her parents Anna and Joachim is being presented to the Temple. At the age of three, Mary climbs the stairs all by herself to the place where she will live until her betrothal to Joseph. Although the story of the Presentation of the Virgin is not found in the Bible, it was a popular subject matter for artists.[9]

The third level of iconography is an examination of the intrinsic meaning of a work of art. On this level the painting becomes a distillation of the ideology of its period through the personality of the artist and shows subject matter universal to humankind. The ideology of the period was humanism and it is here that references to chaos appear in Titian's painting. It is my belief that at its deepest level, Titian's painting symbolizes self-transformation, a concept found throughout civilization and a central idea of Renaissance humanism.[10]

3 Humanism and Initiation

Humanism, the great intellectual movement of the Renaissance, began in 1439 as a Neoplatonic Academy[11,12] housed in a Medici villa north of Florence. The leader of the Academy was Marsilio Ficino, who along with his most famous disciple, Pico della Mirandola, and other scholars, studied disciplines such as Neoplatonism, Graeco-Roman paganism, the ancient mystery traditions and the Jewish Kaballah. In all of these traditions, initiation was the method of personality transformation.

In the Renaissance there was a great interest in classical wisdom.[13,14] Ficino translated Plato, Plotinus, and the Hermetic treatises from Greek into Latin. Pico learned Hebrew and Aramaic, integrating the Kaballah and the Chaldean Oracles into humanist scholarship. The Egyptian mysteries of Isis and Osiris became so popular that Pope Alexander VI had allusions to them painted on the Walls of his Vatican apartments. Osiris, a gentle suffering deity was seen as a precursor to Jesus and Isis who was often portrayed with the baby Horus on her lap was reminiscent of Mary. Another passion of Renaissance humanism was geometry and its application to art.[15] By the third decade of the sixteenth century, Venice eclipsed Florence and Rome as the center of Renaissance humanism. It was the world leader in publishing,[16] producing texts in Hebrew, Greek, and Latin, and also in the vernacular, making humanist scholarship available to people who were only fluent in their native language like Titian, its leading painter. Titian was born in Pieve da Cadore, a town in the mountains north of Venice, in about 1488 and came to the city as a young boy to begin his training in art. In the sixteenth century, Venice was the center of Renaissance humanism and he was its leading painter. He socialized with the intellectuals there [17,30] and kept in contact with scholars at the nearby University of Padua.[19] Like other Renaissance artists of his generation, Titian thought of himself more as a poet than a craftsperson.

Humanists hoped to unify the philosophies of Antiquity with Christianity and through cross-cultural analyses, discover the workings of the human mind. According to Panofsky,[20] humanism's concern with mental processes "achieved a success comparable only to that of psychoanalysis in our own day." (p. 162) Hillman [21,22] says the word "Renaissance" did not only refer to the revival of interest in antiquity but also to the renewal of the self and as such was a focal point for the symbolism of the age. Art throughout Italy was influenced by humanism but its symbolism was often hidden.

4 Humanist Symbolism

Beneath the overt Christian iconography of Titian's painting there appears to be a hidden layer of humanism. Scholars were obsessed with hidden symbolism [23] partly because of the constant political upheavals of the time, but also because humanism was considered a doctrine for the elite. Paintings were supposed to have multiple layers of meaning; the artist and the client both expected it.

Humanists believed that one image could convey many meanings.[49] In Titian's painting, figures suggest multiple associations. For example, the black garments of the high priest in the most elevated position suggest the underworld and may refer to Saturn, to Hades, and to Osiris, all gods of the underworld. By placing the black-robed figure on the highest step, Titian has placed a reference to chaos in a place of honor. Hillman [22] calls the lord of the underworld the "God of the Renaissance" and says he exemplifies the "archetypal principle of the deepest aspect of the soul." (p. 205-07, 210)

I will start the analysis the *Presentation of the Virgin to the Temple* by indicating some of its references to self- transformation and then I will examine two of those references in depth. Titian's painting abounds with the iconography of chaos, initiation, and the underworld. For example, the acanthus leaves on the Corinthian capitals are a sign of life after death.[25] The foliate heads are associated with Osiris and with Dionysus, gods of death and resurrection.[26] The double mountain in the background may refer to the Egyptian double mountain, a symbol of death and rebirth and the pyramid near it to a final resting place.[10]

The Roman armor in the corner of the painting suggests Mithraism, a popular mystery religion for men in the Roman empire, whose initiations took place in underground caves.[27] Because the armor has been taken off and placed on the floor, it implies the end of strife.[28] Its position underneath the Temple, a place of peace, emphasizes this symbolism.

Death followed by rebirth to a higher level of functioning appears to be part of the intrinsic meaning of the *Presentation of the Virgin to the Temple* and can be found in the stairs in the foreground of the painting, and in the painting's hidden the astronomical references. These last two examples will be discussed in greater depth.

5 The Stairs

Looking at the painting as a whole, it appears to be an initiation scenario. Mary as the initiate is leaving profane existence behind by climbing the stairs

Fig. 2 The Schneider analysis of the staircase showing the angles and the triangles of the stairs.

to a new life in the Temple. The psychological nature of her ascent appears to be amplified by the mathematical relationship of the three banks of stairs, 8:5:3, which form the Fibonacci Series. (Fig. 2) The Fibonacci Series is a progression of integers in which each number is the sum of the two numbers proceeding it. The relationship between each successive integer in the Fibonacci Series eventually approximates 1.6181, which is known as the Golden Mean or the Greek letter ϕ. In the beginning of the 16th century, Pacioli published a book on the Golden Mean illustrated by Leonardo da Vinci.[29]

The lower bank of stairs contains 8 steps, the middle bank 5 steps and the upper bank 3 steps. In giving us this proportion, which would continue mathematically as 2, 1, 1, Titian is showing us that Mary is climbing from profane multiplicity into divine unity. Psychologically this would indicate integration of the personality. The mathematics are a reference to humanism; in the Christian tradition, she would be climbing 15 stairs after the first 15 Psalms. The Golden Mean is also shown by the little girl leaning on the stairs whose hands divide the bottom eight steps into groups of five and three. If we count her hands as two we again get the Fibonacci series.[15]

A closer examination of Mary's seemingly straight line of ascension finds reference to a cycle of death and rebirth, making a loop around the lower bank of stairs. Although the stairs appear to form a 3-4-5 triangle, a mathematician, M. Schneider [30] has found that they actually measure 36, 54, and 90 degrees and not the 37, 53, and 90 degrees necessary to form a 3-4-5 triangle. The 36 degree angle of ascent in the stairs is the angle in the point of the pentagram, the five-sided star. Humanism associated the pentad with death and rebirth in the physical world as seen in seasons of vegetation and in the cycles of the moon.[31] It may also refer to the cyclical resurrection of Osiris, a lunar god of vegetation and king of the underworld.

6 The Triple Goddess

The reference to the underworld suggested by the geometry of the lower bank of stairs is amplified by the three female figures surrounding it. (Fig. 3) They appear to be the Maiden, Mother, and Crone of the Triple Goddess, an archetype of transformation through death and rebirth. Mary at the top of the stairs suggests the Maiden, the woman in white and gold at the foot of the stairs suggests the Mother, and the Eggseller beneath the stairs suggests the Crone. They are the most prominent female figures in the painting and form an isosceles triangle with the Eggseller at the pivot. It is a downward pointing triangle, an ancient symbol of the feminine.[32] Dividing the large triangle into eighteen smaller triangles demonstrates their use as spatial organizers in the

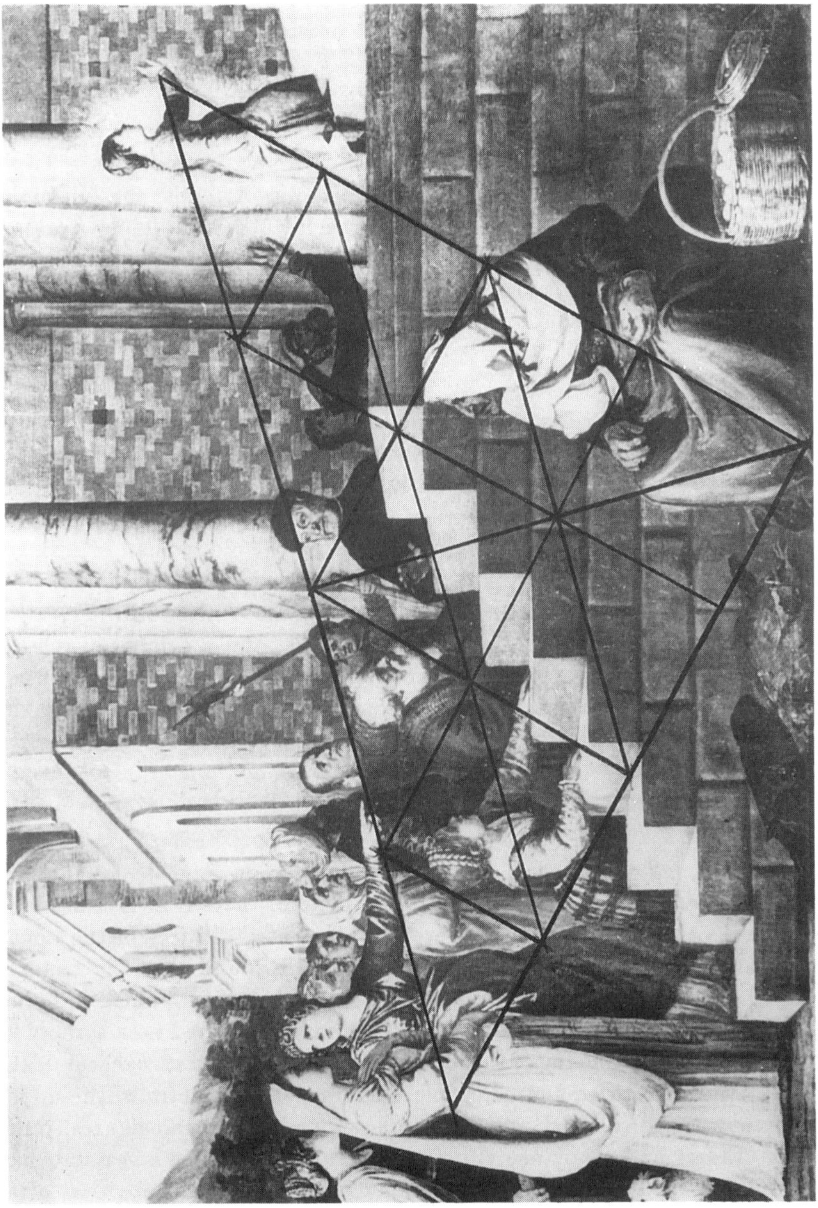

Fig. 3 The isosceles triangle of the Triple Goddess of Mary, the woman in white and gold, and the Eggseller.

painting and transforms the larger triangle into a fractal arrangement. Although fractal geometry is a late twentieth century study its intuitive use by artists has been demonstrated in works of art previous to this time.[33]

The three figures represent new life, maturity, and death as regenerating each other. Like the angle of the stairs they refer to the cycles of vegetation and the cycles of the moon. Although humanists knew the moon as a Triple Goddess in Greek mythology: the crescent moon was Artemis, the full moon was Selene, and the waning or dark moon was Hecate,[2] the Triple Goddess of the moon is also a cross-cultural archetype found in the Arabian, African, Celtic, and Ephesian civilizations.[2,34] The moon symbolizes the cyclical aspect of human nature and the continued struggles of life, which are our deaths and rebirths.

The Triple Goddess rules the sky, the earth, and the Underworld.[35] Mary climbing the stairs suggests the sky, the woman in white and gold at the foot of the stairs suggests the earth, and the Eggseller beneath the stairs suggests the underworld. In all of the cultures of antiquity, a trip to the underworld and its chaos was the beginning of a new life. Near the old woman is a basket of eggs, symbolizing new beginnings. On her other side is a black lamb, an animal which was sacrificed to Hecate, Goddess of the Underworld.[36] Next to the lamb is a rooster, an animal sacrificed to Osiris, and a symbol of rebirth because it announced the dawn.[37] Titian underscores the importance of the Eggseller by making her the largest figure on the canvas. If she stood up she would be the tallest in the crowd. Hillman [22] believes that the Renaissance focus on death and downward imagery in art was a method of introspection and self-transformation.

The upper bank of stairs has a different configuration from the lower stairs and suggests that after a period of death and rebirth, there is an ascension to a higher level of functioning. Schneider [30] noted that the bricks of the upper stairs are longer than those of the lower stairs. He found that the triangle made by the upper bank of stairs is 38, 52, and 90 degrees, which is found in the seven-pointed star, or heptad. The heptad was revered as a symbol of spirituality.[31] As the geometry of the lower bank of stairs suggests cyclical death and rebirth with the three-figured Triple Goddess, the geometry of the upper bank of stairs suggests ascension and constancy with its four male figures: three Synagogue Elders and a boy acolyte. In Jungian theory, three is a number of change and four is a number of completion,[38] which may indicate that after repeated struggles a new self emerges.

7 The Pleides

Interest in astronomy was a part of humanism, and astronomical references have been studied in the art of Michelangelo [15] and of Raphael.[39] In the sixteenth century, Venice was a center for the study of astronomy. It is my belief that in Titian's *Presentation of the Virgin to the Temple*, the group of seven women facing the Temple may refer to the Pleides, one of the most famous constellations of the ancient world.[40,41] In Greek mythology, they were the seven daughters of Atlas and Pleione and temples were oriented in their direction.[42] In Titian's painting, they are the only group of figures where every member is facing the Temple. (Fig. 1)

Myths about the Pleides say there were originally seven stars, but one sister hid her face in shame and now only six stars are visible.[41,42] In Titian's painting, the face of the woman with the small boy is obscured by the woman in white and gold. The constellation has four bright stars and two smaller ones. In the group of women, four are full figures and two are small heads.

The Pleides are only periodically visible in the sky. The ancient Greek writer Hesiod [43] associates them with the seasons of vegetation and warns that the forty days the Pleides dip below the horizon is a time of storms and not for navigation. Because the constellation is symbolically visible in Titian's painting, we may assume that the time is propitious for Mary's journey to the Temple. The periodicity of the constellation may refer to the symbolism of death and rebirth in the *Presentation of the Virgin to the Temple.*

8 Ursa Major

The three high priests and the acolyte at the top of the stairs stand in close enough proximity to a man leaning out of the window to form a group of five figures. Their geometric arrangement makes a square with a line jutting out of it, reminiscent of the constellation of Ursa Major also known to us as the Big Dipper. (Fig. 4a) The angle of line attached to the square and the uppermost corner of the square are in fact, angles found in the constellation, but the body of the constellation is an uneven parallelogram.

Tracing the exact configuration of the stars of Ursa Major onto Titian's painting,(Fig. 4b) finds one star precisely in the center of the hand of the high priest in blue, while the other star falls perfectly on the seam of his garment. Extending the line jutting out of the constellation, the first star falls directly at the edge of the column, while the second fits perfectly into the corner of the painting, and the third continues off the canvas as does the temple. Many ancient civilizations, such as that of Egypt, used Ursa Major in the construction

Fig. 4a The square with a line jutting out from it formed by the five male figures at the top of the stairs.

Fig. 4b The constellation of Ursa Major mapped onto the five male figures at the op of the stairs and onto the architecture of the temple.

of a Temple.[44] It was cited to determine the corners of the building.[45,46]

Unlike the Pleides, the constellation of Ursa Major never sets but circles the pole star in the sky. Because of its constancy, it may refer to the ascended initiate after the trials of death and rebirth. The position of the constellation in Titian's painting enforces its symbolism. According to Suzanne Chippendale, a writer at the Hayden Planetarium,[47] Ursa Major is at the exact angle found in Titian's painting four times a year, but she believes only one of those times is indicated in the painting.

Although the constellation is at that angle on June 20th at 6:00 PM and September 20th at noon, the sky is not dark enough to see it. It is visible on March 20th, but appears after midnight and because ancient civilizations did not have exact time reckonings, midnight was variable and Chippendale does not believe it was the time indicated. She says that archaic peoples made star sightings right after twilight or just before dawn. Chippendale believes the time in Titian's painting to be 6:00 AM on December 20th, the winter solstice.

The Solstice can occur anywhere from December 20th through the 22nd. It refers to the birthday of Jesus, celebrated on December 25, which was the day of the solstice in the ancient Alexandrian calendar. In humanism, it is also a holiday of Osiris, Dionysus, Mithra, and Hercules,[48,49] all gods of death and resurrection, who are celebrated when the days begin to lengthen. In the Christian iconography of the painting, Mary is walking to her destiny as the Mother of Jesus. In the humanist iconography the ascended initiate has attained a new and spiritual life. Psychologically, Mary is all of us, who through the deaths and rebirths of the chaos of life are finally capable of our highest level of functioning.

References

1. H.F. Ellenberger, *The Discovery of the Unconscious*, (Basic Books, New York, 1970).
2. M.E. Harding, *Woman's Mysteries Ancient and Modern*, (Harper and Row, New York, 1971).
3. T. Zausner, *The Society for Chaos Theory in Psychology and the Life Sciences Newsletter* **2**, 6, Jul- Aug, (1995).
4. D. Rosand, *Burlington Magazine*, **115**, 603, (1973).
5. *Webster's New Universal Unabridged Dictionary*, ed. J. L. McKechnie, (Simon and Schuster, New York, 1983).
6. E. Panofsky, *Studies in Iconology*, (Harper and Row, New York, 1972).
7. J. De Voragine, *The Golden Legend*, tr. William Caxton, (AMS Press, New York, 1973).

8. *The Apocryphal New Testament*, tr. M. R. James, (Clarendon Press, Oxford, 1972).
9. D. Rosand, *Painting in Cinquecento Venice: Titian, Veronese, Tintoretto*, (Yale University Press, New Haven, 1982).
10. T. Zausner, (1993), *Transformation of Personality and Emergence of Self in Titian's "Presentation of the Virgin to the Temple"*, (UMI Dissertation Services, Ann Arbor, 1993).
11. J. Hall, *A History of Ideas and Images in Italian Art*, (Harper and Row, New York, 1983).
12. R. Foster and P. Tudor-Craig, *The Secret Life of Paintings*, (St. Martin's Press, New York, 1986).
13. F. Yates, *Giordano Bruno and the Hermetic Tradition*, (University of Chicago Press, Chicago, 1979).
14. F. Yates, in *L'Opera e il Pensiero di Giovanni Pico della Mirandola*, (Istituto Nazionale di Studi sul Rinascimento, Florence, 1965).
15. C. Funck-Hellet, *Composition et Nombre d'Or dans les Oeuvres Peintes de la Renaissance*, (Editions Vincent, Freale and Cie, Paris, 1950).
16. C. Grayson, in *The Age of the Renaissance*, ed. D. Hay, (Thames and Hudson, London, 1967).
17. G. Padoan, in *Titian. Catalog for the exhibition*, (Marsilio Editori, Venice, 1990).
18. J. Schulz, in *Titian, His World and His Legacy*, ed. D. Rosand, (Columbia University Press, New York, 1982).
19. R.W. Kennedy, *Novelty and Tradition in Titian's Art*, (Smith College, Northampton, 1963).
20. E. Panofsky, *Renaissance and Renascences in Western Art*, (Almqvist and Wiskels, Stockholm, 1965).
21. J. Hillman, *Loose Ends: Primary Papers in Archetypal Psychology*, (Spring Publications, Zurich, 1975).
22. J. Hillman, *Re-Visioning Psychology*, (Harper and Row, New York, 1975).
23. P, Burke, *The Italian Renaissance*, (Princeton University Press, Princeton, 1986).
24. A. Chastel, *Marsile Ficin et L'Art*, (Libraire Droz, Geneva, 1975).
25. Vitruvius, *Vitruvius: The Ten Books of Architecture*, tr. M. H. Morgan, (Dover Publications, New York, 1960).
26. W. Anderson, *Green Man*, (Harper Collins Publications, London, 1990).
27. M.J. Vermaseren, *Mithras, the Secret God*, tr. T. and V. Megaw, (Barnes and Noble, New York, 1963).
28. R. Wittkower, *Allegory and the Migration of Symbols*, (Westview Press,

Boulder, 1977).
29. G. Doczi, *The Power of Limits*, (Shambhala, Boulder, 1981).
30. M, Schneider, (personal communications) (1992).
31. Iamblichus (attributed to), *The Theology of Arithmetic*, ed. R. Waterfield, (Phanes Press, Grand Rapids, 1988).
32. M. Gimbutas, *The Goddesses and Gods of Old Europe*, (University of California Press, Berkeley, 1982).
33. B.B. Mandelbrot, *The fractal geometry of nature*, (W.H. Freeman, San Francisco, 1983).
34. B.G. Walker, *The Crone*, (Harper and Row, San Francisco, 1988).
35. R. Graves, *The White Goddess*, (Farrar, Straus and Giroux, New York, 1983).
36. Horace, *Satires, Epistles, Ars Poetica*, tr. H. R. Fairclough, (Harvard University Press, Cambridge, MA, 1991).
37. Plutarch, *Plutarch's Moralia, V.*, tr. F. C. Babbitt, (Harvard University Press, Cambridge, MA, 1969).
38. E.F. Edinger, *Ego and Archetype*, (G. P. Putnam's Sons, New York, 1972).
39. N. Rash-Fabri, *Gazette des Beaux-Arts*, **94**, Oct, 97-104, (1979).
40. R. Graves, *The Greek Myths. Vols. 1 and 2*, (Penguin Books, Harmondsworth, 1969).
41. L. Motz and C. Nathanson, *The Constellations*, (Doubleday, New York, 1988).
42. R.H. Allen, *Star Names, Their Lore and Meaning*, (Dover, New York, 1963).
43. Hesiod, *Theogony, Works and Days, Shield*, tr. A. N. Athanassakis, (Johns Hopkins University Press, Baltimore, 1983).
44. J. Lundquist, (personal communication), (1992).
45. A. Pogo, *Isis*, **14**, Oct, 301-325, (1930).
46. S, Giedion, *The Beginnings of Architecture*, (Princeton University Press, Bollingen Series, Princeton, 1981).
47. S. Chippendale, (personal communication) (1992)
48. H. Bayley, *The Lost Language of Symbolism*, (Citadel Press, Secaucus, 1988).
49. J. Campbell, *The Mythic Image*, (Bollingen Series, Princeton, 1974).

THE DYNAMICS OF CREATIVITY & THE COURAGE TO BE

FREDERICK DAVID ABRAHAM
Blueberry Brain Institute, http://www.pacweb.com/blueberry/

> Every creative act must pass through a moment when it is neither seed nor flower, through the abyss which the mystics call ayin, that nothingness which is the hidden source-spring of everything. Such a passage is fraught with danger, however, for the pull of the abyss, of anarchy, formlessness, and chaos, is strong as death.
>
> Weiner, 1969, pp. 74-75

> *The combination of the experience of meaninglessness and of the courage to be as oneself is the key to the development of visual art since the turn of the century. In expressionism and surrealism the surface structures of reality are disrupted.*
>
> Tillich, 1952, p. 146

1 Preliminary Note

Tobi Zausner and Rik Paar asked my participation in this fascinating subject for which I am exceedingly grateful. Before their bringing it to my attention very recently I was totally unfamiliar with it. As a rank beginner on the subject of creativity, and as only modestly fluent in dynamics, my thoughts on putting the two together are in a most preliminary state of chaotic ferment, not yet sufficiently coherent or developed. What follows is thus not a systematic treatment of this subject, but merely a first collection of some features that I am currently considering, and await fuller development. When Tobi asked me to participate on her panel on this subject for the 1994 Annual Conference of the Society for Chaos Theory in psychology, my first thought was that I had no credentials for participation, and my second thoughts turned to Tillich and to mysticism, and the idea that creativity in the large dealt with self-organizational bifurcations of the deepest existential character. I am not sure I have any more to add to that, but consideration of the topic is proving fascinating. Needless to say, this treatment is not strictly dynamical except in the metaphorical sense; that is, any mathematical formalization of such complex processes are at best a far bit off. My style thus a bit loose; often more poetic, conversational, and sometimes more cryptic than scholarly, especially when discussing the prepsychological domains of myth, mysticism, and literary approaches to creativity. Dynamics are assumed, not explained; a stage upon which to dance.

2 Definitions Of Creativity

2.1 American Heritage (Morris, 1969-78)[55]:

> creative *adj*. 1. Having the ability or power to create things. 2. Creating; productive. 3. Characterized by originality and expressiveness; imaginative.

Some aspects of this definition are mundane and tautological (defs. 1 & 2), but the emphasis on unique ability and capability, and uniqueness in the created product is well expressed (defs. 1 & 3).

2.2 Howard Gardner (1993)[33]:

> The creative individual is a person who regularly solves problems, fashions products, or defines new questions in a domain in a way that is initially considered novel but that ultimately becomes accepted in a particular cultural setting.
>
> (Gardner, 1993, p. 35.)

There is again the emphasis on unique cognitive production, but there is added the context of social evaluation, and a focus on socially significant creations, though they can be viewed as prototypical of more generic creativity.

2.3 Stephen Nachmanovich (1990)[58]:

> Creative process is the struggle for expression of the self, self-creating, self-organizing. Improvisation, the essence of the creative mystery, is the free play of consciousness as it writes, paints, and plays the raw material from unconsciousness).
>
> (Nachmanovich, 1990, paraphrased, quoted, and montaged from the introduction.)

He seeks the joy and the unblocking of this process to develop human potential. His view is personal, spiritual, and psychological.

2.4 Dynamical Metaphor:

> Creativity is self-organizational bifurcation to novel attractors of being.

It is a continuing but saltatory dynamical process requiring energy and concentration to make leaps across the instability of bifurcation points. It is individual and unique. It is the search for, the affirmation of, and the creation of the self. Like all psychological theories, any theory of creativity must prove ultimately dynamical, and like all good psychological-dynamical theories, self-organization, bifurcation, and chaos must play a major role.

3 Literary, Philosophic, and Religious Roots of Ideas of Creativity

Some principal historical roots for me include myth, mysticism, and existentialism.

3.1 Myth

It is easy to see the relevance of myth, for all myths involve catastrophic bifurcations of creation or annihilation, subtle bifurcations—transformations of individuals, societies, or deities, or explosive bifurcations. Jung[44], Freud, and Campbell[22] are but a few of the many investigating the psychological dynamics of myth. These investigations are too well known to elaborate here. But there is an even more relevant feature about myth, so obvious, that it is easily overlooked, and that is that every myth came into existence as a creative act, either to explain something, or to obfuscate an inability to explain something. Thus when Berman[18] points to Neuman's[59] Jungian thesis that mythology is a map to the development of both individual consciousness and the history of human consciousness, from creation to hero myths, and the intermediate dual-role myths (Cain/Abel, Set/Osiris, Duryodhana/Arjuna), a development that contains the internal conflicts that generate creativity, he overlooks the opportunity to highlight myth itself as a creative process obeying the same dynamics as the story it embodies. A self-organizational creative loop.

It would be interesting to speculate about the creative abilities of the snake in Eden, of Prometheus in giving fire to humans. But I choose a different example, a bit more subtle in that there are different facets of creativity involved, especially as much creativity lies in the interpretation of the myth as well as within the myth itself.

Let us look at the trial of Orestes (Aeschylus, *The Oresteian Trilogy*[13]). The synopsis a simple enough, but dramatic story line. The background is the curse of the house of Pelops (Charts 1&2), and a ferment of mixtures of cultures, with both indigenous Greek and Achaean (Northern European, Nordic invaders) roots. Atreus and Thyestes quarrelled over the throne of Argos, Thyestes seduced Atreas' wife,

Chart 1. Cast of Orestein Characters and Some Related Lineage and Events
(Principal characters in Aeschylus emboldened;
early lineage Hesiodan rather than Orphic.)

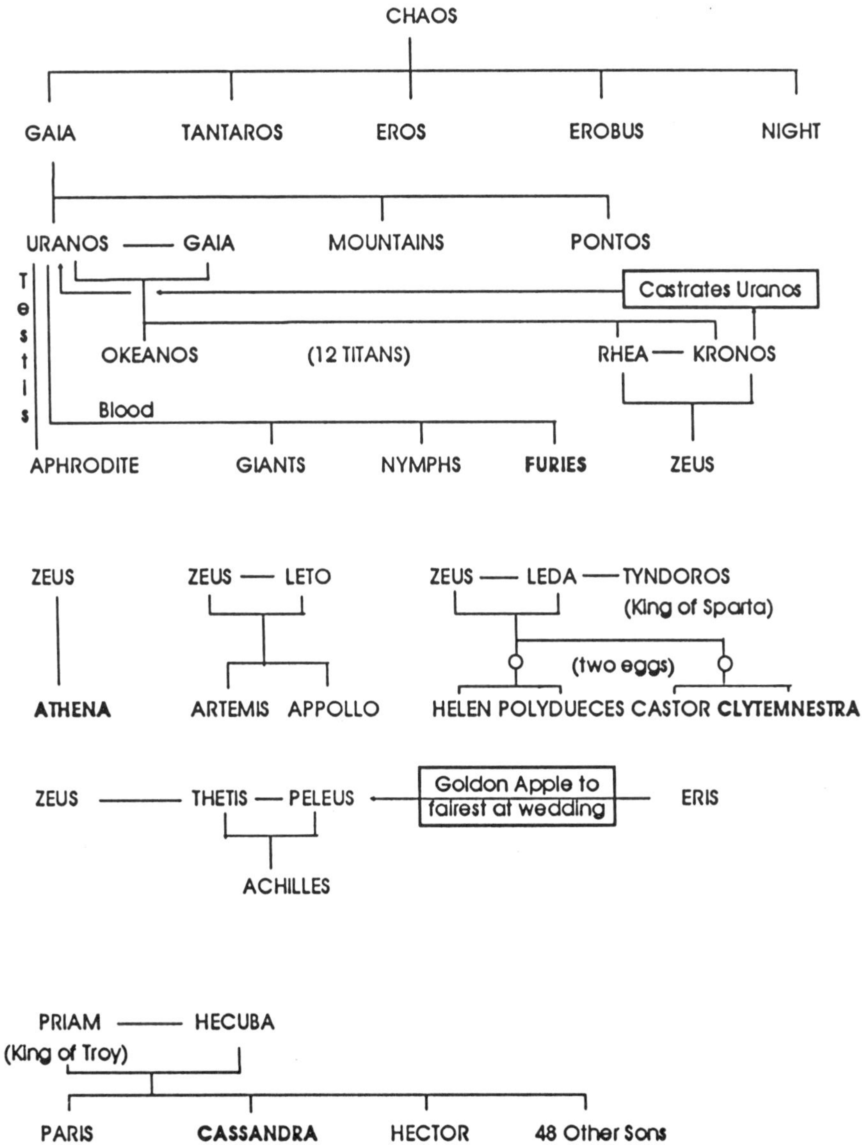

Chart 2. Cast of Orestein Characters and Some Related Lineage and Events
(Continued. Principal characters in Aeschylus emboldened.)

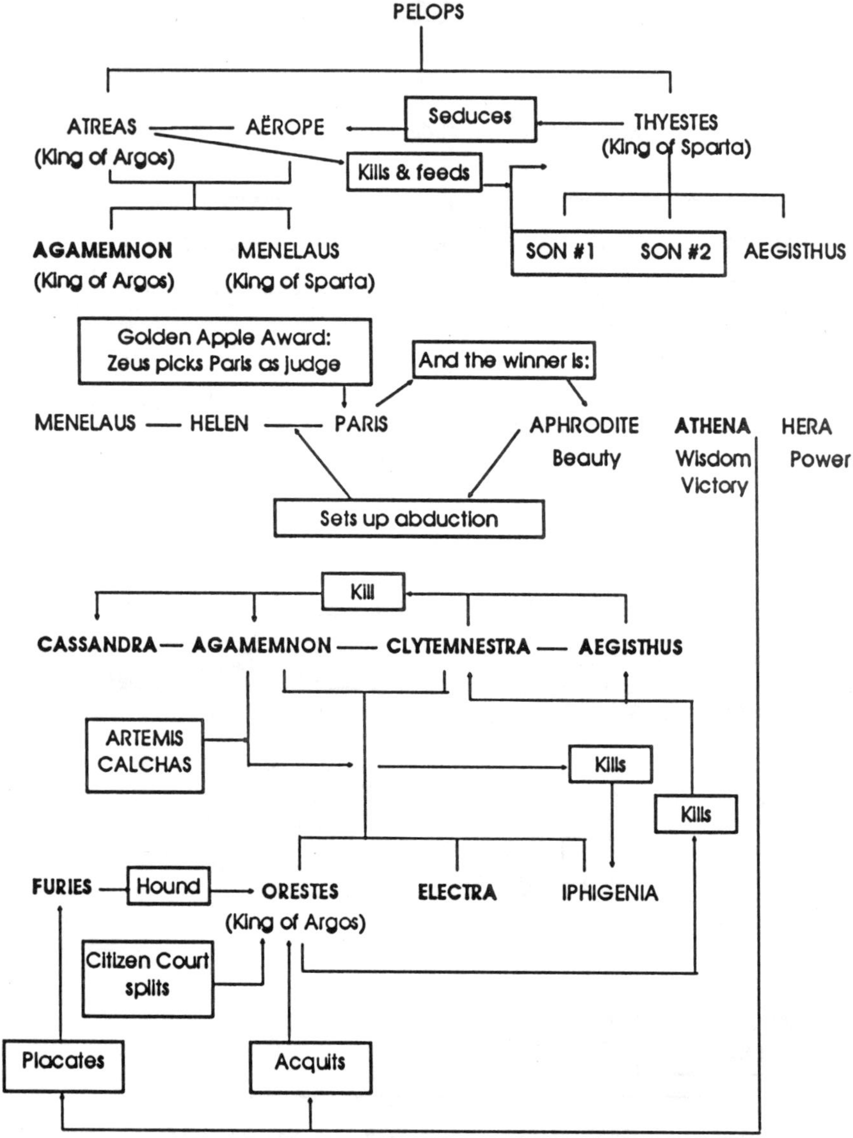

Atreus killed Thyestes' sons and fed them to him which established the downfall of Thyestes as well as the curse. Atreus' son Agamemnon assumed the throne and the curse; his brother Menelaus became king of Sparta. Twin sisters, Helen and Clytemnestra, were daughters of Zeus and Leda. Suitors of Helen had a pact to protect the interests of the winner of her hand. That would be Menelaus (why the clever beauty chose the lout remains a mystery). Agamemnon married Clytemnestra. When Helen ran off to Troy with Paris, the pact comes into play, the suitors suited up and sailed to Troy, but only after Calchas convinced Agamemnon to sacrifice their daughter Iphigenia to obtain favorable winds at Aulis for the trip. Aeschylus's trilogy opens with the arrival in Argos of the news of the fall of Troy. Agamemnon soon returns triumphant and complete with a princess mistress, Cassandra; Clytemnestra slays them both with the help of her lover, Aegisthus (ironically, a surviving son of Thyestes). Apollo orders Orestes, son of Clytemnestra and Agamemnon to slay Clytemnestra and Aegisthus which he does, infuriating the Furies who hound the poor fellow no end. Orestes' trial ends split in the citizen court, so Athena, dual Goddess of both Justice and War casts the deciding vote acquiting Orestes. Athena placates the Furies, again infuriated because of their loyalty to Clytemnestra, or rather their loyalty to the bonds of kinship, by giving the Furies a place of honor despite their demoted status, and stressing their role as the Eumenides (the Kindly Ones).

The emphasis by Aeschylus and Hampden-Turner[39] is upon the cultural bifurcation taking place. The change is from a culture dominated by a traditional Goddess religion and merchant tyrannies and conflicts, a higher-dimensional mix of Gaian and dominator chaos, to a middle-class dominated democracy and modern judicial system, a lower-dimensional, more ordered chaos. A triumph of law and order, civil authority, and due process of law. Ah, but let's back up a bit.

Everybody has their hand on the control stick, that is, participates in affecting movement in the parameter space. Thus this cultural bifurcation is largely self-organizing. The traditional forces of higher-dimensional chaos and the newer forces of democracy which need a lower-dimensional justice system are struggling, are equally represented in the jury, and come to deadlock at the bifurcation point, representing the culture's instability over the struggle at that point. Paradoxically, the resolution has to go to an apparent outsider-of-self-control, using a traditional Goddess, Athena, to complete the motion of the control parameter to pass the bifurcation point to the new order. But note that she embodies the cultural struggle, works out the compromises which leave both the traditional and the new represented, and, as any good therapist, is weaning the society away from its dependence on gods and goddesses and toward democratic self-control. We don't have enough information or explicit enough model to know whether this bifurcation is technically a creative catastrophic, subtle, or explosive bifurcation, but I suspect that the triple blue-sky catastrophe[10] might be a good place to start thinking about the process more explicitly as a dynamical one. Thus in placing chaos in the maneuver to bifurcate, I am giving it a creative cognitive role. [Actually chaos may be best represented by the Goddess Eris, whose darned apple sent Paris' trajectory from Troy to Athens in the first place, but I promised you a simple story line, so we don't pursue that historical trajectory now.]

The especially exciting thing here is that by Ralph's thesis[12], the Furies can be said to represent chaos itself, or I might say, medium-high dimensional chaos. This attractive, feminine chaos, part of the older traditions, is one of the main parties with a hand on the control stick, one of the forces itself struggling for expression. Thus chaos is not only within attractors before and after bifurcation, but a participant in creating the bifurcation. This requires a metacognitional role for chaos in a hierarchical cognitive complex dynamical system (in the technical sense of a network comprised of dynamical schemes whose control parameters are influenced by component systems). I have argued before for a role for dynamics by invoking a dual concept of time, the first where obeying the vectorfield involves a faster, momentary responsiveness; and the second within a context of an overseer with a longer time-integrative memorial function that navigates in the control spaces of these cognitive (or in this case, cultural as well) dynamical schemes/response diagrams and complex dynamical systems/ networks.[3,5,6,10,11]

The popularity and power of Aeschylus' trilogy lies not only in the drama and intrigue of the story line in its portrayal of the cultural struggle, and in explaining it to the citizens, then undergoing that cultural transformation, but also in portraying the creative struggle that goes on at a personal level, as emphasized by Rollo May[53,54]. Does the self stick with past attractors, or is there a mature, ontogenetic, maturing and becoming? Is there creative exploration of new self, and courage to transcend the angst of old self? Self-control keeps a balance of oppositional forces in this development[63] for a medium dimensional chaotic attractor, and prevents a bifurcation to cultural or personal madness, *enantiodromia* (Jung, Vol. 8)[44], or *schismogenesis*[16,17].

> As the cybernetic system revolves, dissent and loyalty, freedom and law, private conscience and public representation 'change places', with one always in the context within which the other occurs. (Hampden-Turner, 1981, Map 58, p. 203.)

Aa a final note on myths and their interpretation itself as a creative act. Aeschylus sees a trilogy of three dramas, whereas Homer was content with a single line using Clytemnestra's act as justification for Odysseus' slaughter of Penelope's suitors. Hampden-Turner sees the dynamical transformation to a just society. Stone[66], Eisler[30], and R. Abraham[12] emphasize that too much is lost when a dominator culture (North European Achaean) bifurcates a chaos-gaian-orphic-cooperative culture into a dominator one. They argue the importance of striving for bifurcations to more gylanic, cooperative cultures. Eisler notes that Athena renounces the role of her own sex in kinship, pronouncing Orestes innocent on those grounds; Clytemnestra was an incubator only, not a DNA donor. Eisler sees the Furies (who argued for the kinship role of the mother and her revenge for Iphegenia) as being returned to the underground (that is, buried, displaced, disgraced). R. Abraham says the underground (which he equates to chaos and godess concepts), although reviled by many, contains metaphors of goodness nonetheless; underground represents "Chaos, Gaia, & Eros"[12]. Hampden-Turner[39] sees the defeat of the Furies as being mollified by Athena; he sees harmony, balance, and compromise. The creation, the myth of the creation, and its interpretation parallel each other. Creative events portray human creativity. Creative interpretation is required to pull the multiple messages out of the story; to co-recreate them.

3.2 Mysticism

Mysticism shares many of the features of myth, with its emphasis on bifurcation, especially of the self and from the attempt to learn and explain the universe. But it has some special characteristics that distinguish it. One is that mysticism focuses on personal experiencing of the cosmos, whereas myth and its successor, institutional religion, place emphasis upon transformation via learning from preexisting institutional and social interpretations of the universe.

Another is that mysticism involves its own brand of paradoxes. One brand of paradox is that while it seeks individual paths to experiential transformation, it nonetheless emphasizes the need for intensive instruction for safe passage through the transformational journey. A second is that the goal of this individual trajectory is often the attainment of loss of self in attaining the unifying experience with the cosmos through meditation. (There are versions in which the persistence of consciousness and productive differentiated cognition exists, obviously different flavors of bifurcaton would be needed to model them.) It is a kind of preparation for death which shows the continuity between life and death, which is often repeated in its emphasis upon the complementarity of creative and annihilative catastrophic bifurcations. A similarity across scale often seen in nearly periodic low-dimensional attractors from breathing and the heart beat and other infradian, to circadian, to ultradian, seasonal, life, generational, and millennial cycles. One can find examples in virtually all forms of religious mysticism. We will mention but a couple of examples.

We might start with Taoist concepts because they are freest from the trappings of institutional religion. They represent the most generic and important features of mysticism, those of the free flowing nature of the universe, and of the importance of freeing oneself to follow that flow (which Nachmanovich[58] emphasizes in his use of improvisation as a primary example of the creative process).

Where does one find creativity in the *I Ching*? Hexagram number one, that's where. Blofeld[20] comments:

> The creative principle functions through Change; accordingly, when we rectify our way of life by conjoining it with the universal harmony, our firm persistence is richly rewarded. This hexagram symbolizes the power of the celestial forces in motion, wherewith Superior [Persons labor] unceasingly to strengthen [their] own character. (Blofeld, 1965, p. 85.)

Dynamical systems stress global pattern through interactive harmony. Thus individual harmony and creativity is achieved by interaction with global harmony and creativity.

Most Eastern religions share this view. It is seen throughout the *Vedas* and the *Mahabarata:*

> Therefore, slay with the sword of knowledge this doubt of yours born of ignorance and rooted in the heart. Be one in self-harmony and resort to Yoga. Stand up, arise Bharata (great warrior), arise and fight. (Bhagavad-Gita, 4:42.)[51]

Krishna, in thus exhorting Arjuna to battle (The spiritual battle of human nature, the battle of the warrior in search of enlightenment). Capra[23], anticipates Tillich and our thesis on several counts. Arjuna's doubt, his inadequate understanding of himself and his universe, and his hesitency to commit to a creative bifurcation to a meaningful self; these set the stage for the teachings of the Gita. These teachings emphasize the requirement of courage as well as enlightenment in order for Arjuna's self-organizational creative growth process to bifurcate successfully.

In one Philippine (Monobo) myth, a folk hero leaped a gap at the horizon (where slashing knives cut the others of his tribe down) to immortality. The leap had to be made at exactly the right moment and was obviously fraught with great danger and required great courage (Magdamo, personal communication)[50]. This universal theme which was also exemplified by the quote of Weiner at the beginning of this essay. Weiner's book makes a dynamicist go tachycardiac because every page of his book on Jewish mysticism reads like a page out of a dynamics text. That quote gets even closer to the dynamics metaphor because it emphasizes (in "the abyss") the instability of the bifurcation point (in dynamics, a point at which an attractor (a) appears, disappears, or undergoes at change but at which the bifurcating attractor does not actually exist; to cross the bifurcation point means abandoning one attractor in order to pass to a new attractor—a subtle bifurcation, or (b) an attractor simply appears or disappears—a catastrophic bifurcation, or (c) an attractor changes size—a plosive bifurcation). The requirement of great courage to cross it comes from the abyss and uncertainty of unknowns about the new attractor. Sometimes the courage to cross a bifurcation is lacking; there is a preference for remaining with an old destructive attractor, an unpleasant aspect of one's own personality or situation. This theme, of course, holds not only for individual self-creativity, but also for that of the creativity of a therapist, or of any one who seeks to assist others to find and to create themselves.

One may note also the great complexity of the Weiner quote[71]. The 'nothingness' of the bifurcation point is well captured, but the inclusion of chaos is not so comfortable to us chaos-crazy folk who love chaos as good. In Jewish mysticism, chaos is often seen as evil, as we have also seen before in the Oresteian myth, and which has been treated extensively by Eisler[30] and in Ralph's new and great book, *Chaos, Gaia, Eros*[12]. That may be because the term is used to mean such high-dimensional chaos, or even beyond chaos, where forces of divergence exceed those of convergence so that trajectories could wander off anywhere. One might consider this latter an annihilation bifurcation to no attractor, though Weiner points out the attractiveness of this condition. Another paradox, whose resolution lies in the yin-yang complementarity of virtual (apparent) rather than real opposites. Dissatisfaction with inhabiting current attractors, and doubt about the attainability or desirability of new attractors, makes death, self-destruction, more desirable. The theme of the self-destructiveness of the creative personality is one which is given much attention in the literature on the psychology of creativity.

Perhaps we can conclude this preliminary rumination on mysticism by anticipating another theme on the dynamical psychology of the cognition of creativity. If we do some more dynamical extrapolation from Weiner[71] and Tillich[67], we might say that not only are the anxiety and courage necessary for being, but there is a medium-dimensional chaos that must prevail as necessary conditions for a creative bifurcation. We leave this conceptual saddle-point until later, and hope that it is not too cryptic left this way. The term *ayin* comes closer to the original meaning of the Greek word for gap (especially Hesiod's creative gap between father heaven—Ouranus, and mother earth—Gaia), which we here liken to bifurcation; while the term *tohu* comes closer the modern notion of chaos, athough mystical lieterture does not distinguish these properties as precisely as modern dynamics (re. the nature of precision in the language of dynamics, see Abraham[7]; R. Abraham[12]; and Goldstein[34]).

Zaehner[73] notes that even medium-dimensional chaos is bad in the eyes of Buddhism since all cognition is impermanent and in flux, and is thus a source of pain, the release from which is the suppression of desire. Even if it is granted that the attainment of Nirvana or Atman is the desirable goal, I find it difficult to label chaos as evil. Big surprise. Chaos is. Chaos can even have its measure of *joie de vivre*. It has both good and evil, both joy and angst (*Abraham's First Protoplatitude*)[10]. There is paradox and ambiguity in much of the mystical literature, as in the Weiner quote where his 'chaos' can be seen either good or bad. This ambiguity could be understood as arising from the dual or multiple forces involved in all of life's vicissitudes.

3.3 Existentialism

Well by now it should be no mystery why I consider Existentialism as practically synonymous with the study of creativity, or at least, as providing a major perspective on the creative process. Existentialism, like mysticism, places emphasis on personal experience rather than cultural and institutional interpretations of the meaning of existence, but goes further in rejecting prescriptive ideologies even at the risk of ambiguity and limited meaning about the human situation when derived from personal experience. It anticipated the depth psychology of psychoanalysis in its perceptions of the personal dynamics of the search for meaning, and was in turn anticipated by Neitzsche[59], Melville, and Dostoevsky, and other writers[45,56,57]. Existentialism especially reacted against the impersonalization of a technological society based on a mechanistic, reductionistic, Laplacian world view. Many modern chaos enthusiasts share this concern[2,6,7,9,10,12,45,56,57].

Kierkegaard emphasized the conflict and frustration arising out of the despair from the disappointed hope generated by scientific-technological promise, and the separation from those hopes caused by human finitude. Denial of this conflict leads to despair. Humanity seems comprised of both the finite and infinite, mortality and immortality, in dynamic interplay. This interplay was also true for many characters in mythology, and could be said to be the dynamic struggle of mysticism as well, of attempting to unite the finite with the infinite. The anxiety generated by this dynamic conflict feeds the development of the self. It is creative. It involves chaotic attractors from the fluid interplay of these forces. It is self-organizational as seen in the struggle of the self for bifurcation to more meaningful attractors.

Camus, unlike mystics, saw absurdity in the human relationship to the universe because of the huge amount of injustice in the world. He especially despised polarized political conflict over ideological abstractions. He saw meaning in the struggle for understanding, within the dynamics. Morality lay in the process of human dialogue. Always a union of opposites. For example, rebellion implies loyalty. We must engage these contradictions, especially the ultimate struggle with the inevitability of our death. Greeley[35], in her dynamical analysis of early Socratic dialogues, also saw creation from dialogue:

> It is my understanding that the intent of the original Socratic dialectical method was to capture and realize, in its 'realized metaphysics', and understanding of the essence of how man proceeds to mediate, or realize, or actualize, his being and his world, for it requires a continual and conscious examination of its own foundations and relationships to others, thus serving as man's mode of access to his world, similar to the Heidiggerian-Tillichian tradition of understanding man's being in the world. It is not just a thinking process; it participates in the creation of reality by being the way reality is transformed, and is primarily dependent on the process of philosophical inquiry, which creates the question, which creates the form of the answer, which is developed and built on in reality, thus creating and bringing an actualized metaphysics into being, into existence. (Greeley, 1991.)

Her use of the term existence need not be considered different from that of the existentialists. It points to the dynamical process of becoming, of the self-organizational creation of reality. The search for truth creates truth. It is motivated by the hopes and anxieties and search for meaning which leads to a self-organizational bifurcation to new attractors of selfhood.

Tillich deals with the dynamics of some binary complementary concepts. The first pair, courage and anxiety:

> Anxiety turns us toward courage. . . Courage resists despair by taking anxiety into itself. (Tillich, 1952, pp. 66.)

Another pair, being and nonbeing:

> For if being is interpreted as life or process or becoming, nonbeing is ontologically as basic as being. (Tillich, 1952, p. 32.)

And these two pairs themselves constitute another higher-level pair:

> Speaking of courage as the key to the interpretation of being-itself, one could say that this key, when it opens the door to being, finds, at the same time, being and the negation of being and their unity. (Tillich, 1952, p. 32.)

Tillich identifies three forms of anxiety stemming from threats to being.

> Nonbeing threatens man's ontic self-affirmation, relatively in terms of fate, absolutely in terms of death. It threatens man's spiritual self-affirmation, relatively in terms of emptiness, absolutely in terms of meaninglessness. It threatens man's moral self-affirmation, relatively in terms of guilt, absolutely in terms of condemnation. (Tillich, 1952, p. 41.)

I find dynamical aspects of such ideas well revealed when he discusses Spinoza's and Leibniz's use of terms like *conatus* and *potentia* as striving for, and the power of, self-realization and as the nature of being, and Nietzsche's "will to power" as

> the self-affirmation of life as life, including self-preservation and growth. Therefore the will . . . wills itself in the double sense of preserving and transcending itself. This is its power, and also its power over itself. Will to power is the self-affirmation of the will as ultimate reality. (Tillich, 1952, pp. 26-27.)

The forces of attraction and convergence within a chaotic attractor should be self-evident here, as well as that of transformation and growth, the transcendence being accomplished with the preservation of being after the bifurcation. From *Thus Spake Zarathustra*[59]:

> Behold, I [Life} am that which must ever surpass itself.
> (Nietzsche, 1883, II 34.)

It is therefore not surprising to find Nietzsche's recognition of the risk of the bifurcation, because he developed considerably from Spinoza, who as a Jewish mystic, shared the insights of the opening quotation of Weiner:

> He hath heart who knoweth fear but vanquisheth it; who seeth the abyss, but with pride. He who seeth the abyss but with eagle's eyes, he who with eagle's talons graspeth the abyss: he hath courage. (Nietzsche, 1883, IV, 73, 4.)

The eagle's eye represents the higher dynamical cognitive view I mentioned earlier, the talons (in part) represent the play on the vectorfield with its domain of the faster time parameter[5,6,8].

A commentary on existentialism should, as it does for Tillich, Camus, Sartre and others, express a concern about authoritarian, collectivist, and totalitarian bifurcations. The swing to a contrary attractor is a type of bifurcation which Jung calls *enantiodromia* (*Vol. 8*)[44], and could be modeled as a triple blue sky bifurcation, perhaps with approach-avoidance or Tomkins ideological dynamics[8,10]. Existentialism extols the divergent forces of individuality and creativity, which are feared by the convergent forces of collectivism and facism. Even the conformist democracy of the Orestein bifurcation was a form of collectivism. Differential equations of conformist/individualist interactions would show dependence on each other as well as themself. Any town meeting, newspaper, or call-in talk show also provides evidence of their interaction.

Tillich[67] also notes, as mentioned earlier, the participatory aspect of mysticism, in polar dynamics with individualization:

> If both poles are accepted and transcended the relation of being-itself has the character of faith.
> Our question is not whether this goal [of participatory identification] can ever be reached by a finite being but whether and how mysticism can be the source of the courage to be. (Tillich, 1952, pp. 156-157.)

The concepts of will and surrender which achieve such full treatment in Schelling, Schopenhauer, and Nietzsche suggest a continuity to William James, though Tillich makes a distinction in his claim that pragmatism overlooks that the "courage to create implies the courage to replace the old by the new" which requires the risk of sensitive dependence on initial conditions, an unpredictability born of finite knowledge of chaotic deterministic systems (well, maybe those are not his exact words).

While existentialism plays on angst and the dark side of finitude for much of its psychological dynamics, it also emphasizes enthusiasm, hope, happiness, and play, which we will see is a repeated theme in creativity. I love the play and humor that is not only revered by Mosca[56,57], but is very nicely practiced by him as literary author, critic, and therapist. Mosca is an explicitly dynamical commentator. In the center of his concern is freedom and choice, although he cautions that dynamics can be pressed not only into service for the play of freedom and choice, but also that it can also be pressed into service

> to capture and reduce these newly discovered exotic cosmic and microcosmic rhythms and resonances and place them in a variety of deterministic cages, all to be sure, for the betterment of humankind. (Mosca, 1995, p. 181.)

After such a caveat, he postulates:

> I assert that Choice/Will/Truth/Happiness are coextensive when they are phase locked, i.e., when they are superposed in the service of implicate order creativity. It is my thesis that this occurs in the commonplace motion of everyday life when we choose to be happy, that is, to be consonant with the basic chirality or spin of the universe which I posit as the 'will to happiness' which is the ultimate strange attractor in human experience. (Mosca, 1995, p. 182.)

He talks of the creative weaving of the fabric of self in everyday experience as aesthetic, truthful, happy, open ended, flowing, unpredictable. He, like Tillich and other existentialists recognize that freedom and choice are perceived as a threat to the hope for "utopian invariance". This threat is not only felt in totalitarian regimes, but reigns within each of us. He mentions Berdyaev's summary of Dostoevsky's concern that people renounce their freedom (e.g., the Grand Inquisitor in *The Brothers Karamozov*). But Frank is drawn to life's uncertainties where others flee from them. He revels in flying the chaotic trajectory of life; that is what gives life meaning. He is a happy Hermes. Flying this trajectory is a process of creative improvisation.

> Play, then, brings us to an awareness of life as an aesthetic process, an ongoing praxis of self-creation and ever changing fictional states, i.e., a process that does not proceed *kronot*ically but *kairo*tically, with major far-from-equilibrium leaps into ever burgeoning states of new potentialities. (Mosca, 1995, p. 187.)

Notice the implication of the complex hierarchical cognitive system and its dependence of the chairos/chronos temporal distinction I have previously mentioned, one in which saltatory bifurcational sequential irreversible progress typical of nonlinear dynamics is made. If there is no fixed point static reality to be discovered and lived by, reality is a process of continual self-creation.

4 Psychological Analysis of Creativity

4.1 Berman (& Freud), The Two Faces of Creativity (1989)[18]

It would be reasonable to segue from existentialism to psychological treatments of creativity through the psychoanalytic disciple of Tillich, Rollo May[53,54], but the stage can also be set by considering Berman's typology of creativity which is also rooted in Freud's analysis of creativity. Berman notes that many creative and performing artists suffer burn-out or self-destruction. Those most intense at self-creation seem to give up and give-in to non-being and self-destruction, or what Bateson[16,17] calls 'schismogenesis', and is another example of enantiodromia. In the existentialist terms, you risk everything to cross the abyss, but you fall in, or give in to an opposing attractor from the despair of the inability to self-create the most ideal self-attractor. Freud's analysis is laid out in his psychobiography of Leonardo da Vinci (1910)[32] and goes something like this. The child expresses sexuality in uninhibited exploration, play, curiosity, and engagement with the environment. Parents suppress/repress this tendency. There are three possible adaptations of the child to this suppression. The *first* is that curiosity and creativity are shut down for the rest of the life. This is *inhibition*. The *third* is that the libido survives intact by successful *sublimation* in a smooth bifurcation right at the outset. The *second* is intermediate. There is some repression and some resistance to suppression. The result is a great deal of compulsion and anxiety. Anxiety continues and eros is transformed through breakthrough or eruption into creative activity. This is *neurotic compulsion*.

Berman classifies creativity according to these categories. Type I is the negation of creativity. Type II is neurotic creativity, typical of most creative work. Type III is a nearly null set; a maturity that still gives free play to spontaneous and joyful creativity that was so well characterized by Mosca, I think.

Type II takes center stage here. It has several characteristics. *First*, it arises from inner conflict occurring at an unconscious level, which becomes manifest in the form of a block to creativity. There is a discrepancy between the hope and a realization or there is conflict between two ideals. *Second*, there is self-creation, that is born of, as Fowles[31] characterizes it, an escape from whatever threatens individuality, a desire to prove one's existence. *Third*, the creative insight emerges from the unconscious. Awareness of the emergent material generates the inner conflict needing healing through self-development. *Fourth*, neurotic creativity can be compulsive. Such intensity can result in the emotional exhaustion that generates the enantiodromic bifurcation, the schismogenesis, especially if the capability for creative novelty wears out early. *Fifth*, there is an erotic component to the creative activity, a lot of passion in obsessive creativity.

Berman observes that the conflict-generational mode of creativity may be an artifact or aberration of modern western culture:

> There is a way in which the very brilliance of Western creativity depends on its instability, its extremely high level of tension and stress. (Berman, 1988, p. 140.)

In terms of dynamical system theory, this brings us to an important idea in which I would slightly modify the sequential nature of these psychodynamics. Freud points out the complexity of dynamics occurring at the unconscious level, so that the conflict is likely preceding the conscious stage. One could characterize eruption into consciousness as an explosive bifurcation[4] occurring under the self-organizational control of the unconscious dynamics. The idea here is that the energy of the libido and these Type II dynamics are responsible for creative bifurcations. Berman's use of the term 'instability' not only merely reminds us of the instability of the bifurcation point (our now familiar *ayin*), but suggests special significance for creativity of an aspect of cognitive navigation in the parameter space. The point that has been made by Kugler (personal communication), Hubler[42] and many others, that the instability which increases as one navigates nearer the bifurcation point on the control parameter, might be a condition that a person seeks. I have discussed this from the viewpoint of its advantage for choice or free will[5,6,7], but its advantage as a place to reside for optimizing opportunities for creative bifurcations is also obvious. For both choice and creativity, there is great psychic cost of residing there so much of the time as well as in the extensive energy expended crossing such bifurcation points. This cost is what may lead to psychic breakdown, self-destruction, as the individual eventually, self-organizationally, retreats to safer, more stable regions of parameter space, into safer stable attractors. In the extreme, the fixed-point attractor of suicide may be chosen.

Perhaps we could leave psychoanalytic psychodynamics for the moment, but not without first mentioning Berman's Type III and a transitional type IIb where anxieties are confronted in consciousness. Type III dynamics are devoid of sexual conflict, there is less investment of self compared to participation in culture or cosmos, the creative process is spontaneous rather than schismogenic. It is 'continuous with life' or as he quotes Coomaraswamy[24]:

> The artist is not a special kind of person; rather, each person is a special kind of artist. (Coomaraswamy, 1934, quoted in Berman, 1988, p. 150.)

4.2 *Stephen Nachmanovich, **Free Play: Improvisation in Life and Art** (1990)*[58]

This continuity with life lies at the heart of the attitude of our next commentator, Stephen Nachmanovich, who plays on the joyful side of creativity. Nachmanovich is a concert quality violinist/violist who loves to play totally improvised concerts, I presume, in classical, jazz, and personal styles. His dissertation in the University of California at Santa Cruz's History of Consciousness Program (which focused heavily on Blake and Bateson) provides a sparkling and brilliant perspective on life and creativity, summarized in his book on the "inner sources of spontaneous creation".

René Dubos[29] in his discussion of creative enthusiasm quotes the poem of the Eskimo sculptor:

> Ringed seal, one carves it
> Hiding now
> Ringed seal, one carves it
> Moving now
> Ringed seal comes to me. (quoted in Dubos, 1972, p. 12.)

And Michaelangelo's sonnet:

> The best of artists has that thought alone
> Which is contained within the marble shell;
> The sculptor's hand can only break the spell
> To free the figures slumbering in the stone.
> (quoted in Dubos, 1972, pp. 10-11.)

Nachmanovich similarly quotes Blake[19]:

> melting apparent surfaces away, and displaying the infinite, which was hid.
> (Blake, 1793; quoted in Nachmanovich, 1990, p. 4.)

If that is all there is to it, why doesn't the muse always come when requested? He answers:

> What we have to express is already with us, is us, so the work of creativity is not a matter of making the material come, but of unblocking the obstacles to its natural flow.
> (Nachmanovich, 1990,p. 10).

Creative breakthroughs come when some impediment or fear is overcome. But the creative process is a continual saltatory parade of provisional bifurcations, not an ultimate, final breakthrough. He finds his music is mystical self-discovery as well as self-expression. And he observes, as did we earlier, that mystics represent the creative side of religion. He considers creativity a mystery, in the full sense of the word. And

> No kind of linear organization can do justice to this subject. Interreflecting themes, the prerequisites of creation, are playfulness, love, concentration, practice, skill, using the power of limits, using the power of mistakes, risk, surrender, patience, courage, and trust.
>
> Creativity is a harmony of opposite tensions, as encapsulated in lila [sanskrit for divine play]. As we ride through the flux of our own creative processes, we hold onto both poles. If we let go of play, our work becomes ponderous and stiff. If we let go of the sacred, our work loses its connection to the ground on which we live.
> (Nachmanovich, 1990, p. 12).

Nachmanovich thus connects to the relationship of individuality and community/ cosmos with which the existentialists as well as the mystics wrestled.

He also highlights the concept of time as have I and Mosca, as mentioned previously.

> Whether we are creating high art or a meal, we improvise when we move with the flow of time and with our own evolving consciousness, rather than with a preordained script or recipe. In composed or scripted art forms, there are two kinds of time: the moment of inspiration in which a direct intuition of beauty or truth comes to the artist; then the often laborious struggle to hold onto it long enough to get it down. In improvisation there is only one kind of time, real time.
> (Nachmanovich, 1990, pp. 17-18.)

I would claim that chairos and chronos are interacting in real time, the play in the vectorfield is under the higher cognitive influence. The jazz musician carries a compositional structure, a melody line, harmonic sequence, rhythmic structure, thematic sequences, historical concepts of performance and of that tune's place in the history of music, etc. while simultaneously interacting with inner sources of inspiration and outer activities of audience and fellow musicians.

I like Nachmanovich's more democratic and pervasive views of creativity, and I note with special interest that his concept of time especially leads him to this position of the generalization of these improvisational qualities to ordinary life. He sees this generalization as residing in the development of the cognitive intersection of varied temporal parametric mental activity derived from developing the ability

> to stretch out those momentary flashes, extend them until they merge into the activity of daily life. (Nachmanovich, 1990, p. 19.)

This is very Heraclitian[10,63]. Heraclitus loved flow, time, and creativity. And the kitchen. A down to earth kind of guy. These two kinds of time are merged in the creativity of teaching as well. Teaching involves the real time performance of interacting students with a body of knowledge and with their own learning and creativity. He notes also the quality of surrender to the present moment that is required, and that surrender includes surrender to the uncertainties of sensitivity to initial conditions, again a recurrent theme of risk, courage, and will.

> Improvisation is acceptance, in a single breath, of both transience and eternity. (Nachmanovich, 1990, p. 21.)

He too notes that the blocks to creative bifurcations come from the retreat of self to safe stable attractors of ego rather than the courage and energy needed for creative expression of a deeper self. And of course, those blocks are working against the explosive bifurcation of intuition from the unconscious into the conscious arena. (p. 29).

4.3 Rollo May ***Man's Search for Himself,*** *Chap 6, The Creative Conscience (1953)*[53].

As is well known, May was a psychoanalytically oriented psychologist and existentialist, a disciple of Tillich's. And also as is well known, one cannot consider

May without returning to myth, for it was one of his favorite playgrounds. His favorite myth for portraying the nature of creativity was that of Prometheus, whose torture represented the inner conflict of creativity, his punishment for the sin of creative rebellion against the proscriptions of the gods. Remember that Prometheus gave the fire of the gods to humans, and Zeus chained him to a mountain top in the Caucuses, where in perpetual torment, an eagle tore his liver by day, and the liver repaired by night. May notes the Babylonian myth of the Garden of Eden is similar but is a myth of self-consciousness more than creativity. While we may admire creative rebellion, May warns that existentialist anxiety can stem from the frustration of attempting to overreach human limitations. The myths also warn against false pride and aspiration. Gods, or their institutional representatives as we saw with the Grand Inquisitor, fight against creativity. The Grand Inquisitor claimed he was saving people from this burden, but projected his own fears of creative rebelliousness onto people as their preferring bread to wrestling with the conflict of differing ideals and aspirations arising from the exercise of freedom and conscience. But there is a constructive side of self-awareness, self-organization, self-creativity,

> which always involves potential conflict with those in power, be they gods or parents. (May, 1953, p. 159.)

And they reveal personal conflicts for psychologically within even the creative person there is not only courage but there is

> a servile side which would prefer comfort to freedom, security to one's own growth. (p. 160.)
>
> Is it not the conflict between every human being's need to struggle toward enlarged self-awareness, maturity, freedom and responsibility, and his tendency to remain a child and cling to the protection of parents or parental substitutes? (May, 1953, p. 166.)

We note the mix of intrapersonal fores of convergence and divergence that make for chaos. May argues that anxiety and guilt are not too great a price to pay for self-knowledge and creativity that comes from participation in this chaotic, productive process (p. 160).

*4.4 Joy P. Guilford, **The Psychometric Approach.** (1950, 1953)*[37,38]

By mid-century psychometry considered that it had successfully analyzed intelligence. It could be argued that this knowledge provided for the better development of individual potential, but it could also be argued that it served the forces of a technological and conformist society, placing round people in round holes, square people in square holes, and black, white, and orange people in very deep holes, and women in ticky-tacky houses. Guilford saw that creativity was not equivalent to intelligence, but thought the psychometric approach could shed some light on creativity. It did. And some very useful ones. Intelligence testing challenged people for solutions to problems with a best solution; intelligent people could solve them with convergent thinking. This enterprise, until the 70's when it was culturally challenged, was considered reliable and valid. Creativity testing challenged people with problems with multiple solutions; creative people found multiple solutions with divergent thinking. This enterprise proved reliable, but not valid. That is, creativity defined by one test might not generalize to other tests or to life.

For our purposes, we will claim there is a mixture of divergent and convergent thinking required for creative cognition, and that there is a range of optimal dimensionality to the chaotic nature of this process in the mid-dimensional range. This process is autopoetic, self-organizational—it uses chaos and it self-perpetuates these chaotic elements it needs for creative progress. We will return to this theme later as Gardner uses a similar concept, and in our summary of some of the most useful dynamical concepts for modeling creativity. But first let's continue to mention other psychologists who studied convergent and divergent thought.

*4.5 Wolfgang Köhler, **The Mentality of Apes** (1925)*[46,47]

Köhler studied creativity in apes. That's right, apes. Only it has been called 'reasoning', 'problem solving', and 'insight'. Well, ok, they are not really the same, and his tasks did not permit multiple solutions, but rather insight to get to the right solution. Food placed out of reach required the use of tools or objects to bring the food and ape together. Köhler analyzed these from a dynamical point of view, using field theory. Organism and environment were interactive, the organism behaving in a force field. But without getting into the details of Gestalt theories of perception and behavior, I mainly wanted to focus on a couple of features for present purposes. In the piling of boxes to reach suspended food, Sultan, after solving a one box problem, was tested with a two box problem. When one box failed to bring him close enough, he

> seizes the other box, and pulling it behind him, gallops about the room, making his usual noise, kicking against the wall and showing his uneasiness in every possible way. He certainly did not seize the second box to put it on the first; it merely helps to give vent to his anger. But all of a sudden his behavior changes completely; he stops making a noise, pulls his box from quite a distance right up to the other one, and stand it upright upon it. He mounts the somewhat shaky construction.
>
> (Köhler, 1925, pp. 135-136.)

This episode contains the essence of creativity. There is conflict and anxiety in the discrepancy between the need for food and the difficulty of attaining it. There is the bifurcation or breakthrough from unconsciousness of the solution, there is compulsion for food. Another chimp, Tschego, gave up trying on a similar problem, until other chimps approached the food. There was a similar incubation before eruption of the solution into consciousness due to the added motivation provided by the threatened loss of the food to the other animals, but also, there was willingness to accept nonbeing (well if she didn't eat for long enough). Both chimps exhibited divergent thinking, interactive cognitive multiplexing over components of their environment with their problem, until a creative result was obtained. But what I wanted to emphasize here is that there is convergent thinking as well, the ability to put a problem center stage and focus on it. When we think of creative products that have stood out historically we see the same thing. Newton was at home in Woolsthrope, away from the distractions of the University which was closed by the plague (1665-6) when he did his early creative work on motion and optics (though the *Principia* came later, 1685-7, after his return to Cambridge), Einstein and Freud were similarly unburdened. Actually, studies of reasoning, problem solving, and insight, in humans and animals, have indicated that intermediate levels of focus of attention and motivation are best for achieving solutions. This is typical of dynamical systems with chaotic dynamics, formalized in the ubiquitous logistic and other single-lobed equations which exhibit most of the features of dynamical systems (or more accurately, their finite incarnations, iterative maps), of the tendency to move outward from center when the trajectory is too close to center, or toward center when orbiting too far from center. This explains the ubiquity of the $\cap$ curve in psychology, which of course survives and thrives on dynamical systems. At any rate, Köhler's apes emphasize both convergent and divergent forces in creativity.

4.6 *Csitszentmihalyi, Humanistic Attributional Approaches; **Flow** (1990)*[25]*; **The Evolving Mind** (1993)*[26]

These approaches emphasize lifetime development, including bifurcational "peak" experiences, intrinsic motivation, and more or less continuous engagement with life. Amabile[14] points out that creative solutions are more frequent in activities done for sheer pleasure. Social judgement seems to favor the conventional. Activities done in more conformist social arenas (school, chruch, the workplace) are less likely to be creative. Popular images seems to focus more on the products of play (writing, arts, music, performance, science) as creative than products of the work place, but such a distinction distinction is easily broken as most of these activities are well paid, and so are the engineers who produce new materials or electronic components. Of course there is much effort to increase creativity in the work place, but in general, we tend to agree with her that freedom from extrinsic reward seems to give rise to more spontaneous creativity.

There is a genre of high-risk sports. Just to engage in them is a form of creativity. Why? The flow of life is intensified in an affirmation of self. One of the sports that most typifies flow could be whitewater kayaking. If that isn't flow, I don't know what is. These folk take drops off 50 foot waterfalls, pass through standing waves towering several feet overhead, dig out of holes capable of sucking them under, and pass rapids and boulders at speed that would terrify one of us mere mortals used to highway speeds and freeway slalom. Csikszentmihalyi has taken this term "flow" and used it to encapsulate the peak experience, the exhilaration of creativity. He notes its similarity to Maslow's[52] 'self-actualization', White's[72] 'competence', deCharm's[27] 'personal causation', Bandura's[15] 'effectance motivation', Deci & Ryan's[28] 'autonomy', Amabile's[14] 'intrinsic motivation' (above), and Hebb's[40] 'optimal arousal'. He notes that his approach "focuses more on what happens in the stream of consciousness". His flow is Mosca's happiness, the mystic's transcendence, the existentialist's engagement in the search-process of meaning and self-creation. Like Nachmanovich he sees spillover into everyday life. For me the exhilaration of peak flow comes from exploring ideas in chaos and psychology, sitting down to write what I think might be a straight forward essay on it and discovering that the trajectory winds through territory more vast and more complex and interesting than I ever thought it might. This is fun and meaningful, even if next week or next year, I find this essay banal or out of date. I love being lost in the turbulence of the continued passage of the trajectory. And as with him and Nachmanovich, I find more flow in everyday life, whether assisting my wife in the development of her work, in the canoe or in backwoods skiing, pruning around the house, fixing supper, or making music alone or with various ensembles; I can cope, laugh at my mistakes, can learn from them and still not let them keep me from enjoying the flow. Csikszentmihalyi deserves more consideration, and my apologies for not reporting more. I thought it worthwhile to mention him anyway, despite my fast and incomplete skimming of him. He does incorporate a complexity approach to a certain extent, but without an explicit dependence on dynamical systems concepts.

4.7 Howard Gardner, Creating Minds (1993)[33]

Gardner criticizes the psychometric approach on grounds of its banality, its superficiality, and its stringent requirements for unambiguity of creative events. He credits Boden[21], Perkins[62], and Sternberg[64] for progress in recognizing that creative minds use problem-solution spaces. I presume that space allows convergent and divergent interactions to occur which gives rise to the optimal range of chaotic dimensionality for creativity I mentioned above. He also likes Gruber's[36] observation that creative scientists such as Darwin and Piaget are involved in networks of projects and individuals and ideas that are highly focused with a sense of will or purpose

> in the entire network, giving direction to their daily and their yearly activities; favor the creation and exploitation of images of wide scope; and display a close and continuing affective tie to the elements, problems, or phenomena that are being studied. (p. 23).

We should not be too surprised then that Gruber assumes an evolving systems approach, since we can also see the elements of a self-organizing complex dynamical system (the network), bifurcations (implied in creativity and the use of the term evolving), and chaotic dynamics that recognizes convergent (focus, determination, and probably a bit of Berman's type II dynamics and a lot of existential angst) and divergent thinking (an expanding search pattern open to new ideas). Gruber uses a developmental approach and studies the multiple component systems comprising the network and their interaction.

Gardner too is basically dynamical without explicit use of system's concepts, recognizing especially the influence of Gruber, Csikszenthmihalyi, and Feldman, and having a very explicit socio-cognitive framework. He describes three organizing themes to his approach. They are developmental, and relational to people, and to domains of professional activity. The life-course developmental perspective examines how childhood patterns of self-discovery establishes patterns for adulthood. It examines the extent to which the continued childhood freedom of curiosity survives and interacts with features of maturity to establish the basis of creativity. In early adulthood he sees a learning period for mastering a discipline or craft of at least 10 years duration, but he says:

> Childhood can be a very powerful ally. Indeed, I contend that the creator is an individual who manages a most formidable challenge: to wed the most advanced understandings achieved in a domain with the kinds of problems, questions, issues, and sensibilities that most characterized his or her life as a wonder-filled child. It is in this sense that the adult creator draws repeatedly on the capital of childhood.
>
> (p. 32).

He emphasized the ecstasy, enthusiasm, the intoxication, and romance of this wonder, exploration, and discovery. He suggests that certain personality characteristics are necessary for the bifurcation to an innovative life following skill development. He also focuses on the instability of creative breakthrough, of that bifurcation point where he sees the creation of a first major work:

> At this point in the general trajectory of development, I examine human creativity most intensively. I attempt to re-create the mental model, the representation map, that each individual formed about the chosen creative task.
>
> (p. 33).

He uses a historiometric psychobiographical approach of Creativity with a big C, examining seven great creative individuals from basically the early part of the 20th Century (Freud, Einstein, Picasso, Stravinsky, Eliot, Graham, and Gandhi). In these folks he finds a pattern of dissatisfaction with past solutions to problems and a persistence in finding new directions:

> A seemingly local solution needs to be abandoned in favor of a far more extensive reorientation or reconceptualization.
>
> These are the times that try the mettle of the creator. No longer do the conventional symbol systems suffice; the creator must begin, at first largely in isolation, to work out a new, more adequate form of symbolic expression, one equal to the problem or product in all of its complexity. . . In this pursuit, there are no guarantees or even reliable guides; the creator must trust his or her own intuition and must be braced for repeated and unrequited failures. (pp. 33-34.)

His summary of the developmental features of creativity includes: (1) childhood, universals and specifics; (2) conversion of initial interest into sustained mastery; (3) subsequent discovery of novel or discrepant elements after mastery of a domain; (4) how initial novelty leads to a program of exploration; (5) supportive and inhibitory roles of other people; (6) how the new symbol system, language, or mode of expression is gradually worked out; (7) reactions to critics; and (8) a second, mid-life, more comprehensive innovation or creative bifurcation.

One might make a comment on his choosing individuals. Quantum theory was a creative scientific breakthrough of the same era, but was more of a cooperative than individual creative effort, just as cooperativity marks the creative breakthrough of 'chaos theory' (dynamics) today. Dynamics represents a kind of ultimate 'strange loop' or self-awareness bifurcation. A meta-hermeneutical cycle. The subject matter is self-organizationally making us aware of itself, it is self-creating by its emergence in our consciousness, a new paradigm shift (A theme recurrent in my brother Ralph's thinking). It is almost as if we were both the state spaces in which this story plays itself out, and at the same time nodes of dynamical systems forming a complex dynamical system or network with new emergent properties. This raises the question of whether the same might be said for Freud, Einstein, Picasso, Stravinsky, Eliot, Graham, and Gandhi. Their creations may have been inevitable, though with a unique flavor and finding an easier way through their particular personalities. We see in all of this both the risk, the unpredictability of the sensitivity to initial conditions, and the stability and inevitability of what I have called 'insensitivity to initial conditions', the repeatable, generalizable, and invariant aspects of the chaotic dynamic (Abarbanel, Brown, Sidorowich, & Tsimring, 1993, also use the phrase "insensitivty to initial conditions" to reference the invariant properties of chaotic attractors). As with quantum phenomena, a trajectory chooses from its potential future states a new state with the unfolding of the present moment. Martha Graham said (gleaned from Nachmanovich):

> There is a vitality, a life force, an energy, a quickening, that is translated through you into action, and because there is only one of you in all time, this expression is unique. And if you block it, it will never exist through any other medium and will be lost. (Quoted in Nachmanovich, 1990, p. 25.)

If that doesn't capture the essence of the courage, the power, the will and the surrender to be, I don't know what does. Well Frost does, actually, in his greatest of poems "The Trajectory Not Taken" which ends:

> I took the one less traveled by,
> And that has made all the difference. (1915)

Frost also talked about the *Courage To Be New* (1946) using the familiar warrior image (which is also involved in *The Road Not Taken* by the context of its generation) used by Heraclitus, Plato, the *Gita*[51], Trungpa[69], Shakespeare, Nietzsche[59], and many others, in which he seems to be saying, as in the *Gita*, that creative affirmation, both personal and cultural, comes from the violence and conflict inherent in a major bifurcation. He sees that the bigger the abyss, the better the attractor on the other side. These two poems of Frost convey much the same idea despite the different fractal comfort of the images of their setting, a feature which emphasizes the creativity in all our lives, and in the whole fabric of our lives, not in just special instances. But enough commentary, now to return to Gardner and his most important thoughts.

Besides the developmental perspective, his organizing framework includes an interactive perspective. Its essentials he develops from Csiklszentmihalyi who looks to interaction of three nodes of a complex dynamical system. Those nodes are the individual, the domain or discipline, and the judgmental social milieu. He says:

> In Csikszentmihalyi's persuasive account, creativity does not inhere in any single node, nor, indeed, in any pair of nodes. Rather, creativity is best viewed as a dialectical or interactive process, in which all three of these elements participate.
> (Gardner, 1993, p. 38.)

For visualization he diagrams these on a triangle (Fig. 3), very reminiscent of Torre's[68] triadic view of cognitive development of students in an academic institution. The holistic nature of a process residing in the interactive process and not in a product or component of the machinery is also explicit in the neurological positivism and ecological psychology of Vandervert[70], Kugler *et al.*[48] and Abraham[6]. Gardner refers to the diagram as seeing 'its operation in dynamic form'. Actually, if the triangle were unfolded into the orthogonal dimensions of a 3D state space, the trajectories could spin out in a complete geometric visualization of the dynamics of the interaction of individual, domain, and society (a pseudo-simulation is shown in Fig. 4).

In a given domain there is a Darwinian selection of a fit few, and of their products, only a few are judged creative or 'highly novel yet appropriate for the domain'. The domain bifurcates as a result of the individual bifurcation and the interaction of the individual and the domain's knowledge base and the judgmental community continues. Working with the big C this way, he claims:

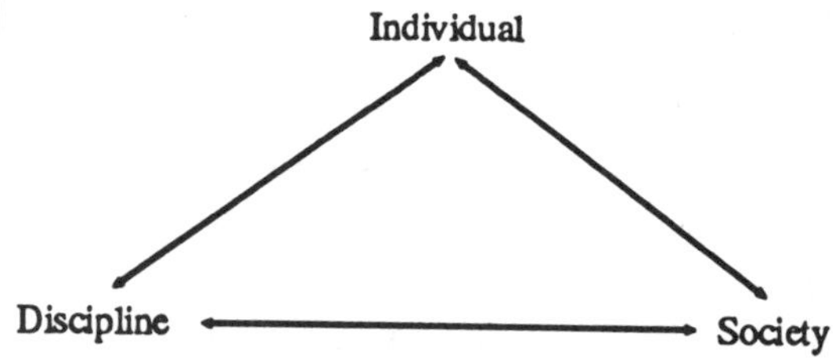

Figure 3.

The Csikszentmihalyian-Gardnerian Triangle of Systems Creativity

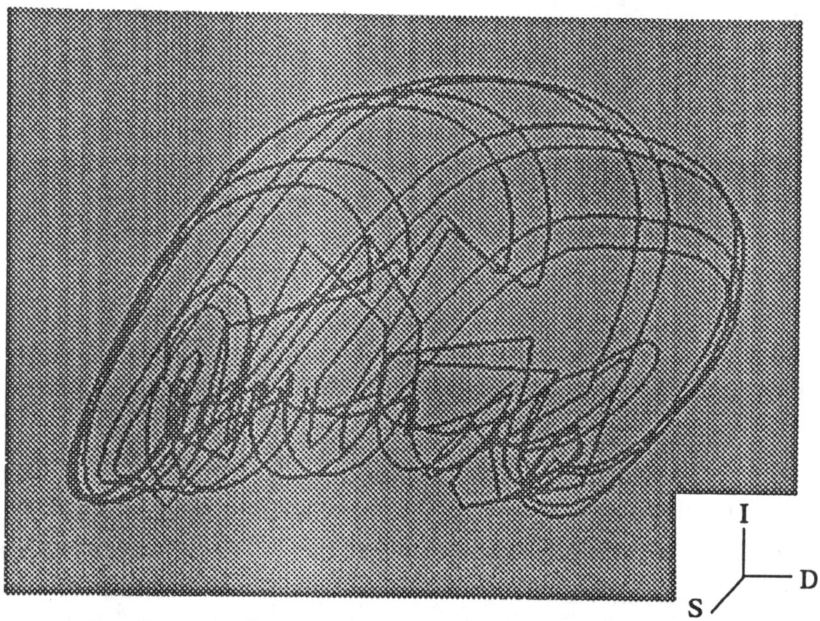

Figure 4. Pseudosimulation of a 3D Chaotic Attractor for the Csikszentmihalyian-Gardnerian Creativity System

The asynchrony is between the three aspects of the individual (1), the discipline (D), and the judgmental social milieu (S). Each would require many differential equations to represent it, but we pretend that three would do, and further imagine that there are convergent and divergent forces within each (positive and negative Lyapunov exponents on each dimension representing them) making for the chaotic nature of the attractor.

> In the absence of a judgment by a competent field, one simply cannot determine whether an individual merits the epithet "creative". (Gardner, 1993, p. 40.)

Throwing my lot in with Nachmanovich and Mosca, my claim is that all we ordinary mortals also participate in creativity, realizing differing degrees of satisfaction, and ups and downs, in pursuit of fulfilling our potential for it. That is not to deny validity of Gardner's view of creativity, that it has its own real playing field, I just want to emphasize that his dynamics are applicable to creativity with a smaller c as well. That is, they are useful outside his domain of publically acclaimed great achievements as well.

So far, Gardner is offering a nice organizational scheme, its developmental and interactive perspectives nicely dynamical, and modestly creative and highly useful in their own right. His third organizational perspective excites the dynamicist even more. He calls it *Fruitful Asynchrony*.

> Specifically, I claim that there exist certain kinds of asynchrony within or across these nodes and that these may well enhance the likelihood of creativity. (Gardner, 1993, p. 40).

Yesssss! Chaosssss! Thank you Howard!

> I refer to a lack of fit, an unusual pattern, or an irregularity within the creativity triangle. (Gardner, 1993, p. 41.)

Chaos within a node he sees typified by Picasso's peculiar balance of talents. Chaos across nodes he sees typified by Freud's intelligences being at odds with that of Freud's domain and with science. He talks about not too much chaos, and not too little, but just the right amount; thus the term 'fruitful asynchrony'. Or what I earlier called an optimal range of dimensionality.

5 Summary of Dynamical Concepts Involved In Creativity

5.1 Self-Organizational Bifurcations are Creativity

When I first embarked on these beginning explorations, as I said, I had brief and rather obvious insights that self-organizational bifurcations were involved, and that creativity seemed driven by attempts to self-create a meaningful self, that is, that cognitive navigation of the relevant parameter maps was driven by an existential anxiety. We have explored in miniature this meaning and some of its expressions in the literature of myth, mysticism, existentialism, and various psychological views of creativity. I have emphasized an old theme of mine, that such cognitive navigation, while over-stated, involves a hierarchical complex dynamical system network approach, in which a faster, lower cognitive navigation in the state space interacts with a slower, higher cognitive navigation in the parameter space.

5.2 Chaos and Instability Facilitate Creativity

Soon there was a turn of the trajectory I had not anticipated; further hidden treasures as the onion was unpeeled; a more creative and exhilarating aspect of my exploratons and discoveries. Actually there is not any real originality or novelty here in any other sense, that is, from the viewpoint of the 'domain' of dynamics; and in retrospect these discoveries can be seen as an inevitable conclusions. We have mentioned that systems, in order to bifurcate, must first get near the bifurcation point. Individuals who can either enjoy the intoxication (and the toxification, the stress) of living near the abyss, who surf on the instability of the bifurcational wave, or who face, with the eagle's courage and strength of grasp of the talons, the existential challenge of *ayin*, are more likely to have a life process that gives them meaning. They create themselves.

The other facet of living on the edge is *chaos management*. Managing chaos within reasonable mid-dimensional fractal levels depends on skill levels, ability to focus and sustain an effort, and cognitive and emotional forces of convergence and divergence that are within acceptable ranges in order for creative and useful bifurcations to occur. As mentioned earlier, there are three generic types of bifurcation: (1) subtle bifurcatons, (2) catastrophic bifurcations, both creative and annihilative, and (3) explosive bifurcations (especially from unconsciousness to consciousness, even as some of the trajectory consciously multiplexes across cognitive components). An alternative model to the explosive unconscious-to-conscousness bifurcation is a complex dynamical system where unconscious and conscious are conceived to interact, similar to our model of the triple blue-sky bifurcation of Jungian enantiodromia, individuation, and archetypal expression[2,8,10]. The point here, however, is not which way to model the process exactly, several candidates are obvious, but to recognize that positioning near the bifurcation point, accepting its instability, and optimizing chaos while there, are important aspects of creative life. This implies that the higher cognitive navigation, with or without awareness, makes dimensional choices (actually dimensional choices are mediated by position in parameter space, but here we only try to recognize higher cognitive control) about component dynamical schemes and the entire complex dynamical network, in order to successfully and creatively navigate. In other words, the management of chaos is a part of the imaginative sculpting of the whole dynamical domain.

5.3 Chaos at the Controls

This management of optimal levels of chaos and the inherent hierarchical cognitive control of it implies yet another role for chaos. We saw, in the Orestein myth, as the cultural representatives of chaos, the Furies and the older culture, not only had a hand on the controls, but also were considered an important aspect of their culture. What does this mean? It means that not only is chaos a rather ubiquitous feature of the behavior of complex dyanamical systems, schemes, and networks, but it is a desirable goal in itself. In the evolution of the complexification of the human mind and human society, with their bifurcations and also with their gradual changes in between bifurcations, there is an intentional dynamic, managing the chaos within and between the interactions in emotional and intellectual activity, which reads and feedbacks to control chaos to those optimal levels. Societies and individuals who have stronger convergent tendencies have lower preferred levels of dimensionality; they tend to lower dimensionality and are threatened by creativity and bifurcations to more complexity. Totalitarian regimes are almost completely convergent to fixed point attractors and ruthlessly squash emergent and divergent tendencies. Ultra-creative artists, like Van Gogh, are never satisfied with occupation of their current attractor, and bifurcate to ultimate complexity which also collapses into a fixed point. That is, infinite information is also zero information. The alpha and omega, zero and infinite dimensionality collapse into the unity of zero-dimensionality. Meanwhile, we shall prefer to live our lives with a mixture of joy, contentment, courage, and meaning, and with optimal levels of creativity and chaos and humor. We prefer to resider near the bifurcation point. The creative potential is worth the risk. We hope sharing this desire will encourage others to find and share these pleasures.

Anaxagoras gave a chaos creation myth for the creation of the universe, claiming there was an original homogeneous random universe in which an accidental local perturbation occurred creating initial conditions for the evolution of a complex universe. Chardin's alpha and omega for human culture may represent similar extreme dimensionalities, and Locke made similar claims for the ontology of the human mind using Aristotle's *tabula rasa.* We have not, and are not, concerned with cosmic creation here, and we don't know whether to characterize random and independent initial conditions as zero dimensionality or infinite dimensionality, but we can see that somewhere between these two extremes a happy universe or a happy mind can exist in creative self-organizational growth.

6 Summary

Creativity is driven by conflict/cooperation, the yin/yang of a two-sided coin called interaction, born of dual motivations to surf the positive emotions, especially joy, play, curiosity, and to escape the negative emotions, especially the angst of existential anxiety and fear. We can even say these can be viewed as a creativity motivation, an intentional system that evolves to greater complexity within a range of optimal degrees of chaotic character, of optimal dimensional complexity, and by residing close to the abyss, the bifurcation point, with all its attendant risks, and with the requirement of making the leap of the creative effort. Perhaps this would be better stated as a drive to chaos surfing on creativity. It rides on intentional self-organized bifurcations, dwelling only briefly in stable chaotic attractors, residing more often near bifurcation points as chaotic strategies are worked out enabling bifurcations to greater complexity.

Acknowledgement

Many friends embody the spirit of free play, curiosity, and creativity in their life styles and cognitive explorations. They have shared these styles and ideas, and have inspired me by being themselves. Their dancing on happiness is far more important than my literary, scientific, and philosophic searches. They include Precy Magdamo, Connie Abraham, Ralph Abraham, Lew Abraham, Esther Abrams, Ted Melnechuk, Allan Combs, David Loye, Frank Mosca, Rick Paar, Dave Adams, Steve Guastello, Patricia

Tuccio, Tobi Zausner, Steve Bernstein, Igor Gamov, and the late Virginia Voeks, to name but a few of the precious many.

References

Most authors have other works whose redundancy renders them acceptable as substitutes.
*** indicates special relevance.**

1. H. Abarbanel *et al.*, *Reviews of Modern Physics,* **65**, 1331 (1993).
2. F.D. Abraham, *Psychological Perspectives,* **20**, 156 (1989).
3. F.D. Abraham in McGraw-Hill Yearbook of Science and Technology, ed. S.P. Parker *et al* (McGraw-Hill, New York 1992).
4. F.D. Abraham, *World Futures,* **37**, 41 (1993).
5. F.D. Abraham in *Encyclopedia of Time,* ed. S.L. Macey (Garland. Hamden, 1994
6. F.D. Abraham, *Psychoscience,* **1**, 85 (1994).
7. F.D. Abraham in *Chaos Theory in Psychology,* ed. F. D. Abraham & A. R. Gilgen (Praeger/Greenwood, Westport, 1995).
8. F.D. Abraham, *Psychological Perspectives,* **31**, 70 (1995).
9. F.D. Abraham in *Chaos Theory in Psychology and the Life Sciences.* ed. R. Robertson & A. Combs (Erlbaum. Hillsdale, 1995).

*10. F.D. Abraham, R.H. Abraham, & C.D. Shaw, *A Visual Introduction to Dynamical Systems Theory for Psychology* (Aerial Press, Santa Cruz, 1990).

11. F.D. Abraham & A.R. Gilgen, eds., *Chaos Theory in Psychology* (Praeger/Greenwood, Westport, 1995).

*12. R.H. Abraham, *Chaos, Gaia, Eros* (Harper, San Francisco, 1994).

13. Aeschylus, *Agamemnon, The Choephori, The Eumenides* (Ατηενσ Φρεε Πρεσσ, Athens, 466BCE). And in Aeschylus, *The Orestein Trilogy,* ed. & trans. P. Vellacott (Penguin, Harmondsworth & New York, 1956).
14. T.M. Amabile, *The Social Psychology of Creativity* (Springer-Verlag, New York, 1983).
15. A. Bandura, *Psychological Review,* **84**, 1919 (1977).

16. G. Bateson, *Steps to an Ecology of Mind* (Ballantine, New York, 1972).)
17. G. Bateson, *Mind and Nature: A Necessary Unity* (Dutton, New York, 1979).
*18. M. Berman, *Coming to Our Senses* (Simon & Schuster, New York, 1989). Also in *Speculations, the Reality Club*, ed. J. Brockman (Prentice Hall, New York, 1988)
19. W. Blake, *The Marriage of Heaven and Hell* (1793).
20. J. Blofeld, transl. & ed., *I Ching* (George Allen & Unwin. London, 1965/1968).
21. M. Boden, *The Creative Mind* (Basic Books, New York, 1990).
22. J. Campbell, *Transformations of Myth through Time* (Harper & Row, New York, 1990).
23. F. Capra, *The Tao of Physics* (Shambala, Boulder, 1983).
24. A.K. Coomaraswamy, *The Transformation of Nature in Art* (Dover, New York, 1934/1956).
25. M. Csikszentmihalyi, *Flow* (Harper & Row, New York, (1990).
26. M. Csikszentmihalyi, *The Evolving Self: A Psychology for the Third Millennium* (HarperCollins, New York, 1993).
27. R. deCharms, *Personal Causation: The Internal Affective Determinants of Behavior*, (Academic., New York, 1968).
28. E.L. Deci & R.M. Ryan, *Intrinsic Motivation and Self-determination in Human Behavior* (Plenum, New York, 1985).
29. R. Dubos, *A God Within* (Charles Scribner's Sons, New York, 1972).
30. R. Eisler, *The Chalice and the Blade* (Harper & Row, San Francisco, 1987).
31. J. Fowles, *The Artistos* (Pan, London, 1969).
32. S. Freud, S. *Leonardo da Vinci and a Memory of his Childhood*, trans. A. Tyson (Norton, New York, 1910/1964).
*33. H. Gardner, *Creating Minds* (Basic Books, New York, 1993.
34. J. Goldstein, in *Chaos Theory in Psychology and the Life Sciences*, ed. R. Robertson & A. Combs (Erlbaum, , 1995).
35. L. Greeley, *Philosophical Spacing: Key to the Nonlinear Complex Dynamics of the Attentional System of the Cognitive Learning Process in the Philosophical Dialectical Method* (Unpublished dissertation, Harvard University, 1991).

36. H.E. Gruber, & S.N. Davis, in *The Nature of Creativity*, ed. R. J. Sternberg (Cambridge University Press, New York, 1988).
37. J.P. Guilford, J.P. *American Psychologist*, **5**, 444 (1950).
38. J.P. Guilford, J.P. *Creativity and its Cultivation* (Harper & Row, New York, 1953).
*39. C. Hampden-Turner, *Maps of the Mind* (Beazley, London, 1981).
40. D.O. Hebb, *Psychological Review*, **62**, 232 (1955).
41. K. Horney, *The Neurotic Personality of our Time* (Norton, New York, 1937).
42. A. Hubler, in *Modeling Complex Systems*, ed. L. Lam (Springer, New York, 1992).
43. W. James, *Principles of Psychology* (Holt, New York, 1890).
44. C. Jung, Autwort auf Hiob/Answer to Job in *The portable Jung*, ed. J. Campbell (Penguin, New York, 1957). Also in *Psychology and Religion: West and East, Vol. 11. The Collected Works of C.G. Jung (2nd ed.)*, ed. & trans. R.F.C. Hull (Rascher, Zurich, 1952; Routledge & Kegan Paul, London, 1954; Princeton: Princeton, Bollingen Series XX, Princeton, 1969).
45. A. Koestler, *The Act of Creation* (Hutchinson, London, 1964).
46. W. Köhler, *The Mentality of Apes*, trans. E. Winter (Harcourt, Brace, New York, 1925).
47. W. Köhler, *Dynamics in Psychology*. Liverright, New York, 1940).
48. P.N. Kugler *et al.*, *Psychological Research*, **52**, 98 (1990).
49. Lao-tze, *Tao Te Ching*, ed. & trans. G-F. Feng & J. English, (orig. 6th CBCE; Random House, New York, 1972).
50. P. Magdamo, *Palaas and Salarung: A Folktale of Self-referential Justice of the Philippine Monobo* (unpublished notes, and concert presentations; see F. Abraham[6] for comments).
51. Maharishi MaheshYogi, *On the Bhagavad-Gita: A New Translation and Commentary, Chapters 1-6* (Penguin, Harmondsworth, 1969).
52. A. Maslow, *Toward a Psychology of Being* (Van Nostrand, Princeton, 1968).
53. R. May, *Man's Search for Himself* (Norton, New York, 1953).
*54. R. May, *The Courage to Create* (Norton, New York, 1975).
55. W. Morris, ed., *The American Heritage Dictionary of the English Language* (Houghton Mifflin, Boston, 1969-78).

*56. F. Mosca, *The Unbearable Wrongness of Being* (Options for Living Press, Thornwood, 1994).

57. F. Mosca, in *Chaos Theory in Psychology*, ed. F.D. Abraham & A.R. Gilgen (Greenwood/Praeger, Westport, 1995).

*58. S. Nachmanovitch, *Free play: Improvisation in Life and Art* (Tarcher, Los Angeles, 1990).

59. F. Neitzsche, *Thus Spake Zarasthrustra* (1883).

60. E. Neumann, The Origins and History of Consciousness, trans. R.F.C. Hull (Princeton, Princeton, 1949).

*61. D. Paar, in *Psychotherapy and the Promiscuous Patient*, ed. E.M. Stern (Haworth Press, New York, 1992).

62. D.N. Perkins, *Creativity: Beyond the Darwinian Paradigm*. Unpublished manuscript for the Achievement Project Symposium, Kent, UK, December 13-15, 1991).

63. H.C. Sabelli, *Union of opposites. A Comprehensive Theory of Natural and Human Processes* (Brunswick, Lawrenceville, 1989).

*64. R.J. Sternberg, ed., *The Nature of Creativity* (Cambridge University Press, New York, 1988).

*65. I. Stewart, *Does God Play Dice? The Mathematics of Chaos* (Basil Blackwell, Cambridge, 1989).

66. M. Stone, *When God Was a Woman* (Harcourt, Brace, Jovanovich, New York, 1978).

*67. P. Tillich, *The Courage To Be* (Yale University Press, New Haven, 1952).

68. C.A. Torre, in *Chaos Theory in Psychology*, ed. F.D. Abraham & A.R. Gilgen (Greenwood/Praeger, Westport, 1995).

69. C. Trungpa, *Shambhala: The Sacred Path of the Warrior* (Shambhala, Boulder, 1984).

70. L.R. Vandervert, *The Journal of Mind and Behavior,* **14**, 277 (1993).

71. H. Weiner, *9 1/2 Mystics: The Kabbala Today* (Macmillan, New York, 1969).

72. R.W. White, *Psychological Review, 66*, 297 (1959).

73. R.C. Zaehner, *Mysticism: Sacred & Profane* (Clarendon, New York, 1957).

CODA
CONSCIOUSNESS: CHAOTIC AND STRANGELY ATTRACTIVE[a]

ALLAN COMBS

University of North Carolina at Asheville,
1 University Heights, Asheville, NC
28804-3299, USA
and
The Saybrook Institute
San Francisco, CA, USA

Consciousness is viewed from an ecological perspective in which the ongoing cognitive, perceptual, and emotional events that structure it are seen as a rich complex of interacting processes, analogous to the interactive metabolism of a living cell. The result is an organic, self-generating, or autopoietic, system, continuously in the act of creating itself.

1 Process

> Rather than discrete things and independent events, there are but ripples upon ripples upon waves upon waves in this universe, propagating in a seamless sea.
> Ervin Laszlo[1]

There is evidently a strong human tendency to conceptualize reality in terms of objects. This may in part be due to the influence of language. For example when infants acquire names for things they immediately begin to form hard and fast object categories from which accurate predictions about real physical objects are made.[2] Members of each such category seem to acquire an underlying "essence." That the formation of firm linguistic-bound classifications occurs so early in life suggests that the tendency to form them is rooted in the biology of the brain itself. Evidently there has been an evolutionary advantage to seeing the world in terms of crisply defined objects. Such an advantage is not difficult to imagine if one thinks about the early history of our species. There are obvious benefits to living in a world which yields clear distinctions, for example, between lions and gazelle, or between ripe red apples and bitter green ones. Monkey's of course can do this too. But monkeys cannot distinguish similar but distinct types of flint rock, or variations in the appearance of grain that indicate resistance to blight–and certainly cannot

[a]This paper originally appeared in *Psychoscience* and is reprinted with permission

pass these on to offspring.

Thus the world presents itself in terms of objects and categories of objects. Processes and patterns of activity, on the other hand, are seen only as through a glass darkly. This is especially true where it comes to the processual aspects of our own inner lives. Emotion, memory, and thought are such processes, and can only awkwardly be construed as having fixed form, shape, or permanence. Consciousness itself is such a process, or at least those aspects or reflections of it that can be sensed. If one thinks of consciousness in terms of metaphors such as light or clear water, then it has no discernible substance other than that which is disclosed by it–such as stones seen at the bottom of a brook, to use Sartre's analogy. A more process oriented image, however, would be that of a shaft of light falling across a vortex of airborne dust. The light itself is both emptiness and illumination, but the vortex of dust that gives it definition is awhirl with motion. In the following paragraphs I will try to show that consciousness is similarly awhirl with motion, both at the level of experience, and at the level of the neurological events which undergird experience.

2 States of consciousness

In the mid 1970s psychologist Charles Tart [3,4] showed that consciousness can be usefully understood from a systems perspective. In particular he argued that ordinary waking consciousness, or "ordinary reality" as it has come to be called, can be seen as a discrete state of consciousness surrounded by a potentially large number of alternative or "altered" states, as suggested by William James' phrase:

> Our normal waking consciousness...is but one special type of consciousness, whilst all about it, parted from it by the filmiest of screens, there lie potential forms of consciousness entirely different. We may go through life without suspecting their existence; but apply the requisite stimulus, and at a touch they are all there in all their completeness.[5]

Tart's basic notion is that a state of consciousness is comprised of a harmonious set of psychological functions. The latter include memory, cognition, sense of humor, sense of self, external perception (exteroception), perception of internal body states (interoception), and so on. Together they form a gestalt-like whole, or in other words a working system. Ordinary waking consciousness is one such system. Dream sleep is another–though in fact several states may be accessible in dreams.[6,7,8] Others include an unknown number of ecstatic states that can erupt spontaneously into ordinary consciousness, plus states that are accessible through meditation, the shamanic trance, hypnosis, and myriad drug induced states, as well as ordinary non-dream sleep.

Tart conceptualized states of consciousness as discrete. At first glance this idea may seem questionable, but it is useful in contrasting different states, and in fact seems to have some empirical justification. For instance falling asleep is accompanied by a slow vertical eye roll which marks an abrupt transition of awareness away from exteroceptive stimulation and the outside world and into an internal state of reverie. One abruptly looses awareness of external stimulation such as light or sound.[9] The onset of dream sleep is likewise abrupt. Drug intoxication, brought on, for example, by alcohol or marijuana, seems gradual, but often is heralded by a distinct moment when one first feels intoxicated.

Tart described several types of processes that stabilize a state of consciousness. One of the most important of these is loading stabilization, or in plain English keeping a person busy with activities that support the desired state. For certain ecstatic states this might include chanting, or the repetition of a prayer or mantrum, but in the case of ordinary reality it usually means staying productively busy. Anything from doing dishes to mowing lawns, paying bills, building houses, or writing books will do it. Such productive business is the passion of our civilization, and we tend to look poorly on anyone or anything that is not so.

Transitions between states of consciousness can be brought about by disrupting the stabilizing processes and bringing to bear positive patterning forces in the direction of the new desired state. For example, to go to sleep we must quit doing dishes, mowing lawns, and writing bills, and move to a dark and relatively quiet place where natural biological tendencies toward rest and sleep can have their play. The technologies of consciousness, found for example in yoga or shamanism, involve many patterning techniques designed both to disrupt ordinary consciousness and to move the practitioner toward extraordinary states.[10,11]

To my mind Tart's ideas are of the first order, but can benefit from more recent advances in the sciences of complexity, which yield more dynamic and fluid conceptions of the nature of systems. For example, a state of consciousness can be reconceptualized as an attractor. Speaking informally, an attractor is a condition to which a system is drawn by its own nature. If a cup is placed slightly tilted on a table, it will roll about in a spiral till it comes to rest standing up. This latter condition is termed a static attractor, because it represents the static position to which the cup is disposed. More interesting are cyclic or fixed cycle attractors. The human heart, for instance, runs through its cycle many times each minute. The moon passes through its various phases each month. These, and many others, are instances of systems that naturally settle into predictable cyclic routines. Most interesting, however, are the class of attractors that are neither fixed nor precisely predictable. These are termed strange or chaotic attractors.

On close inspection the cyclic rhythm of the human heart is found not to be precise, like the motions of a clock, but only approximately so. It's global form is well known and easily recognized, but the precise action of an individual heart differs from beat to beat, thus defying exact prediction. Moreover, it is unlikely that the heart ever, in the strictest possible sense, repeats itself the same way twice. This situation of global familiarity but non-predictability, along with the idea that the system never exactly repeats itself, is exactly what defines a chaotic system, one whose action is described by a strange attractor. Though it is often difficult to completely satisfy these criteria in particular instances, 12 systems of this general type are found abundantly in nature, including biological systems such as the human brain.[13,14] Here I will refer to them as chaotic or chaos like.

3 States of consciousness as chaotic attractors

I want to suggest that a state of consciousness is a chaotic attractor, one that brings together the many elements that form it in a way that coalesces them into a unified pattern of activity. The tendency of an attractor to draw its various constituents into a coherent configuration can be seen in the tendency to "fall" into a state of consciousness such as dream or nondream sleep once we tip into its basin (to use systems terms). This tendency of the attractor to capture consciousness in a particular state goes beyond Tart's original patterning forces and represents an intrinsic dynamic of the system itself. When sleep begins to overtake consciousness we drift off swiftly, carried away into its attractor basin. The swiftness of the transition depends on the steepness of the basin's slope. If we have gone without sleep for a long time the passage may be quick, but if we are worried, or have had too much coffee, it may be slow and troubled.

Why view consciousness as a chaotic attractor? The answer is that consciousness as a total event is, as William James[15] pointed out a hundred years ago, a constantly changing process, clearly not static or even following a fixed cycle, but nevertheless one that has an identifiable global character, at least for each individual. Memories come and go, thoughts pass through the mind only to disappear and return again later, moods are continually changing, and alertness and energy levels vary from hour to hour. These are the elements of a kind of mental soup, or more accurately a kind of mental weather, with the equivalent to the latter's constantly fluctuating temperature, humidity, wind, barometric pressure, and so on.

It is not surprising that weather is chaotic. Indeed, the elements that comprise it, such as temperature, oscillate in an identifiable cycle from day to day, but cannot be predicted with precision. What is more, it is unlikely that temperature fluctuations ever follow exactly the same course on any two days. Much the same can be said about mental weather. It is formed of the interaction of elements such as moods, thoughts, memories, and so on. These are Tart's original psychological functions. For some, such as moods, there is already empirical evidence that they are chaotic [16,17,18] while virtually all are consistent with the general description, above, of chaotic processes. As a group, their interaction, like the interaction of the elements of the weather, yield an exquisitely complex process fabric that we know as consciousness. This fabric is far too complicated to describe in detail, but efforts have been made to mathematically conceptualize it as a grand chaotic attractor. Chaos mathematician Ben Goertzel [19] recently developed the broad conception of a mathematical expression, which he calls the cognitive equation, that represents the entire process structure of an individual's mental life.

Goertzel imagines this structure as operating on two levels, that of the mind and that of the brain. He observes that, "the brain, like other extremely complex systems, is unpredictable on the level of detail but roughly predictable on the level of structure. This means that the dynamics of its physical variables display a strange attractor with a complex structure of "wings" or "compartments.""[20] These wings or compartments are in effect small attractors that reside in the larger attractor of the overall neurological activity of the brain. They might, for example, be associated with individual states of consciousness. For the sake of clarity I will continue to speak of them as separate, understanding that they are always part of the larger event which is the total process structure of the brain.

Goertzel views mental activity as running on top of the brain process, creating the second level of the system. The mental level is somewhat less finely detailed, however, and more generalized than the neurological level of activity. "If physical level attractors are drawn in ball-point pen, process [mental] level attractors are in magic marker."[21] Nevertheless, the same overall process structure is apparent at both levels.

Conceptualizing states of consciousness as chaotic attractors takes us a good way down the road toward understanding something about their internal dynamics. For one thing, many complex chaotic systems are self-organizing. For example, a living cell is composed of a rich and complex matrix of chemical cycles which self-organize in such a way as to regulate the overall activity of the cell. In 1974 biologists Maturana, Varela, and Uribe,[22] carried this notion further, suggesting that the total ongoing product of this matrix of activity is no less than the cell itself. In other words the principle activity of a living cell, when all its complex metabolic activities are summed up, is the continuing creation of itself. The above authors termed this process autopoiesis, or self-creation. Living cells are autopoietic systems. So are ecologies, as it turns out, as are many other complex systems, such as the international economy, and even human societies.

I believe that consciousness, and in particular a state of consciousness, is also an autopoietic event.[10,23,11] Like a living cell, it is made up of complex processes that interact in such a fashion as to create as a net result the state itself. It is not hard to see how this works. When we are depressed, for instance, we tend to selectively recall unhappy events,[24] while these in their turn contribute to the mood of depression. On the other hand, a good mood facilitates happy memories, which support a positive disposition. Beyond this, the very context created by a particular state of consciousness conspires to support each of its own elements. Consider, for example, marijuana intoxication.[25] A decline in short term memory contributes to an inability to sustain concentration for significant periods of time, while also supporting a style of cognition that relies more on intuition and imagination than discursive thought. This in turn fosters the style of inane yet imaginative humor so characteristic of that state.

It seems that this kind of self-regulating, essentially autopoietic activity, is seen at all levels of the system. For instance, beliefs rush to support each other even if they have to be invented on the spot, and even if they fail to form a logically coherent cloth. It is well known in social psychology, for instance, that disproving a valued belief, say, that the Second Coming is at hand, does not eliminate it. Rather, it is modified, extended, and reissued with new force.[26]

Interestingly, psychological stages of development, like states of consciousness, also appear to be autopoietic. Each is constructed and supported by a process fabric. In the case of Piaget's developmental epistemology, this fabric is woven of behavioral and mental schemata.[27,28] These wind together in a mutually supportive fashion. For instance the schema or concept of conservation, which specifies that matter does not appear or disappear out of nowhere, is supported by the schema of reversibility, the ability to perform mental operations in reverse. If a child is presented with the puzzling observation that a thin glass of water poured into a wide glass does not fill it to the level of the original, he or she can imagine the water poured back again into the tall glass and note that it's level is at the original height. The point here is simply that the mental processes which undergird consciousness, whether they be states of consciousness or developmental stages, seem to represent a single common regimen, that of an autopoietic process matrix.

4 The edge of chaos

The basic idea here is that consciousness comes in units, termed states, each woven of psychological processes, or functions, such as memory, emotion, cognition, one's sense of humor, one's sense of self, etc. Here, I want to momentarily consider these elements in a more formal fashion, as simply forming a self-organizing assembly of constituents that interact with each other in a rich variety of ways. European systems theorist George Kampis [29] has pondered such systems in detail. He terms them component-systems, and argues that they produce new and creative combinations that cannot, even in principal, be predicted by computational procedures such as performed by computers. The reason for this fundamental creativity is that the interactions of the component parts of these systems (processes to be more accurate) tend during their ordinary activity to build novel new structures (processes) while at the same time destroying some of those already present. These new elements in turn interact with each other and with the previously existing ones to produce new components not foreseeable from the original constituents. The net result is a rolling autopoietic event in which old structures are destroyed and original new ones routinely come into existence.

It is not hard to imagine how this works in day to day reality. The thoughts, memories, and feelings of one moment combine to create the thoughts, memories, and feelings of the next moment, and so on for each moment of our lives. Thus, like Heracleides' river, our inner lives are constantly in flux, never the same twice. But wait a moment; this is precisely what one would expect of a chaotic system. In this regard, it turns out that Goertzel's notion of the cognitive equation, a chaotic attractor, has a great deal in common with Kampis' idea of a component-system. Indeed, Goertzel was both influenced and encouraged by Kampis' thinking in the development of his own work.

In an interesting twist on this whole line of thought, Goertzel considers the problem of how such a network of interacting processes could find the limited though real constancy that characterizes our inner lives. To this end he offers the "productivity hypothesis," according to which "within a high degree of approximation, every mental process X which is not a pattern in some other mental process, can be produced by applying some mental process Y to some mental process Z, where Y and Z are patterns in some other mental process."[30] In other words, any mental process that stands alone can be produced by the interaction of other mental processes, themselves the products of still other mental processes. Thus, for example, particular ideas and beliefs are the products of the interaction of other ideas and beliefs, themselves dependent upon still other interactions. In a different example, a particular emotion may result from the interaction of certain memories with certain thoughts. Alternatively, the interaction of the emotion with the memories might create the thoughts. Whether the combination of the emotion and the thoughts will actually recreate the memory is perhaps more questionable, but there is considerable evidence that memory is much more the product of construction that would have been guessed a few decades ago.[31,32]

Goertzel is a computationalist, and does not agree with Kampis that the creativity of such a system as the human mind cannot, in principal, be modelled by a computer program. The question of whether the properties of consciousness can be represented computationally is the source of considerable debate today [33,34,35] but need not distract us at the moment. Here I want to return to Goertzel's notion that the cognitive equation, which represents the complex process fabric consciousness, is chaotic. The idea that the processes which configure consciousness are chaotic is broad, and does not mean that they exhibit noticeable chaotic properties at every moment. Indeed, they do not. Even a relatively simple chaotic function such as the Lorenz attractor does not necessarily appear chaotic on brief observation. More to the point, psychological processes [15,10] as well as the neurological events that undergird them [12,13] seem to move back and forth between relatively chaotic patterns of activity and more cyclic or even static ones. For example, moods can oscillate in a regular and predictable rhythm over the course of the day. Such a pattern, however, could not last indefinitely without perturbations contributing to more chaotic periods of activity. Moods are an obvious example, but there is no reason not to suspect that functions such as memory, thought, dreams, and general arousal do not also exhibit periods of constant or rhythmatic activity, and other periods of more chaotic change. If this is correct, then the overall process fabric of consciousness would likewise be expected to exhibit periods of calm, periods of more or less regular oscillation, and pronounced of periods chaotic activity. The long view of such a regimen would be that of a very complex mathematical attractor such as suggested by the cognitive equation. A shorter view, however, would disclose a system moving in and out of chaos.

Such systems on the edge of chaos are not unknown in the sciences of complexity. They may in fact turn out to be common if not universal among self-regulating systems such as living cells, ecologies, and brains. Discussing the evolution of biological systems, Stuart Kauffman recently observed that "selection achieves and maintains complex systems poised on the boundary or edge between order and chaos. These systems are best able to coordinate complex tasks and evolve in a complex environment."[36]

The ability of a system to move in and out of chaos gives it a creative advantage. It is capable of shifting from a steady or cyclic routine to one that generates novel emergent properties, whether those be original ideas or perceptions, new patterns of behavior, or novel emotional responses. Moreover, there is a tenacious resilience to a chaotic regimen that is absent in routinized behavior. This is seen in the excited dance of both prize fighters and neurons. We meet challenge effectively, not by assuming a fixed posture, but by moving in a dynamic cross-step that absorbs blows, counterpoises against shifting ground, and yields new and unexpected rejoinders. From a theoretical point of view, chaos protects a system from sticking in small grooves or attractors, and thus failing to find larger, more effective outcomes. For instance, a memory search can be thought of as a trip through neural "state space" in search of the correct memory attractor.[37,38,39,40] If the system gets stuck in the attractor of a wrong solution, subsequent recall will be incorrect. What is needed is a process that keeps it from settling down too quickly in the first attractor basin that comes along. This process is chaos. One can easily think of it as operating in a similar fashion during the search for a solution to a mathematical or linguistic problem, or a quest for the right artistic expression. Chaos is the antidote to stasis and stagnation.

References

1. E. Laszlo, *The creative cosmos: A unified science of matter, life, and mind*, (Floris, Edinburgh, 1993).
2. S.A. Gelman and E.M. Markman, *Cognition*, **23**, 183-209 (1986).
3. C.T. Tart, *Science*, **176**, 1203-1210 (1972).
4. C.T. Tart, *States of consciousness*, (E.P. Dutton, New York, 1975).
5. W. James, *The varieties of religious experience*, (Modern Library, New York, 1902/1929).
6. S. Krippner, *Anthropology of Consciousness*, **5**(3), 17-23 (1994).
7. C.T. Tart in *Altered States of Consciousness*, ed. C.T. Tart (Doubleday, New York, 1969).
8. S. LaBarge in *Conscious mind, sleeping brain*, ed. J. Gackenbach and S. LaBarge (Plenum, New York, 1988).
9. W.C. Dement, *Some must watch while some must sleep: Exploring the world of sleep*, (W.W. Norton, New York, 1972).
10. A. Combs, *World Futures: The Journal of General Evolution*, **38**, 43-62 (1993).
11. A. Combs, *The radiance of being*, (Floris, Edinburgh, 1995).
12. P. Rapp, *Biologist*, **40**(2), 89-94 (1993).

13. E. Basar, ed., *Chaos in Brain Function*, (Springer-Verlag, Berlin, 1990).
14. K.H. Pribram, ed., *Origins: Brain and self organization*, (Lawrence Erlbaum, Hillsdale, New Jersey, 1994).
15. W. James, *The principles of psychology*, (Harvard University Press, Cambridge, Mass, 1890/1981).
16. A. Combs, M. Winkler and C. Daley, *The Psychological Record*, **44**, 359-368 (1994).
17. O. Sacks, *Awakenings*, (Harper, New York, 1973/1990).
18. M. Winkler and A. Combs, *A chaotic systems analysis of individual differences in affect*. Paper presented at the 24th Interamerican Congress of Psychology, Santiago, Chile (1993, July).
19. B. Goertzel, *Chaotic logic*, (Plenum, New York, 1994).
20. B. Goertzel, ibid., p.157.
21. B. Goertzel, ibid., p.158.
22. H.R. Maturana, F.J. Varela, and R. Uribe, *Biosystems*, **5**, 187-196 (1974).
23. A. Combs, in *Chaos Theory in Psychology*, ed. F. Abraham and A. Gilgen, (Greenwood Pub, Westport, CT, 1994).
24. G.H. Bower, *American Psychologist*, **36**, 129-148 (1981).
25. C.T. Tart, *On being stoned*, (Science and Behavioral Books, Palo Alto, CA., 1971).
26. L. Festinger, H.W., Riecken, Jr., and S. Schacter, *When prophecy fails*, (University of Minnesota Press, Minneapolis, 1956).
27. J.H. Flavell, *The developmental psychology of Jean Piaget*, (Van Nostrand, New York, 1963).
28. Gruber and Voneche, eds., *The essential Piaget*, (Basic Books, New York, 1977).
29. G. Kampis, *Self-modifying systems in biology and cognitive science*, (Pergamon, New York, 1991).
30. B. Goertzel, ibid., p.157.
31. E.F. Loftus and H.G. Hoffman, *Journal of Experimental Psychology: General*, **118**, 100-104 (1989).
32. E. Winograd, and U. Neisser, *Affect and accuracy in recall*, (Cambridge University Press, Cambridge, 1992).
33. P.S. Churchland, *Matter and consciousness*, (MIT Press, Cambridge, MA, Bradford Books, 1984).
34. D.C. Dennett, *Consciousness explained*, (Little, Brown, Boston, 1991).
35. R. Penrose, *Shadows of the mind*, (Oxford University Press, Oxford, 1994).
36. S.A. Kauffman, *The origins of order*, (Oxford University Press, New

York, 1993).

37. F.D. Abraham in *Proceedings of The Society for Chaos Theory and the Life Sciences*, ed. R. Robertson and A. Combs (Lawrence Erlbaum, Hillsdale, N.J., 1995).
38. W.J. Freeman, it Scientific American, **Feb**, 78-85 (1991).
39. W.J. Freeman, *Societies of brains: A study in the neuroscience of love and hate*, (Lawrence Erlbaum Associates, Hillsdale, N.J., 1995).
40. C.A., Skarda and W.J. Freeman, *Behavioral and Brain Sciences*, **10**(2), 161-195 (1987).